IEE CONTROL ENGINEERING SERIES 34

Series Editors: Prof. D. P. Atherton
                Dr. K. Warwick

Singular perturbation methodology

IN

CONTROL

SYSTEMS

Other volumes in this series

Singular
perturbation
methodology
IN
CONTROL
SYSTEMS

DS Naidu

Peter Peregrinus Ltd. on behalf of the Institution of Electrical Engineers

Published by: Peter Peregrinus Ltd., London, United Kingdom

© 1988 Peter Peregrinus Ltd.

British Library Cataloguing in Publication Data.

Naidu, D. S.
  Singular perturbation methodology in
  control systems.—(IEE control
  engineering series; 34)
  1. Automatic control—Mathematical models
  I. Title   II. Institution of Electrical
  Engineers   III. Series
  629.8'312   TJ213

ISBN 0 86341 107 X

Printed in England by The Bath Press

Dedicated to my
wife Sita
and
daughters Radhika and Kiranmai

Contents

Preface

A fundamental problem in the theory of systems and control is the mathematical modelling of a physical system. The realistic representation of many systems calls for high-order dynamic equations. The presence of some 'parasitic' parameters such as small time constants, masses, moments of inertia, resistances, inductances and capacitances is often the cause for the increased order and 'stiffness' of these systems. The stiffness, attributed to the simultaneous occurrence of 'slow' and 'fast' phenomena, gives rise to time scales. The systems, in which the suppression of a small parameter is responsible for the degeneration of dimension, are labelled as 'singularly perturbed' systems, which are a special representation of the general class of time-scale systems.

The 'curse' of dimensionality coupled with stiffness poses formidable computational complexities for the analysis and control of these systems. The singular perturbation and time-scale methods, 'gifted' with the two remedial features of dimensional reduction and stiffness relief, are considered as a 'boon' to control engineers. These techniques have attained a certain level of maturity in the theory of continuous control systems described by ordinary differential equations. The methodology has an impressive record of applications in a wide spectrum of fields such as circuits, networks, power systems, electromagnetics, fluid mechanics, structural mechanics, soil mechanics, flight mechanics, celestial mechanics, geophysics, nuclear-reactor dynamics, oceanography, biology and ecology.

Discrete systems are very prevalent in science and engineering. There are mainly three sources of discrete models described by difference equations containing several small parameters. The first source is digital simulation of continuous systems where the differential equations are approximated by the corresponding difference equations. The study of sampled-data systems and computer control systems leads in a natural way to another source of discrete models. Finally, many economic, biological and sociological systems are represented by discrete models. In spite of the fact that the control of physical systems with a digital computer is of paramount importance, the field of singular perturbations and time scales in discrete control systems is only of recent origin. Of late, there has been considerable interest in this field including the optimal and adaptive control.

This book deals mainly with the singular perturbations and time scales in continuous and discrete control systems. The book is composed of seven chapters. The first introductory chapter contains chapter preview, nature of different problems and the spectrum of applications.

Chapter 2 presents the singular perturbation method for continuous and discrete control systems. We start with the definition of singular perturbation and discuss its various characteristic features such as order reduction, boundary-layer phenomena, degeneration, loss of auxiliary conditions, stretching transformations as applicable to differential and difference equations. Then we present Vasileva's method of obtaining asymptotic expansions for differential equations in terms of outer, inner and intermediate series. The boundary-layer method is also discussed where the approximate solution is expressed as an outer series and a boundary-layer correction which is equivalent to the difference between the inner and intermediate series. The structural properties of stability, controllability and observability are discussed. A singular perturbation method is given for initial- and boundary-value problems arising in discrete systems.

In Chapter 3, we focus our attention on the time-scale analysis of continuous and discrete systems. First, we describe a block diagonalisation procedure to decouple a two-time-scale continuous system into a slow and a fast subsystem. The aspects of permutation and scaling are discussed. It is also seen that the singularly perturbed continuous systems are represented as two-time-scale systems. The feedback design for eigenvalue displacement is given. The aspects of block diagonalisation and feedback design are discussed also with reference to discrete systems.

Chapter 4 deals with singular perturbation method to obtain asymptotic power-series expansions for the two-point boundary-value problem (TPBVP) arising in the open-loop optimal control of linear and nonlinear continuous systems. The expansions are constructed using an outer solution and the initial and final boundary-layer corrections. The fixed-endpoint linear optimal control problem is also discussed.

The open-loop discrete optimal control problem is the content of Chapter 5. Here we consider the open-loop optimal control of a linear, shift-invariant, singularly perturbed discrete system leading to a TPBVP, cast in the singularly perturbed structure. A singular perturbation method is discussed to obtain approximate solutions composed of outer series and initial and terminal correction series.

Chapter 6 considers the closed-loop optimal control of singularly perturbed, linear, time-invariant continuous systems. This leads us to the solution of matrix Riccati differential equations in singularly perturbed form. A method is given to obtain Riccati solutions consisting of outer series and inner and intermediate series. The steady-state regulator problem is discussed. The fixed-endpoint problem and the associated special matrix Riccati equation are given. A subsystem regulator design based on time-scale decomposition is also provided.

Finally, Chapter 7 addresses the closed-loop optimal control of a linear, shift-invariant, singularly perturbed discrete system. The resulting matrix Riccati difference equation is formulated in a structure amenable for singular perturbation analysis. The approximate solution is sought in the form of an outer series and a final correction series. The time-scale analysis of the closed-loop discrete optimal control problem is discussed.

There are many other important topics such as multi-modelling, differential games, stochastic filtering and control, adaptive control and a wide spectrum of applications which do not form a part of this book, the main reason for omission being the space limitation.

The salient features of this book are:

(i) Both singular perturbation technique and time-scale analysis are presented to problems in systems and control.
(ii) The book deals with both continuous systems (differential equations) and discrete systems (difference equations).
(iii) The material of the book is based largely on the past experience of the author in teaching and research. As such, most of the material is not found in any of the books published in this field.
(iv) Wherever necessary, examples are given to illustrate the methodology.
(v) A unique feature of the book is the extensive list of references which will be of great help for further study and research.

The material of the book falls in the areas of modelling and control of large-scale systems, perturbation techniques and model simplification. The book is aimed at both engineers and applied mathematicians interested in studying the field of singular perturbations and time scales. The book can also be used for teaching graduate students in systems and control engineering. The background required for understanding the book is elementary theory of differential and difference equations, matrices, control systems and optimal control.

Many people have influenced me in the field of singular perturbations and time scales. First among them is Professor P. K. Rajagopalan, former Dean of Planning and Co-ordination, and former Head of the Department of Electrical Engineering, Indian Institute of Technology, Kharagpur, to whom I wish to express my deep sense of gratitude for his inspiration and direction.

It is with a sense of great joy and pride that I would like to express my deep gratitude to my mother Subbamma and my father Ramaiah who made several cheerful sacrifices throughout their life to make me what I am today. Also, I like to record a high appreciation of my wife Sita and my daughters

Radhika and Kiranmai, who gave me a tremendous amount of love, under-standing, co-operation and encouragement in spite of the fact that they missed me a lot during the entire period of my research work and the preparation of this book.

NASA Langley   Desineni S. Naidu
Hampton, USA
January 1988

Acknowledgements

The author would like to acknowledge the permission given by IEEE (USA), Springer–Verlag (West Germany), Taylor & Francis Ltd. (UK), John Wiley & Sons (UK), Peter Peregrinus Ltd. (UK), and Elsevier–Sequoia SA (Switzerland) to use the material published in the various journals, proceedings, and/or research monographs.

Thanks are due to National Research Council, Washington, DC, USA, which offered the author a Senior Research Associateship Award tenable at NASA Langley Research Center, Hampton, USA, where the book took the final shape. The author would like to express his sincere thanks and high appreciation to Dr. Douglas B. Price, Assistant Head, Spacecraft Control Branch, for his constant interest, encouragement, and support. The final stages of this book were completed under the support from NASA Langley Research Center, Grant No. NAG-1-736, administered through ODU Research Foundation.

Chapter 1

Introduction

1.1 Introduction

A fundamental problem in the theory of systems and control is the mathematical modelling of a physical system. The realistic representation of many systems calls for high-order dynamic equations. The presence of some 'parasitic' parameters such as small time constants, masses, moments of inertia, resistances, inductances and capacitances is often the cause for the increased order and 'stiffness' of these systems. The stiffness, attributed to the simultaneous occurrence of 'slow' and 'fast' phenomena, gives rise to time scales. The systems in which the suppression of a small parameter is responsible for the degeneration of dimension are labelled as 'singularly perturbed' systems, which are a special representation of the general class of time-scale systems.

The 'curse' of dimensionality coupled with stiffness poses formidable computational complexities for the analysis and control of these systems. The singular perturbation and time-scale methods, 'gifted' with the two remedial features of dimensional reduction and stiffness relief, are considered as a 'boon' to control engineers. These techniques have attained a certain level of maturity in the theory of continuous control systems described by ordinary differential equations (Vasileva, 1963; Van Dyke, 1964; Wasow, 1965; Cole, 1968; Butuzov et al., 1970; Nayfeh, 1973; O'Malley, 1974a; Kokotovic et al., 1976; Eckhaus, 1979, Eckhaus and De Jager, 1982; Ardema, 1983a; Saksena et al., 1984). The methodology has an impressive record of applications in a wide spectrum of fields such as circuits (Sastry and Desoer, 1981), networks (Sannuti, 1981), power systems (Chow, 1982), electromagnetics (Seshadri, 1976), fluid mechanics (Van Dyke, 1964), structural mechanics (Flaherty and O'Malley, 1982), soil mechanics (Dicker and Babu, 1974a,b), flight mechanics (Ardema, 1977), celestial mechanics (Verhulst, 1975), geophysics (Carrier, 1970), nuclear reactor dynamics (Reddy and Sannuti, 1975), oceanography (Ruijter, 1979), biology (Carpenter, 1977) and ecology (Naidu and Rajagopalan, 1979).

Discrete systems are very prevalent in science and engineering. There are mainly three sources of discrete models described by difference equations containing several small parameters. The first source is digital simulation of continuous systems where the differential equations are approximated by the corresponding difference equations (Hildebrand, 1968). The study of sampled-data systems and computer-control systems leads in a natural way to another source of discrete models. Finally, many economic, biological and sociological systems are represented by discrete models. In spite of the fact that the control of physical systems with a digital computer is of paramount importance, the field of singular perturbations and time scales in discrete control systems is only of recent origin (Comstock and Hsiao, 1976; Locatelli and Schiavoni, 1976). Recently there has been considerable interest in this field including optimal and adaptive control (Phillips, 1980a; Rajagopalan and Naidu, 1980a; Atluri and Kao, 1981; Blankenship, 1981; Naidu and Rao, 1982, 1984, 1985a,b; Mahmoud, 1982a,b,c; Ioannou and Kokotovic, 1983; Kando and Iwazumi, 1983a,b).

The fast growth of research activity in singular perturbations and time scales is seen by the publication of excellent survey papers (Vasileva, 1963, 1976; O'Malley, 1968; Bjurel *et al.*, 1970; Butuzov *et. al.*, 1970; Lagerstrom and Casten, 1972; Smith, 1975; Genesio and Milanese, 1976; Kokotovic *et al.*, 1976; Lions, 1978; Sandell *et al.*, 1978; Blankenship, 1979; Jamshidi, 1980; Utkin, 1984; Kokotovic, 1984, 1985; Saksena *et al.*, 1984), reports and proceedings of special conferences (Kokotovic and Perkins, 1972; Brauner *et al.*, 1977; Eckhaus and de Jager, 1978, 1982; Hemker and Miller, 1979; Verhulst, 1979; Meyer and Parter, 1980; Ardema, 1983a), research monographs and books (Van Dyke, 1964; Wasow, 1965; Kaplun, 1967; Anderson and Moore, 1971; Cruz, 1972; Eckhaus 1973, 1979; Vasileva and Butuzov 1973, 1978; Nayfeh, 1973, 1981, 1985; O'Malley, 1974a; Grabmuller, 1978; Vidyasagar, 1978; Nayfeh and Mook, 1979; Pervozvanskii and Gaitsgori, 1979; Campbell, 1980, 1982; Miranker, 1980; Mishchenko and Rozov, 1980; Schuss, 1980; Bensoussan *et al.*, 1981; Kevorkion and Cole, 1981; Lions, 1981; Lomov, 1981; Mahmoud and Singh, 1981a, 1984; Chow, 1982; Jeffrey and Kawahara, 1982; Dontchev, 1983; Ioannou and Kokotovic, 1983; Bensoussan *et al.*, 1984; Chang and Howes, 1984; Kushner, 1984; Tikhnov *et al.*, 1984; Mahmoud *et al.*, 1985b; Naidu and Rao, 1985a; Smith, 1985; Kokotovic *et al.*, 1986).

1.2 Chapter preview

This book mainly deals with the singular perturbations and time scales in continuous and discrete control systems. The book is composed of seven chapters. The first introductory chapter contains a chapter preview, the nature of different problems and the spectrum of applications.

Chapter 2 presents the singular perturbation method for continuous and discrete control systems. We start with the definition of singular perturbation and discuss its various characteristic features such as order reduction, boundary-layer phenomena, degeneration, loss of auxiliary conditions, stretching transformations as applicable to differential and difference equations. Then we present Vasileva's method of obtaining asymptotic expansions for differential equations in terms of outer, inner and intermediate series. The boundary-layer method is also discussed where the approximate solution is expressed as an outer series and a boundary-layer correction which is equivalent to the difference between the inner and intermediate series. The structural properties of stability, controllability and observability are discussed. A singular perturbation method is given for initial- and boundary-value problems arising in discrete systems.

In Chapter 3, we focus our attention on the time-scale analysis of continuous and discrete systems. First, we describe a block diagonalisation procedure to decouple a two-time-scale continuous system into a slow and a fast subsystem. The aspects of permutation and scaling are discussed. It is also seen that the singularly perturbed continuous systems are represented as two-time-scale systems. The feedback design for eigenvalue displacement is given. The aspects of block diagonalisation and feedback design are discussed also with reference to discrete systems.

Chapter 4 deals with the development of a singular perturbation method to obtain asymptotic power-series expansions for two-point boundary-value problem (TPBVP) arising in open-loop optimal control of linear and nonlinear systems. The expansions are constructed using an outer solution and an initial and a final boundary-layer correction. The fixed-endpoint linear optimal control problem is also discussed.

The open-loop discrete optimal control problem is the content of Chapter 5. Here we consider the open-loop optimal control of a linear, shift-invariant singularly perturbed discrete system leading to a TPBVP cast in the singularly perturbed structure. A singular perturbation method is discussed to obtain approximate solutions composed of outer series and initial and terminal correction series.

Chapter 6 considers the closed-loop optimal control of singularly perturbed, linear, time-invariant continuous systems. This leads us to the solution of matrix Riccati differential equations in singularly perturbed form. A method is given to obtain Riccati solutions consisting of outer series and inner and intermediate series. The steady-state regulator problem is discussed. The fixed-endpoint problem and the associated special matrix Riccati equation are given. A subsystem regulator design based on time-scale decomposition is also provided.

Finally, Chapter 7 addresses the closed-loop optimal control of a linear, shift-invariant, singularly perturbed discrete system. The resulting matrix Riccati difference equation is formulated in a structure amenable for singular perturbation analysis. The approximate solution is sought in the form of an outer series and a final correction series. The time-scale analysis of the closed-loop discrete optimal control problem is presented.

There are many other important topics such as multi-modelling, differential games, stochastic filtering and control, adaptive control and a wide spectrum of applications which do not form a part of this book, the main reason for omission being the space limitation.

1.3   Classification of problems

There are various types of problems or systems often discussed in the theory of singular perturbations and time scales. Some of these problems are dealt with in detail in this book and other problems are merely mentioned in this Section with appropriate references.

1.3.1   Initial-value problems

The singularly perturbed initial-value problem (IVP) is one of the extensively investigated topics in this field (Vasileva, 1963; Wasow, 1965; O'Malley, 1974a; Kokotovic *et al.*, 1976). In this we consider a linear system

$$\frac{dx}{dt} = A_{11}x + A_{12}z + B_1u \tag{1.1a}$$

$$\varepsilon\frac{dz}{dt} = A_{21}x + A_{22}z + B_2u \tag{1.1b}$$

or a nonlinear system

$$\frac{dx}{dt} = f(x,z,u,\varepsilon,t) \tag{1.2a}$$

$$\varepsilon\frac{dz}{dt} = g(x,z,u,\varepsilon,t) \tag{1.2b}$$

with initial conditions

$$x(t=0) = x(0), \quad z(t=0) = z(0) \tag{1.3}$$

where x and z are m- and n-dimensional state vectors, u is an r-dimensional control vector, As and Bs are matrices of appropriate dimensions and ε is a small scalar positive parameter responsible for causing singular perturbation in the sense that when ε is neglected (made zero) the order of the system (1.1) or (1.2) is reduced. A boundary layer is said to exist at the initial point, thus needing only one correction (O'Malley, 1974a).

This problem is studied in detail in Chapter 2.

1.3.2   Boundary-value problems

In the case of singularly perturbed boundary-value problems (BVP), we have the system (1.1) or (1.2) with boundary conditions such as

$$x(t = 0) = x(0), \quad z(t = T) = z(T) \tag{1.4}$$

The full impact of the theory of singular perturbations is felt mainly in BVP. All the essential features are clearly brought out in this case. Here, the boundary layers or layers of rapid transition occur at the initial and final (terminal) points, thus necessitating two corrections. The BVPs are thoroughly investigated in the literature (Harris, 1960; Vasileva and Butuzov, 1973; O'Malley, 1974a; Kokotovic *et al.*, 1976).

   Chapter 2 considers details of this problem.

1.3.3   Two-time-scale systems

The concept of time scales is a relatively recent notion associated with a general class of systems where a small parameter need not explicitly be present like in singularly perturbed systems (Chow, 1977). Thus, a two-time-scale system can be modelled as

$$\frac{dx}{dt} = f(x,z,u,t) \tag{1.5a}$$

$$\frac{dz}{dt} = g(x,z,u,t) \tag{1.5b}$$

in the nonlinear case, and

$$\frac{dx}{dt} = A_{11}x + A_{12}z + B_1u \tag{1.6a}$$

$$\frac{dz}{dt} = A_{21}x + A_{22}z + B_2u \tag{1.6b}$$

in the linear case, with appropriate initial or boundary conditions. These systems possess two widely separated clusters of eigenvalues giving rise to slow and fast phenomena and leading to the formation of boundary layers. The two-time-scale property is exhibited in a natural way in systems with singularly perturbed structure (1.1) or (1.2).

   Further details of time-scale analysis are presented in Chapter 3.

1.3.4   Discrete time systems

It is surprising that the discrete systems described by difference equations have received attention only recently in the theory of singular perturbations and time scales (Comstock and Hsiao, 1976; Rajagopalan and Naidu, 1980a; Phillips, 1980a). Here, we model a singularly perturbed discrete system as

$$\begin{bmatrix} x(k+1) \\ z(k+1) \end{bmatrix} = \begin{bmatrix} A_{11} & \varepsilon A_{12} \\ A_{21} & \varepsilon A_{22} \end{bmatrix} \begin{bmatrix} x(k) \\ z(k) \end{bmatrix} + \begin{bmatrix} B_1 \\ B_2 \end{bmatrix} u(k) \tag{1.7}$$

and a two-time-scale discrete system as

$$\begin{bmatrix} x(k+1) \\ z(k+1) \end{bmatrix} = \begin{bmatrix} A_{11} & A_{12} \\ A_{21} & A_{22} \end{bmatrix} \begin{bmatrix} x(k) \\ z(k) \end{bmatrix} + \begin{bmatrix} B_1 \\ B_2 \end{bmatrix} u(k) \tag{1.8}$$

with appropriate auxiliary conditions.

Details of discrete systems form parts of Chapters 2, 3, 5 and 7.

1.3.5   Open-loop optimal control problems

The real need for order reduction associated with singularly perturbed systems is most acutely felt in optimal control design which demands the solution of higher-order equations (Kokotovic and Sannuti, 1968). The performance index to be minimised is usually taken as

$$J = \frac{1}{2} \int_0^T (y'Dy + u'Ru)\, dt \tag{1.9}$$

where $y' = (x', z')$, D and R are positive semidefinite and definite symmetric matrices, respectively. For the minimisation of the system (1.1) with respect to (1.9), we arrive at the open-loop optimal control which requires the solution of a singularly perturbed two-point boundary-value problem (TPBVP) in terms of the state and costate variables.

In the case of discrete control systems, we arrive at the same situation for the system (1.7) and the cost functional

$$J = \frac{1}{2} \sum_{k=0}^{N-1} [y'(k)Dy(k) + u'(k)Ru(k)] \tag{1.10}$$

The development of appropriate methodology for continuous and discrete systems is detailed in Chapters 4 and 5, respectively.

1.3.6   Closed-loop optimal control problems

The closed loop optimal control of (1.1) and (1.9) leads to the solution of a singularly perturbed matrix Riccati differential equation (Sannuti and Kokotovic, 1969*a*). Similarly, for the discrete systems (1.7) and (1.10), we face the matrix Riccati difference equation (Naidu and Rao, 1984).

Chapters 6 and 7 describe the continuous and discrete problems, respectively.

The above problems are discussed in detail in this book whereas the following are the further problems which are of importance in the theory of singular perturbations and time scales.

1.3.7 *Classical control problems*

Under classical control problems, we include analysis in the frequency domain, state feedback, composite control, multivariable systems, eigenvalue problems and other general problems dealing with order reduction.

In frequency-domain analysis, we consider the Laplace transform of (1.1) and the output relation

$$y = C_1 x + C_2 z \tag{1.11}$$

to obtain

$$\begin{bmatrix} A_{11}-pI_m & A_{12} & B_1 \\ A_{11}/\varepsilon & A_{11}/\varepsilon-pI_n & B_2/\varepsilon \\ C_1 & C_2 & 0 \end{bmatrix} \begin{bmatrix} x(p) \\ z(p) \\ u(p) \end{bmatrix} = \begin{bmatrix} -x(0) \\ -z(0) \\ y(p) \end{bmatrix} \tag{1.12}$$

where p is assumed to be a Laplace variable.

The matrix (1.12) denoted by $A(p)$ is known as the system matrix. The invariant zeros of the system are those complex frequencies $p = p_0$ for which $A(p_0)$ loses rank and the corresponding invariant-zero directions are those vectors that lie in the null space of $A(p_0)$. The asymptotic values, the invariant zeros and associated invariant-zero directions as $\varepsilon \to 0$ are the values computed from the system matrices

$$A^{(s)}(p) = \begin{bmatrix} A_{11}-pI_m & A_{12} & B_1 \\ A_{21} & A_{22} & B_2 \\ C_1 & C_2 & 0 \end{bmatrix} ; A^{(f)}(p) = \begin{bmatrix} I_m & 0 & 0 \\ 0 & A_{22}-pI_n & B_2 \\ 0 & C_2 & 0 \end{bmatrix} \tag{1.13}$$

associated with the slow and fast subsystems, respectively (Porter, 1974). Similar relations are obtained between the input $u(p)$ and output $y(p)$ (Porter and Shenton, 1975*b*; Porter, 1982; Luse and Khalil, 1985).

In state feedback, we consider the composite control u_c, consisting of slow and fast controls as

$$u_s = K_s x_s; \quad u_f = K_f z_f \tag{1.14}$$

$$u_c = u_s + u_f = K_s x_s + K_f z_f$$

where the slow and fast subsystems are defined by

$$\dot{x}_s = A_0 x_s + B_0 u_s, \quad x_s(t{=}0) = x(0) \tag{1.15a}$$

$$y_s = C_0 x_s + D_0 u_s \tag{1.15b}$$

and

$$\varepsilon \dot{z}_f = A_{22} z_f + B_2 u_f, \quad z_f(t{=}0) = z(0) - z^{(0)}(0) \tag{1.16a}$$

$$y_f = C_2 z_f \tag{1.16b}$$

with

$$z_f = z - z_s; \quad u_f = u - u_s; \quad y_f = y - y_s$$

$$A_0 = A_{11} - A_{12} A_{22}^{-1} A_{21}; \quad B_0 = B_1 - A_{12} A_{22}^{-1} B_2$$

$$C_0 = C_1 - C_2 A_{22}^{-1} A_{21}; \quad D_0 = - C_2 A_{22}^{-1} B_2$$

The complete results on the state feedback with composite control of multivariable systems are obtained by several authors (Porter, 1977, 1982; Suzuki and Muira, 1976; Fossard and Magni, 1980; Hickin, 1980; Bradshaw and Porter, 1981; Doraiswami, 1982; Young and Kokotovic, 1982; Shieh and Tsay, 1984). Other related problems are concerned with eigenvalues (Porter and Shenton, 1975a; Moore, 1976; Chow, 1978a; Boglaev, 1979; Allemong and Kokotovic, 1980; Anderson and Hallauer, 1981; Ardema, 1983a; Freiling, 1984; Fu and Sawan, 1986), root loci (Sannuti and Wason, 1983; Sastry and Desoer, 1983), Routh approximations (Hutton and Friedland, 1975) and order reduction in control systems (Davison, 1966; Kuppurajulu and Elangavon, 1970; Aoki, 1978; Campbell, 1980, 1982; Tse *et al.*, 1978; Hartwig, 1979; Hickin and Sinha, 1980; Elrazaz and Sinha, 1981; Newcomb, 1981; Verghese *et al.*, 1981; Wasynozuk and Decarlo; 1981, Duc *et al.*, 1983; Hoppenstead, 1983; Siljak, 1983; Tzafestas and Anagnostou, 1983, Lastman *et al.*, 1984) and balanced systems (Fernando and Nicholson, 1982a,b, 1983a,b; Dauphin-Tanguy *et. al.*, 1985).

1.3.8 Time-delay systems

Pure time delays are commonly encountered in many engineering systems. A general time-delay system in the singularly perturbed structure is represented as

$$\frac{\mathrm{d}x}{\mathrm{d}t} = f(x, \bar{x}, z, \bar{z}, u, \varepsilon, t) \tag{1.17a}$$

$$\varepsilon \frac{\mathrm{d}z}{\mathrm{d}t} = g(x, \bar{x}, z, \bar{z}, u, \varepsilon, t) \tag{1.17b}$$

where $\bar{x}$ and $\bar{z}$ are the delayed states $x(t-\varepsilon)$ and $z(t-\varepsilon)$, respectively. Physically, the parameter ε could represent pure time delays, coupling coefficients, moments of inertia, inductances and other small parameters whose presence is responsible for increased order of the system (Reddy and Sannuti, 1976). The theoretical aspects of the differential equations with small delay have been considered by Vasileva (1963), Halanay (1966), O'Malley (1971) Gichev and Dontchev (1979), Malek-Zavarei (1980).

There are applications of the singular perturbation method for time-delay systems occurring in nuclear reactors (Sannuti and Reddy, 1973; Reddy and Sannuti, 1975), power systems (Reddy and Sannuti, 1976) and biological oscillations (Grasman and Jansen, 1979).

1.3.9 Observers

An observer is a dynamic system which reconstructs the state of the original system on the basis of known system inputs and measurements (O'Reilly 1983a). For the singularly perturbed system described by system (1.1) and (1.11), the full-order observer is governed by

$$\dot{\bar{x}} = A_{11}\bar{x} + A_{12}\bar{z} + B_1 u + L_1(y - C_1\bar{x} - C_2\bar{z}) \tag{1.18a}$$

$$\varepsilon\dot{\bar{z}} = A_{21}\bar{x} + A_{22}\bar{z} + B_2 u + L_2(y - C_1\bar{x} - C_2\bar{z}) \tag{1.18b}$$

where $\bar{x}$ and $\bar{z}$ are the estimates of x and z and the state reconstruction errors are $x_e = x - \bar{x}$; $z_e = z - \bar{z}$. Then

$$\dot{x}_e = (A_{11} - L_1 C_1) x_e + (A_{12} - L_1 C_2) z_e \tag{1.19a}$$

$$\varepsilon\dot{z}_e = (A_{21} - L_2 C_1) x_e + (A_{22} - L_2 C_2) z_e \tag{1.19b}$$

System (1.18) functions as an observer for system (1.1) and (1.11), if matrices L_1 and L_2 of appropriate dimensionality are chosen such that the equilibrium state $x_e = 0$, $z_e = 0$ of system (1.19) is asymptotically stable (Porter, 1977).

Full details of observer design are found in Porter (1974), Balas (1978), O'Reilly (1979a, 1980, 1983a), Javid (1980, 1982), Mahmoud (1982a), Dragan (1985), Fu and Sawan (1986), Kando and Iwazumi (1985), Silva-Madriz (1986).

1.3.10 High-gain feedback systems

A fundamental property of high-gain systems is their relationship with singularly perturbed systems. It has been observed that all singularly perturbed systems can be represented as high-gain systems and all high-gain systems can be analysed as singularly perturbed systems (Kokotovic and Sannuti, 1968). High-gain methodologies can be used to analyse 'cheap' control problems where small penalties are made on control variables and a class of so-called variable structure systems.

(a) High-gain systems

A high-gain feedback system is represented as

$$\dot{x} = Ax + Bu, \quad x \in R^n, \quad u \in R^p \tag{1.20a}$$

$$u = \frac{1}{\varepsilon}Cx = \frac{1}{\varepsilon}y, \quad y \in R^p, \quad g = \frac{1}{\varepsilon} \tag{1.20b}$$

where g is the large scalar gain factor.

We can rewrite system (1.20) as

$$\varepsilon\dot{y} = CBy + \varepsilon CAx \tag{1.21}$$

Here y is the fast variable provided CB is nonsingular. After the fast transient near the range space of B decays, the slow motion is continuous near the null space of C (Young *et al.*, 1977). From the theory of asymptotic root loci, the p fast eigenvalues of the high-gain system (1.20) tend to infinity (the infinite zeros) along the asymptotes defined by the directions of $\lambda_i(CB)$, while the $(n-p)$ slow eigenvalues tend to the transmission zeros (finite zeros) of the open loop system (1.20a) with the output $y = Cx$. Thus when CB is nonsingular, system (1.20) can be expressed in the standard singularly perturbed form of system (1.1).

If CB is singular the situation is more complicated. For example, under the condition

$$CA^iB = 0, \quad i = 0, 1, \dots, q-2$$

$CA^{q-1}B$ is nonsingular.

This corresponds to cheap control and singular-arc problems and q fast time scales

$$\frac{t}{\varepsilon}, \quad \frac{t}{\varepsilon^2}, \dots \frac{t}{\varepsilon^q}$$

Studies on high-gain feedback systems and singular perturbations are found in Shaked (1976), Young *et al.* (1977), Young (1978a, 1982a,b), Bradshaw and Porter (1979), Khalil (1981a,b), Porter (1982), Willems (1981, 1982), Khalil and Saberi (1982), Kimura (1982), Sannuti (1983) and Schumacher (1984).

(b) Cheap control

A 'cheap' control problem arises when the system (1.20) is regulated with respect to a quadratic performance index having small (cheap) penalty on u (Young *et al.*, 1977) as

$$J = \frac{1}{2} \int_0^\infty (x'Qx + \varepsilon^2 u'Ru) \, dt \tag{1.22}$$

When $B'QB>0$, a composite control is obtained with state feedback gain matrices K_s and K_f obtained from the solutions of slow and fast regulator problems. Further discussion on singular perturbation analysis of 'cheap control' problems is found in O'Malley (1974a, 1976b), Jameson and O'Malley (1975), Francis and Glover (1978), Murthy (1978), Francis (1979, 1981), Grasman (1979a), Dragan and Halanay (1982), Kimura (1981), O'Reilly (1983b), Sannuti (1983), Sannuti and Wason (1983, 1985) Priel and Shaked (1984).

(c) Variable-structure systems
Variable-structure systems are systems (1.20a) with discontinuous feedback control for which the so-called sliding mode occurs on the switching surface $s(x) = 0$ (Utkin, 1977a). While in the sliding mode, the system remains insensitive to parameter variations and disturbances, similar to a high-gain system. In sliding mode, the discontinuous feedback control, componentwise, is

$$u_i(x) = u_i^+(x), \quad s_i(x) > 0$$
$$u_i^-(x), \quad s_i(x) < 0 \tag{1.23}$$

and system (1.20a) is governed by the 'equivalent control'

$$u_{eq} = -(CB)^{-1}CAx \tag{1.24}$$

obtaining by requiring that

$$\dot{s} = CAx + CBu = 0 \tag{1.25}$$

Detailed treatment is available from Utkin (1977a,b, 1984), Young *et al.* (1977), Young (1978b), Young and Kwatny (1982), Zinober *et al.* (1982).

A seemingly related problem is concerned with singular systems attracting considerable attention (O'Malley, 1976b; Schaar, 1976; Vasileva, 1976; Flaherty and O'Malley, 1977; Campbell, 1980, 1982; Campbell and Rose, 1982; Ardema, 1979, 1980, 1983a; Verghese *et al.*, 1979; Utkin, 1984; Paraskevopoulos and Christoodoulou, 1984) including descriptive variable systems (Cobb, 1981, 1983).

1.3.11 Multimodelling

Consider a large-scale system consisting of one slow and N fast subsystems as (Kokotovic, 1981)

$$\dot{x} = A_{i0}x + \sum_{i=1}^{N} A_{0i}z_i + \sum_{i=1}^{N} B_{0i}u_i \tag{1.26a}$$

$$\varepsilon_i \dot{z}_i = A_{i0}x + A_{ii}z_i + \sum_{j=1}^{N} \varepsilon_{ij} A_{ij} z_i + B_{ii}u_i, \quad j \neq i \tag{1.26b}$$

where each fast subsystem is associated with a different singular perturbation parameter ε_i and is weakly coupled to other fast subsystems through ε_{ij}.

In a situation like this it is rational for a fast system controller to neglect all other fast subsystems and to concentrate on its own subsystem, plus the interaction with others through the slow core. For the ith controller, this is simply effected by setting all the ε-parameters to zero except for ε_i. The ith-controller simplified model is then

$$\dot{x}^i = A_i x^i + A_{0i} z_i + B_{0i} u_i + \sum_{j=1}^{N} B_{ij} u_j, \quad j \neq i \tag{1.27a}$$

$$\varepsilon_i \dot{z}_i = A_{i0} x^i + A_{ii} z_i + B_{ii} u_i \tag{1.27b}$$

We denote x^i with a superscript rather than a subscript to stress the fact that x^i is not a component of x, but the ith controller's view of x. In reality, the model (1.27) is often all the ith controller knows about the whole system. The jth controller, on the other hand, has a different model of the large-scale system. This situation is called 'multimodelling'. These problems have been investigated for deterministic and stochastic systems (Khalil, 1978, 1979, 1980a,b, 1981a,b; Khalil and Kokotovic, 1978, 1979a,b,c, 1980; Kokotovic, 1981; Saksena and Cruz, 1981a,b, 1982, 1985a,b; Saksena and Basar, 1982, 1986; Saksena et al., 1984; Abed, 1985b; Ladde and Rajyalakshmi, 1985; Silva-Madriz, 1986; Silva-Madriz and Sastry, 1984, 1986; Gajic and Khalil, 1986; Ladde and Sirisaengataksin, 1986). We note that in multi-parameter systems we treat the systems in which the parameters are of the same order and do not allow multi-time-scale assumption (Khalil, 1978).

1.3.12   *Differential games*

In the design of multi-input control problems, the objectives in the optimal policy may be met by formulating the control problem as a differential game. In a general differential game there are several players, each trying to minimise his individual cost functional. Each player controls a different set of inputs to a single system, described by a differential equation. Three kinds of strategies are considered, Pareto optimal, Nash equilibrium and Stackelberg.

In the case of a two-player Nash game, we have

$$\dot{x} = f(x,z,u_1,u_2,t), \quad x(t=0) = x(0) \tag{1.28a}$$

$$\varepsilon \dot{z} = g(x,z,u_1,u_2,t), \quad z(t=0) = z(0) \tag{1.28b}$$

and

$$J_i = \int_0^T L_i(x,z,u_1,u_2,t)dt, \quad i = 1,2 \tag{1.29}$$

Necessary conditions for closed-loop solution are

$$\dot{p}_i = -\frac{\partial H_i}{\partial x} - \left(\frac{\partial \Psi_j}{\partial x}(t,x,z)\right)' \frac{\partial H_i}{\partial u_j}, \quad p_i(T) = 0 \tag{1.30a}$$

$$\varepsilon \dot{q}_i = -\frac{\partial H_i}{\partial z} - \left(\frac{\partial \Psi_j}{\partial z}(t,x,z)\right)' \frac{\partial H_i}{\partial u_j}, \quad q_i(T) = 0 \tag{1.30b}$$

and $u_i = \Psi_i(t,x,z)$ minimises the Hamiltonian $H(x,z,t,u_i,\Psi_j)$, $i,j = 1,2$; $i \neq j$. For open-loop, expressions (1.30) are replaced by

$$\dot{p}_i = -\frac{\partial H_i}{\partial x}, \quad p_i(T) = 0 \tag{1.31a}$$

$$\varepsilon \dot{q}_i = -\frac{\partial H_i}{\partial z}, \quad q_i(T) = 0 \tag{1.31b}$$

The main question investigated is one of well-posedness whereby the limit of the performance using the exact strategies is compared to the limit of the performance using the simplified strategies. The simplified solution is said to be well posed if the two limits are equal. This and other aspects of differential games have been treated by Khalil (1978, 1980a,b), Gardner and Cruz (1978), Khalil and Kokotovic (1978, 1979a,b), Salman and Cruz (1979, 1983), Khalil and Medanic (1980), Saksena and Cruz (1981a,b, 1982, 1984, 1985a,b), Ozguner (1982), Saksena and Basar (1982, 1986), Saksena *et al.* (1984).

A very useful application of singular perturbations and differential games is to pursuit-evasion problems in aerospace systems (Farber and Shinar, 1980; Shinar *et al.*, 1980; Shinar, 1981, 1983, 1985; Shinar and Farber, 1984).

1.3.13 *Time-optimal control problems*

The majority of results in singularly perturbed control problems is concerned with unconstrained control (Kokotovic *et al.*, 1976; Saksena *et al.*, 1984). The time-optimal control of linear singularly perturbed systems is an interesting and important problem (Collins, 1973; Kokotovic and Haddad, 1975a,b). Here we consider the problem of steering system (1.1) from any state to origin in minimum time subject to the constraint on u as $|u_j| \leqslant 1$, $j = 1, \ldots, r$.

We note that system (1.1) can be decoupled into the slow and fast subsystems (1.15) and (1.16). Thus the control is composed of a slow control, primarily dependent on slow dynamics, followed by a fast control primarily dependent on fast dynamics. The implication of the two-time-scale property of the system is that the time-optimal control strategy should first concentrate on steering the slow states near to their origin and then concentrate on rapidly steering the fast states to their desired final state while steering the slow states the remaining small distance to their final state.

The results dealing with time-optimal control of singularly perturbed systems are Collins (1973), Akulenko (1975), Freedman and Kaplan (1975, 1977), Kokotovic and Haddad (1975a,b), Javid (1978a), Berkey and Freedman (1976, 1979a,b), Freedman (1976), Freedman and Granoff (1976), Gichev and Dontchev (1979), Halanay and Mirica (1979), Javid and Kokotovic (1977), Berkey *et al.* (1978), Dontchev and Veliov (1985).

1.3.14   Stochastic systems

The filtering and stochastic control of singularly perturbed systems requires cautious treatment owing to the fact that the input white noise process fluctuates faster than the fast dynamic variables, no matter how small ε is. In the limit as $\varepsilon \to 0$, the fast variables themselves tend to white-noise processes, thus losing their significance as physically meaningful dynamic variables (Haddad and Kokotivic, 1977).

The singular perturbation formulation of systems with white-noise inputs appears as

$$\dot{x} = A_{11}x + A_{12}z + G_1 w \tag{1.32a}$$

$$\varepsilon \dot{z} = A_{21}x + A_{22}z + G_2 w \tag{1.32b}$$

where $w(t)$ is white Gaussian noise. The main desire in singular perturbation analysis is the simplification in the solution of (1.32). In this case, setting $\varepsilon = 0$ in (1.32) is inadequate for achieving the desired simplification, since

$$z^{(0)} = -A_{22}^{-1}(A_{21}x^{(0)} + G_2 w) \tag{1.33}$$

has a white-noise component, and, therefore has, infinite variance. Thus $z^{(0)}$ cannot serve as an approximation to z in the mean-square sense. Under the assumption

$$R_e\lambda(A_{22}) < 0, \quad R_e\lambda(A_{11} - A_{12}A_{22}^{-1}A_{21}) < 0 \tag{1.34}$$

the mean-square convergence

$$\lim_{\varepsilon \to 0} E\{(x - x_d)(x - x_d)'\} = 0 \tag{1.35a}$$

$$\lim_{\varepsilon \to 0} E\{(z - z_d)(z - z_d)'\} = 0 \tag{1.35b}$$

is demonstrated for x_d and z_d defined by

$$\dot{x}_d = (A_{11} - A_{12}A_{22}^{-1}A_{21})x_d + (G_1 - A_{12}A_{22}^{-1}G_2)w \tag{1.36a}$$

$$\varepsilon\dot{z}_d = A_{21}x_d + A_{22}z_d + G_2w \tag{1.36b}$$

For the linear filtering of system (1.32) with respect to the observations

$$y = C_1x + C_2z + v \tag{1.37}$$

where $v(t)$ is a white Gaussian noise independent of the process noise $w(t)$, the Kalman filter can be approximately decomposed into two filters in different time scales, thereby yielding estimates of slow and fast states (Haddad, 1976).

Several other aspects of the singularly perturbed stochastic problems and nonlinear stochastic control are found in many works (Kushner, 1965, 1967, 1984; Dorato *et al.*, 1967; Fleming, 1971, 1974; Haddad and Kokotovic, 1971, 1977; Holland, 1976, 1981; Haddad, 1976; Teneketzis and Sandell, 1977; Altshuler and Haddad, 1978; Blankenship, 1979; Khalil, 1978; Sebald and Haddad, 1978, 1979; Tsai, 1978; Price 1979; Kortum, 1979; Soliman and Ray, 1979a,b; Schuss, 1980; Bensoussan 1981; Hopkins and Blankenship, 1981; Sastry and Hijab, 1981; Shaked and Bobrovsky, 1981; Bobrovsky and Schuss, 1982; Saksena and Basar, 1982, 1986; Singh, 1982; Willsky *et al.*, 1982; Ardema, 1983a; Sastry, 1983; Gajic, 1986; Khalil and Gajic, 1984; Krtolica, 1984; El-Ansary and Khalil, 1986; Gajic and Khalil, 1986).

A related problem deals with Markov chains. Here finite-state continuous Markov processes with weak interactions are modelled as singularly perturbed systems. Two-time-scale expansions simplify the cost equations and lead to decentralised optimisation algorithms. These are discussed in Gaitsgori and Pervozvanskii (1975, 1980), Delebeque and Quadrat (1978, 1981), Phillips and Kokotovic (1981), Coderch *et al.* (1983a,b), Delebeque (1983), Delebeque *et al.* (1984), Saksena and Cruz (1985a), Dygas *et al.* (1986).

1.3.15 *Adaptive control systems*

The study of robustness of model-reference adaptive schemes in the presence of singular perturbations has recently been initiated (Ioannou, 1981). The general formulation examines situations when the order of the model is equal to the order of the slow part of the unknown plant and the model-plant 'mismatch' is due to the fast part of the plant. A fundamental requirement for the feasibility of an adaptive scheme is that it be robust, that is, tolerates a certain model-plant mismatch. The singular perturbation parameter ε is a convenient parameterisation of this mismatch because it allows an asymptotic analysis using the limit as $\varepsilon \to 0$. In our formulation adaptive schemes are designed for the reduced-order plants and then applied to the actual plants. They are considered robust if the error in their performance, due to model-plant mismatch is $0(\varepsilon)$. The details are given in the works of Ioannou (1981, 1986), Anderson and Johnson (1982), Ioannou and Kokotovic (1982, 1983, 1984a,b, 1985), Ioannou and Johnson (1983), Reidle and Kokotovic (1985, 1986), Rohrs *et al.* (1985).

1.3.16   *Distributed parameter systems*

Many engineering systems exhibit a distributed-parameter nature and, in order to be accurately modelled, they must be described by partial differential equations. The state space for a distributed parameter system (DPS) has infinite dimension and reduced-order modelling is a welcome feature in controller synthesis. The design of finite-dimensional controllers for DPS and the analysis of their closed-loop stability by singular perturbation techniques are the important issues in DPS.

The singularly perturbed Cauchy problem is given as (Asatani, 1974)

$$\frac{\partial x(w,\varepsilon,t)}{\partial t} = A_{11}x(w,\varepsilon,t) + A_{12}z(w,\varepsilon,t) \tag{1.38a}$$

$$\varepsilon\frac{\partial z(w,\varepsilon,t)}{\partial t} = A_{21}x(w,\varepsilon,t) + A_{22}z(w,\varepsilon,t) \tag{1.38b}$$

with appropriate auxiliary data and $w \in \Omega$.

The anaysis of singularly perturbed DPS depends on that of partial differential equations, which has rich literature. However, a few only are mentioned here (Bobisud, 1967; Hoppensteadt, 1971; Eckhuas, 1973; Nayfeh, 1973; Comstock, 1975; Fife, 1976; Grabmuller, 1978; Schuss, 1980).

The details regarding singular perturbations in DPS can be found in Asatani (1974, 1976), Asatani *et al.* (1977), Lions (1973a,b, 1978, 1981), Soliman and Roy (1979b), Balas (1982a,b, 1984), Harten (1984), Stavroulakis and Tzafestas (1982) and Tzafestas (1984).

Once again we note that the problems mentioned in Sections 1.3.1–1.3.6 only are discussed in detail in this book whereas the rest of the problems are not dealt with in detail owing to shortage of space and other considerations. The reader is encouraged to refer to the works listed for further details.

1.4   Applications

Singular perturbation problems arise in a natural way in many fields of applied mathematics, engineering and biological sciences such as fluid dynamics, electrical and electronic circuits and systems, electrical power systems, aerospace systems, nuclear reactors, biology and ecology. Here we shall briefly mention this wide application and the associated references (Verhulst, 1979; Eckhaus and de Jager, 1982).

1.4.1   *Fluid dynamics*

Fluid dynamics played an important role in developing singular perturbation methodology. Prandtl (1905) pointed out that, for high Reynolds number, the tangential velocity in incompressible viscous flow past an object changes very rapidly from its zero value to the value given by the solution of the Navier–Stokes equation. This change takes place in a layer near the boundary, called the boundary layer, of thickness ε proportional to the inverse of the square root of Reynolds number. Friedrichs (1955) developed Prandtl's boundary-layer theory in fluid mechanics.

In studying singular perturbation problems in fluid dynamics, Kaplun (1967) introduced several notions such as degenerate solution, limit process, nonuniform convergence, boundary layer, inner and outer expansions and matching. Fluid dynamics is still an abundant source of many challenging perturbation problems. Attention is drawn to the following important works on singular perturbations in fluid dynamics (Prandtl, 1905; Friedrichs, 1955; Kaplun, 1967; Reiss, 1961; Van Dyke, 1964; Cole, 1968; Fraenkel, 1969a,b,c; Bush, 1971; Lagerstrom and Casten, 1972; Nayfeh, 1973; McLeod and Porter, 1974; Rudraiah *et el.*, 1974; James, 1975; Lagerstrom, 1975; Matkowsky and Siegmann, 1976; Stewartson, 1976; Shepherd, 1978a,b; Callegari and Ting, 1978; Cohen *et al.*, 1978; Cook and Cole, 1978; MacGillivray, 1983; Tavantzis *et al.*, 1978; Skinner, 1981; Eckhaus and de Jager, 1982; Howes, 1982; Hsiao and MacCamy, 1982; Fisher *et al.*, 1985).

1.4.2   *Electrical circuits and machines*

In dynamic modelling of most of the electrical and electronic circuits and electrical machines, we often neglect small parasitic elements such as time constants, resistances, inductances, capacitances, masses and moments of inertia, thereby paving the way for singular perturbation analysis. Several authors have studied the various aspects of electrical circuits including the van der Pol equation arising from a simple *RC* circuit (Bjurel *et al.*, 1970; Desoer, 1970, 1977; Desoer and Shensa, 1970; Shensa, 1971; Wilde and Kokotovic, 1972a; Sannuti, 1976, 1978, 1981; Khalil and Kokotovic, 1978; Campbell, 1981; Matsumoto *et al.*, 1981; Newcomb, 1981; Sastry and Desoer, 1981; Chow, 1982; Sastry, 1983; Eitelberg, 1985), electronic circuits consisting of transistors, diodes, tunnel diodes, and so on (Miranker, 1962; Brayton *et al.*, 1966; Bjurel *et al.*, 1970; Miranker and Hoppensteadt, 1974; Markowich and Ringhoffer, 1984; Markowich, 1986), electrostatics (Abraham-Sharuner, 1974; Dickmann *et al.*, 1980; Chew and Kong, 1982), electromagnetics (Fraenkel, 1969c; Seshadri, 1976; 1977a,b; Asfar and Nayfeh, 1983; Magnan and Goldstien, 1983) and electrical machines consisting of DC motors, amplidynes, synchronous generators and transformers (Vasileva, 1963; Sannuti and Kokotovic, 1969a,b; Bjurel *et al.*, 1970; Jamshidi, 1972, 1974, 1976; Kokotovic and Yackel, 1972; Yackel and Kokotovic, 1973; Chow and Kokotovic, 1978a; Dragan and Halanay, 1982; Naidu and Sen, 1982; Zaid *et al.*, 1982; Haller and Iung, 1983; Sauer *et al.*, 1984).

1.4.3   Electrical power systems

In general, power system models are of high dimensionality and are stiff owing to the interacting dynamic phenomena of widely differing speeds. For example, voltage and frequency transients range from intervals of seconds (voltage regulator, sppeed governor and shaft-energy storage) to several minutes (prime mover, fuel transfer times and thermal-energy storage). Thus, in power-system dynamics, we see a strong motivation for the application of time-scale and singular perturbation methodologies which are known for their features of order reduction and stiffness relief (time-scale separation). Several authors have contributed to the study of dynamics, control and optimisation of power systems including the very useful concept of 'coherency' (Kokotovic, 1975; Alden and Nolan, 1976; Chow and Kokotovic, 1976; Reddy and Sannuti, 1976; Chow *et al.*, 1978; Delebeque and Quadrat, 1978; Khalil and Kokotovic, 1978; Pai and Adgoanker, 1983; Avramovic *et al.*, 1980; Kokotovic *et al.*, 1980; Sastry and Varaiya, 1980, 1981; Winkelman *et al.*, 1980, 1981; Arapostathis *et al.*, 1982; Campbell and Rose, 1982; Chow, 1982; DiCaprio, 1982; Dillon, 1982; Khalil and Saberi, 1982; Peponides *et al.*, 1982; Fossard *et al.*, 1983; Cori and Maffezoni, 1984; Dorsey and Schlueter, 1984; Chow and Kokotovic, 1985; Rao *et al.*, 1985).

1.4.4   Aerospace systems

In most of the aerospace problems, no singular perturbation parameter appears explicitly on physical considerations. In such cases a parameter may be artificially inserted to suppress the variables in the equations which are expected to have relatively negligible effects. For example, in a flight dynamics problem for a manned vehicle, a complete set of equations of motion would consist of the coupled system of the six equations of rigid-body motion of the vehicle as a whole, the equations describing the dynamics of the control systems, the pilot's arm and foot, etc. It is obvious that many of these effects can be neglected if, say, the vehicle trajectory is the only thing of interest (Ardema, 1977). In particular, in a minimum time-to-climb (MTC) problem, it has been found in practical problems for supersonic aircraft that the flight-path angle is relatively fast as compared with altitude, which, in turn, is fast compared to specific energy. It is this separation of the 'speed' of the variables that motivates the formulation of singular perturbation problems by the artificial (forced) insertion of the singular perturbation parameter. This is often referred to as the forced singular perturbation (FSP) technique (Sesak *et al.*, 1979).

In some aircraft problems, we consider the vertical-plane MTC problem, where the drag D is much less than the lift L and we define a natural perturbation parameter $\varepsilon = D/L$ (Mehra *et al.*, 1979).

Optimization of aircraft trajectories has been a challenging problem due mainly to the high dimensional nonlinear TPBVP requiring excessive amount of computation. The ultimate goal is to obtain optimal feedback control which can be computed on line and stored on board. The singular perturbation and time-scale techniques have been successfully applied to a variety of aerospace problems, like trajectory optimisation problems, on-line optimal control problems, ramjet-powered air-to-air missiles and pursuit-evasion problems (Kelley, 1964, 1970*a,b*, 1971*a,b*, 1972, 1973, 1978; Spriggs *et al.*, 1969; Bjurel *et al.*, 1970; Kelley and Edelbaum, 1970; Ardema, 1976, 1977, 1979, 1980, 1983*a,b*; Balachandra, 1975; Calise, 1976, 1977, 1978, 1979, 1980, 1981, 1984, 1985; Teneketzis and Sandell, 1977; Balas, 1978; Mehra *et al.*, 1979; Farber and Shinar, 1980; Shinar *et al.*, 1980; Sridhar and Gupta, 1980; Chakravarty, 1985; Shinar, 1981, 1983, 1985, Cliff *et al.*, 1982; Gracey *et al.*, 1982; Houlihan *et al.*, 1982, Shinar and Negrin, 1983; Price *et al.*, 1984; Rajan and Ardema, 1984, 1985; Shinar and and Farber, 1984; Ardema and Rajan, 1985*a,b*; Breakwell *et al.*, 1985, Moerder and Calise, 1985; Xu, 1985; Gajic, 1986).

1.4.5   *Chemical reaction, diffusion and reactor control systems*

Here we include problems in chemical reactions, diffusion problems and reactor control problems. Mixtures of substances which diffuse and react are described by systems of coupled nonlinear diffusion equations. For example, the behaviour of a single reactant species inside a porous catalyst pellet can be described by a singularly perturbed boundary-value problem, where the small parameter is the inverse of the Thiele modulus (which measures the effect of diffusion as opposed to reaction) (Howes, 1977).

Nuclear-reactor kinetics is concerned with problems including oscillations caused by perturbed coolant flow or control-rod movements, excursions due to loss of coolant, accident studies and so on. The duration of transients extends from a fraction of a second to a few minutes, thus leading in a natural way to time-scale analysis. In a simple case of reactor dynamics we assume the lumped parameters and consider a 'point reactor' or 'one-point' approximation. In error estimation of prompt jump approximation, if the relative rate of change of the reactor power in the mean prompt generation time is sufficiently small, the 'point reactor' model can be fitted into the singularly perturbed structure (Asatani, 1974).

Notable contributions towards the application of singular perturbation and time-scale methods to problems in chemical reactions, diffusion problems and reactor dynamics and control problems are Hirschfelder *et al.* (1953), Birkhoff (1966), Burghardt and Zaleski (1968), O'Malley (1968), Hendry and Bell (1969), Bjurel *et al.* (1970), Hendry (1970, 1971), Bentwich (1971), Cohen (1971, 1973, 1974, 1976), Chen and O'Malley (1972, 1974), Keller (1973), Mika (1976), Reddy and Sannuti (1975), Sannuti and Reddy (1973), Asatani (1974, 1976), Kao and Bankoff (1974), Aris (1975), Parter *et al.* (1975), Auchmuty and Nicholis (1976), Fife (1976, 1979), Kando and Iwazumi (1978, 1981), Cohen *et al.* (1977), Come (1977), Ho and Hsiao (1977), Howard and Koppel (1977), Howes (1977), Butuzov (1978), Aris (1979), Bobisud (1979), Neu (1979, 1980), Soliman and Ray (1979a,b), Bobisud and Christenson (1980), Cope (1980), Kuruoglu *et al.* (1981), Maginu (1981), Kath and Cohen (1982), Rinzel and Terman (1982), Stavroulakis and Tzafestas (1982), Harten (1984), Shimizu and Matsubara (1985).

Stiff differential equations are obtained in theoretical studies of reaction kinetic models of systems where the rate constants of the reactions involved are widely separated. Such systems occur in many fields of chemistry like analytical chemistry (Christensen, 1969; Mathuna, 1971; Fowler, 1977; Frank and Wendt, 1982; Renardy, 1984).

1.4.6   Biology and biochemistry

An interesting application of singular perturbation analysis is in the fields of biology and biochemistry. For example, in the study of Michaelis–Menten kinetics of enzyme reactions, the small initial ratio of enzyme to substrate concentration is taken as the singular perturbation parameter to analyse a simple enzymatic reaction having the irreversible conversion of the substrate into the product (Heineken *et al.*, 1967).

Singular perturbation results have also been used to study a general model of a biological process (e.g. nerve impulse, heart beat, muscle contraction) consisting of slow and fast phenomena. The slow (fast) equations correspond to subprocesses whose rates are slow (fast) relative to the rate of the primary phenomenon. For example, the Hodgkin–Huxley model of nerve impulse transmission consists of a nonlinear diffusion equation coupled with one fast equation and two slow equations (Carpenter, 1977).

Works concerning the application of singular perturbation theory to various problems in biology and biochemistry include Heineken *et al.* (1967), Aris (1968), Barcilon *et al.* (1971), Grasman and Veling (1973), Cohen (1974, 1976), Casten *et al.* (1975), Hoppensteadt (1975), Rubinow (1975), Grasman and Jansen (1979), Carpenter (1977), Cronin (1977), Fowler (1977), Murray (1977), Bell and Cook (1978), Meiske (1978), Wollkind and Logan (1978), Argemi *et al.* (1979), Fletcher (1980), Hunt (1980), Mottani and Rothe (1980), Perelson (1980), Perelson and Delisi (1980), Salateh *et al.* (1980), Do and Greenfield (1981), Ignetik and Deakin (1981), Frank and Wendt (1982), Tanyi (1982), Holden and Winlow (1983), Keener (1983), Lehman and Stark (1983), Schauer and Heinrich (1983), Grasman (1984), and Nipp (1985).

1.4.7   Ecology

Ecology is concerned with the study of the relationships between population and environment. The growth of any population in a restricted environment must eventually be limited by a shortage of resources (Hoppensteadt, 1975). For example, the species-resource logistic model is described by singularly perturbed differential equations (Naidu and Rajagopalan, 1979). The study of the singular perturbation method in ecology is due to Grasman and Veling (1973), Hoppensteadt (1975), Grasman *et al.* (1976), Logan *et al.* (1976), Ludwig (1976), Wollkind and Logan (1978), Hoppensteadt and Miranker (1977), Lakin *et al.* (1977), Wollkind (1977), Wollkind *et al.* (1978), Grasman (1979*a,b*), Naidu and Rajagoplan (1979), Gardner and Smoller (1983).

1.4.8   Other applications

There are a number of other interesting and challenging applications of singular perturbation and time-scale methodologies in a variety of fields (Verhulst, 1979; Eckhaus and de Jager, 1982). These include soil mechanics (Li, 1972; Dicker and Babu, 1974*a,b*), structural mechanics (Crespo da Silva and Davis, 1977; Anderson and Hallauer, 1981; Flaherty and O'Malley, 1982), celestial mechanics (Nayfeh, 1965; Verhulst, 1975, 1979; Smith, 1975; Sarlet, 1978), thermodynamics (Cooper, 1971, 1975; Harten, 1979*a*; Polk, 1976; Kassoy, 1976; Rudraiah and Musuoka, 1982), plates and shells (Knowles and Messick, 1964; Schuss, 1980; Verhulst, 1979), elasticity (Matkowsky and Reiss, 1977); lubrication (DiPrima, 1968; Steinmetz, 1974; Capriz and Cimmatti, 1978; Shepherd, 1978*b*), vibration (Smith, 1975), renewal processes (Hoppensteadt, 1983), magnetohydrodynamics (MHD) (Cook *et al.*, 1972; Harten, 1979*b*; Gustafsson, 1980), oceanography (Carrier, 1970; Ruijter, 1979; Hermans, 1982), acoustics (Einaudi, 1969; Solomon and Comstock, 1973), welding (Andrews and Atthey, 1975, 1976), wave equation (Davis and Reiss, 1970; Comstock, 1975; Greenberg, 1976), production inventory system (Bradshaw *et al.*, 1982), quantum mechanics (Fattorini, 1985), communication theory (Schuss, 1980), wave propogation (Barker, 1984; Eckhaus and de Jager, 1982; Kriegsman and Reiss, 1983), ionisation of gases (Hilhorst, 1982), loading bridge (Litz and Roth, 1981), biped locomotion (Miyazaki and Arimoto, 1979, 1980), freight-car hunting (Whitman, 1980), reliability (Schuss, 1980; Krtolica, 1984), laser (Eckhaus *et al.*, 1985), flexible manipulators and robotics (Chernousko, 1983; Chernousko and Shamaev, 1983; Marino and Nicosia, 1984; Khorasani and Kokotovic, 1985, 1986).

Singular perturbation method

In this Chapter we present the singular perturbation method for continuous and discrete control systems. Firstly, in Section 2.1 we start with the definition of singular perturbation and discuss its various characteristic features such as order reduction, boundary layer, degeneration and loss of auxiliary conditions as applicable to continuous systems described by ordinary differential equations. The same characteristic features are discussed for difference equations describing discrete systems in Section 2.2. We then describe in Section 2.3 Vasileva's method of obtaining asymptotic expansions in terms of outer, inner and intermediate series for singularly perturbed initial-value problems of continuous nature. The boundary-layer method is also discussed where the approximate solution is given by the outer series and a boundary-layer correction which is equivalent to the difference between the inner and intermediate series. The boundary-value problem is treated and the structural properties—stability, controllability and observability—are discussed. Section 2.4. deals with the development of the singular perturbation method for initial- and boundary-value problems arising in discrete systems.

Sufficient number of examples are presented to illustrate the various features and techniques. The final Section is devoted to conclusions and discussion. In particular, we try to scan the literature and other types of problems in control. We conclude by bringing out various characteristic features of singular perturbations in differential and difference equations describing continuous and discrete control systems, respectively.

2.1 Singular perturbations in continuous systems

2.1.1 Definition and features

A basic problem in control system theory is the mathematical modelling of

* The portions on discrete systems in this chapter are based on the research monograph, NAIDU, D. S., and RAO, A. K. (1985): 'Singular perturbation analysis of discrete control systems'. Lecture Notes in Mathematics, Vol. 1154 (Springer–Verlag, Berlin). The permission given by Springer–Verlag is hereby acknowledged.

a physical system. The realistic representation of many systems calls for high-order differential equations. The presence of some 'parasitic' parameters such as small time constants, masses, moments of inertia, resistances, inductances and capacitances is often the cause of the increased order of these systems. Broadly speaking, a system in which the suppression of small parameters is responsible for the degeneration of dimension is termed a 'singularly per-turbed' system. Alternatively, a problem described by a differential equation involving a small parameter ε is called a singular perturbation problem if the order of the differential equation becomes lower for $\varepsilon = 0$ than for $\varepsilon \neq 0$ (Wasow, 1965). Obviously, the small parameter ε multiplies the highest deriva-tives of the differential equation. On the contrary, in a normal or regular perturbation problem, the small parameter is not responsible for the reduction in the order of the problem. In order to have a precise definition of singular perturbation, let us consider the following two simple examples:

Example 2.1
Consider a second-order boundary value problem (BVP)

$$\varepsilon \frac{d^2x}{dt^2} + \frac{dx}{dt} = 0 \tag{2.1}$$

with boundary conditions

$$x(0) = x_i; \quad x(1) = x_f \tag{2.2}$$

and the small parameter ε multiplying the highest derivative.
   The exact solution of (2.1) along with (2.2) is

$$x(t,\varepsilon) = \frac{(x_i - x_f)\exp(-t/\varepsilon) + x_f - x_i\exp(-1/\varepsilon)}{1 - \exp(-1/\varepsilon)} \tag{2.3}$$

For positive values of ε,

$$\lim_{\varepsilon \to 0_+} [x(t,\varepsilon)] = x_f, \quad 0 < t \leq 1 \tag{2.4}$$

The convergence is uniform in every closed interval $0 < \delta \leq t \leq 1$, but not in the whole interval $0 \leq t \leq 1$. The solution $x^{(0)}(t) = x_f$ is that of the reduced or degenerate problem

$$\frac{dx^{(0)}}{dt} = 0, \quad x^{(0)}(1) = x_f \tag{2.5}$$

obtained by letting $\varepsilon = 0$ in (2.1).
   In a narrow interval of width $0(\varepsilon)$, the solution (2.3) changes rapidly from $x(0,\varepsilon) = x_i$ to a value neighbouring $x^{(0)}(t) = x_f$ with difference of the order of ε. Such an interval is often called a 'boundary layer', because of a mathema-

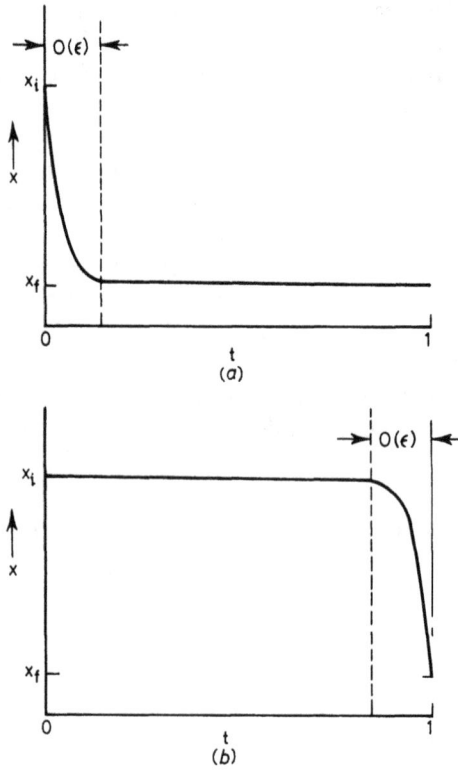

Fig. 2.1   *Boundary layer of Example 2.1*
   a ε > 0
   b ε < 0

tical analogy with the boundary layers of fluid dynamics (Van Dyke, 1964). It is also termed 'layer of rapid transition' (Lagerstrom and Casten, 1972). This situation of a boundary layer occurring at $t = 0$ is shown in Fig. 2.1.

For negative values of ε, we have

$$\lim_{\varepsilon \to 0_-} [x(t,\varepsilon)] = x_i, \quad 0 \leqslant t < 1 \tag{2.6}$$

and the boundary layer occurs at $t = 1$ (Fig. 2.1).

The reduced problem (2.5), being of first order, cannot be expected to satisfy both the given boundary conditions. This 'loss' of some of the boundary conditions is a typical characteristic feature of singular perturbations.

Example 2.2
Consider another BVP,

$$\varepsilon \frac{d^2 x}{dt^2} + \frac{dx}{dt} + x = 0 \tag{2.7}$$

with boundary conditions

$$x(t = 0) = x_i, \quad x(t = 1) = x_f \tag{2.8}$$

and the small parameter ε multiplying the highest derivative.
The solution of (2.7) and (2.8) is

$$x(t,\varepsilon) = \frac{[(x_f - x_i \exp(m_2)) \exp(m_1 t) + (x_i \exp(m_1) - x_f) \exp(m_2 t)]}{\exp(m_1) - \exp(m_2)} \tag{2.9}$$

where

$$m_1 = -1/2\varepsilon + (1 - 4\varepsilon)^{0.5}/2\varepsilon$$
$$= -1 + 0(\varepsilon) \tag{2.10}$$

and

$$m_2 = -1/2\varepsilon - (1 - 4\varepsilon)^{0.5}/2\varepsilon$$
$$= -1/\varepsilon + 1 + 0(\varepsilon) \tag{2.11}$$

are the roots of the characteristic equation

$$\varepsilon m^2 + m + 1 = 0 \tag{2.12}$$

As ε tends to zero either from positive or negative values, we have

$$\lim_{\varepsilon \to 0_+} [x(t,\varepsilon)] = x_f \exp(1 - t), \quad 0 < t \leq 1 \tag{2.13}$$

$$\lim_{\varepsilon \to 0_-} [x(t,\varepsilon)] = x_i \exp(-t), \quad 0 \leq t < 1 \tag{2.14}$$

This situation is depicted in Fig. 2.2 which brings out clearly the fact that the solution has a nonuniform convergence and changes rapidly inside the boundary layers.
The degenerate (or reduced order) problem,

$$\frac{dx^{(0)}}{dt} + x^{(0)} = 0 \tag{2.15}$$

obtained by suppressing the small parameter ε in (2.7), has the boundary condition $x^{(0)}(t = 1) = x_f$ if ε tends to 0_+ and $x^{(0)}(t = 0) = x_i$ if ε tends to 0_-. In either case, one boundary condition has to be sacrificed in the process of degeneration.

$$\lim_{t \to 0} \lim_{\varepsilon \to 0} [x(t,\varepsilon)] = x^{(0)}(0) = x_f e \tag{2.16}$$

$$\lim_{\varepsilon \to 0} \lim_{t \to 0} [x(t,\varepsilon)] = x_i \tag{2.17}$$

From (2.10) and (2.11), it is clear that, for smaller values of ε, one root is large compared to the other root. This means that the solution consists

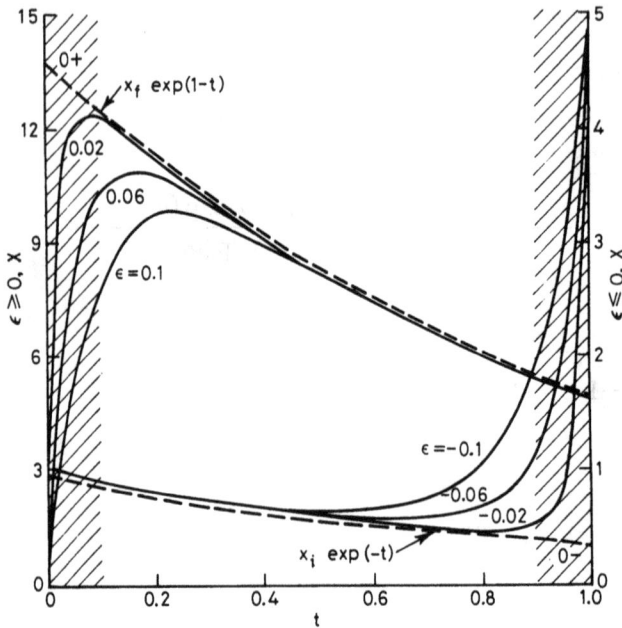

Fig. 2.2 *Degeneration of solution of Example 2.2*
——— solution for $\varepsilon \neq 0$
- - - - - solution for $\varepsilon = 0$
Shaded regions are the boundary layers
Note: $\varepsilon = 0_+$ will satisfy the final condition, while $\varepsilon = 0_-$ will satisfy the initial condition

of 'slow' and 'fast' components (modes) corresponding to 'small' and 'large' eigenvalues, respectively.

The above two classical examples bring out clearly the following important features of singular perturbations:

(i) The perturbation is from $\varepsilon = 0$ to $\varepsilon \neq 0$ and hence the problem (2.1) or (2.7) is called a 'singularly perturbed (singular perturbation) problem,' described by a differential equation with the small parameter ε multiplying the highest derivative.
(ii) The solution has nonuniform convergence.
(iii) There exists a boundary layer where the solution changes rapidly.
(iv) The perturbation process (from $\varepsilon = 0$ to $\varepsilon \neq 0$) and degeneration process (from $\varepsilon \neq 0$ to $\varepsilon = 0$) are opposite in action and effect.
(v) The degenerate problem, also called the 'unperturbed' problem, is of reduced order and cannot satisfy all the boundary conditions given for the original, full, or perturbed problem.
(vi) The singularly perturbed problem possesses a two-time-scale property

due to the simultaneous presence of slow and fast phenomena and the problem is said to be 'stiff'.

2.1.2 *Linear control systems*

From the systems and control point of view, we now introduce the concept of singular perturbations. Using the state-variable representation, a linear time-invariant system becomes

$$\dot{x} = A_{11}x + A_{12}z + B_1u, \quad x(t=0) = x(0) \tag{2.18a}$$

$$\varepsilon\dot{z} = A_{21}x + A_{22}z + B_2u, \quad z(t=0) = z(0) \tag{2.18b}$$

where x and z are m- and n-dimensional state vectors, u is an r-dimensional control vector. The matrices A_{ij} and B_i are of appropriate dimensions. The scalar, positive parameter ε represents all small parameters to be ignored. The system (2.18) is in the singularly perturbed form in the sense that by making $\varepsilon = 0$ in system (2.18) the degenerate system becomes

$$\dot{x}^{(0)} = A_{11}x^{(0)} + A_{12}z^{(0)} + B_1u, \quad x^{(0)}(t=0) = x(0) \tag{2.19a}$$

$$0 = A_{21}x^{(0)} + A_{22}z^{(0)} + B_2u, \quad z^{(0)}(t=0) \neq z(0) \tag{2.19b}$$

or

$$\dot{x}^{(0)} = (A_{11} - A_{12}A_{22}^{-1}A_{21})x^{(0)} + (B_1 - A_{12}A_{22}^{-1}B_2)u \tag{2.20a}$$

$$z^{(0)}(t) = -A_{22}^{-1}A_{21}x^{(0)} - A_{22}^{-1}B_2u \tag{2.20b}$$

The block diagram of the full, high-order, or perturbed system (2.18) and the degenerate, low-order, or unperturbed system (2.19) is shown in Fig. 2.3. Here we assume that the input u is independent of ε. Otherwise u becomes $u^{(0)}$. The effect of degeneration is not only to 'cripple' the order of the system from $(m+n)$ to m by 'dethroning' z from its original state variable status, but also to 'desert' its initial condition $z(0)$. This is a 'harsh punishment' on z for having a close association (multiplication) with the singular perturbation parameter ε. The fact that $z^{(0)}(0)$ need not, in general, be equal to $z(0)$, is readily seen from (2.20b). We assume that the matrix A_{22} is nonsingular.

From Fig. 2.3, we can also view the degeneration as equivalent to letting the forward gain of the system go to infinity.

Example 2.3
Consider an armature-controlled DC motor described by

$$\dot{x} = a_{12}z \tag{2.21a}$$

$$L\dot{z} = -a_{21}x - Rz + u \tag{2.21b}$$

where x, z and u are, respectively, speed, current and voltage, L and R are armature inductance and resistance, and a_{12} and a_{21} are some motor constants.

(a)

(b)

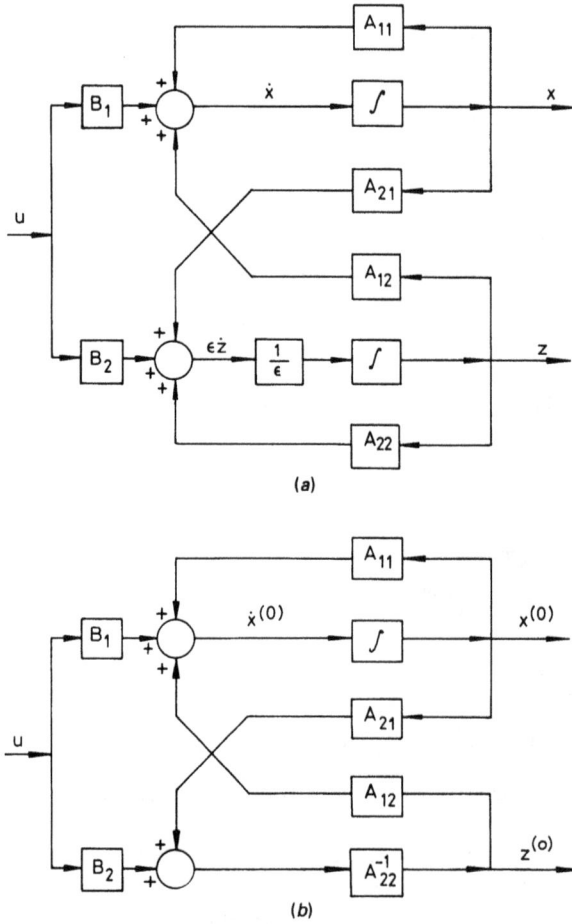

Fig. 2.3   *Singularly perturbed continuous control system*
   a Original system
   b Degenerate system

In many DC motors, the armature inductance L is the small parameter ($L = \varepsilon$) which is often neglected. Hence the simplified model becomes

$$\dot{x}^{(0)} = \frac{a_{12}}{R}(u^{(0)} - a_{21}x^{(0)}) \tag{2.22a}$$

$$z^{(0)} = (u^{(0)} - a_{21}x^{(0)})/R \tag{2.22b}$$

2.1.3   Nonlinear control systems

A singularly perturbed nonlinear system is of the form

$$\dot{x} = f(x,z,u,\varepsilon,t), \quad x(t=0) = x(0) \tag{2.23a}$$

$$\varepsilon\dot{z} = g(x,z,u,\varepsilon,t), \quad z(t=0) = z(0) \tag{2.23b}$$

We assume that, in a domain of interest, f and g are twice continuous differentiable functions of their arguments x, z, u, ε, t. By setting $\varepsilon = 0$, the $(m+n)$th-order differential equation (2.23) degenerates to an mth order differential equation and an nth order algebraic or transcendental equation. That is

$$\dot{x}^{(0)} = f(x^{(0)},z^{(0)},u^{(0)},0,t), \quad x^{(0)}(0) = x(0) \tag{2.24a}$$

$$0 = g(x^{(0)},z^{(0)},u^{(0)},0,t) \tag{2.24b}$$

We assume that one of the several solutions of (2.24b) is

$$z^{(0)} = p(x^{(0)},u^{(0)},t), \quad z^{(0)}(0) \neq z(0) \tag{2.24c}$$

Substituting (2.24c) in (2.24a), we get

$$\dot{x}^{(0)} = f[x^{(0)},p(x^{(0)},u^{(0)},t),u^{(0)},t]$$

$$= f^0(x^{(0)},u^{(0)},t) \tag{2.25}$$

Example 2.4
Consider a nonlinear system shown in Fig. 2.4. The small parameter $\varepsilon = 1/K$,

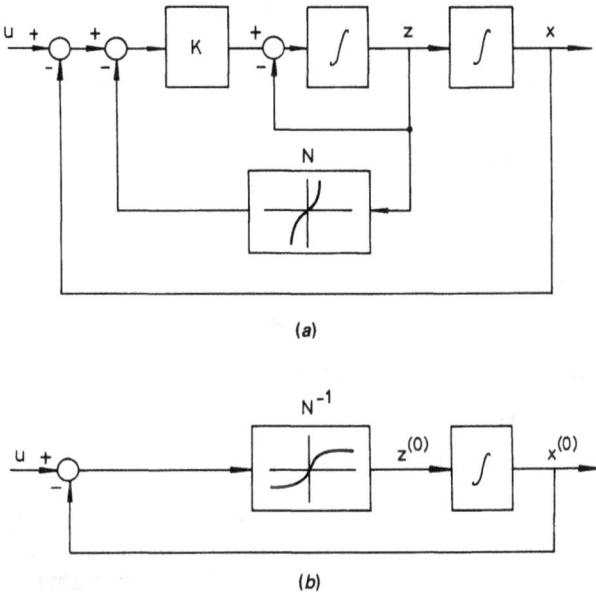

(a)

(b)

Fig. 2.4 *Nonlinear system with high gain of Example 2.4*
  a Full system
  b Degenerate system

where K is the amplifier gain and the nonlinear element N is tan z. The system equations are (Sannuti, 1968)

$$\dot{x} = z \tag{2.26a}$$

$$\varepsilon \dot{z} = -x - \varepsilon z - \tan z + u \tag{2.26b}$$

If the gain K is sufficiently large, $K = \infty$, then $\varepsilon = 0$. The degenerate system becomes

$$\dot{x}^{(0)} = z^{(0)} \tag{2.27a}$$

$$0 = -x^{(0)} - \tan z^{(0)} + u \tag{2.27b}$$

or $\quad \dot{x}^{(0)} = \tan^{-1}(u - x^{(0)}) = z^{(0)} \tag{2.28}$

A useful concept in singular perturbations is the speed of a variable. Consider the singularly perturbed system (2.23) or (2.18). We have $\dot{z} = g/\varepsilon$, if ε is very small and $g \neq 0$, then $\dot{z}$ is very large. We call x the 'slow' variable and z the 'fast' variable. These slow and fast variables make the system (2.18) or (2.23) stiff, which in a computational sense means that there are unreasonable restrictions on the discretising interval, requiring a larger number of mesh points for small values of ε in the region of the boundary layer. On the other hand, the degenerate system (2.19) or (2.24) is relieved of the stiffness.

The degenerate system has the advantages of order reduction and stiffness removal. But, it is unable to satisfy the given auxiliary conditions, and above all its performance may be far from satisfactory compared to the high-order system.

2.2 Singular perturbations in discrete systems

2.2.1 Definition and features

A discrete system is described by a difference equation. In order to get a clear idea of singular perturbations in difference equations let us consider

$$y(k + 2) + ay(k + 1) + \varepsilon y(k) = 0 \tag{2.29a}$$

$$\varepsilon y(k + 2) + ay(k + 1) + y(k) = 0 \tag{2.29b}$$

$$\varepsilon y(k + 3) + ay(k + 2) + by(k + 1) + \varepsilon y(k) = 0 \tag{2.29c}$$

with appropriate boundary conditions, where the parameter ε is small compared to a and b. All the above three types of difference equations, with the small parameter ε multiplying (appearing) the lowest (right), highest (left), and highest and lowest functions, are said to be in the singularly perturbed form, in the sense that

$$y^{(0)}(k + 2) + ay^{(0)}(k + 1) = 0 \tag{2.30a}$$

$$ay^{(0)}(k+1) + y^{(0)}(k) = 0 \tag{2.30b}$$

$$ay^{(0)}(k+2) + by^{(0)}(k+1) = 0 \tag{2.30c}$$

obtained by letting $\varepsilon = 0$ in (2.29) are of reduced order (Comstock and Hsiao, 1976; Rajagopalan and Naidu, 1980a). Note that this situation is different from that in the differential equations where the small parameter multiplying the highest derivative (appearing at the left end) only provides the singular perturbation character.

(i) *Initial-value problem (IVP)*
In particular, consider (2.29a) as an initial-value problem (IVP) with $y(k = 0) = y(0)$ and $y(k = 1) = y(1)$. The complete solution is

$$y(k) = \frac{[y(0)m_2 - y(1)]m_1{}^k + [y(1) - y(0)m_1]m_2{}^k}{m_2 - m_1} \tag{2.31}$$

where m_1 and m_2 are given by

$$m_1 = -\frac{a}{2} + \frac{a}{2}(1 - 4\varepsilon/a^2)^{0.5}$$

$$= -\varepsilon/a + 0(\varepsilon^2) \tag{2.32}$$

$$m_2 = -\frac{a}{2} - \frac{a}{2}(1 - 4\varepsilon/a^2)^{0.5}$$

$$= -a + \varepsilon/a + 0(\varepsilon^2) \tag{2.33}$$

Using (2.32) and (2.33) in (2.31) and neglecting terms with coefficients of ε we get the approximate solution as

$$y^{(0)}(k) = y(1)(-a)^{k-1} \tag{2.34}$$

Now consider (2.29a) with the small parameter ε made zero. Then the resulting degenerate problem is

$$y^{(0)}(k+2) + ay^{(0)}(k+1) = 0 \tag{2.35}$$

and can satisfy only one of the given two initial conditions. If (2.35) is solved with $y^{(0)}(1) = y(1)$, the solution is the same as given by (2.34). This also shows that in general $y^{(0)}(0) \neq y(0)$ and the boundary layer occurs at $k = 0$ (Comstock and Hsiao, 1976).

It is seen from (2.32) that, as ε is made smaller, the roots m_1 and m_2 tend to 0 and $-a$, respectively. This means that the solution (2.31) possesses 'fast' and 'slow' components, thus giving rise to a two-time-scale character.

Example 2.5
Consider

$$y(k+2) - 0.8\,y(k+1) - \varepsilon y(k) = 0$$

with $y(0) = 2.0$; $y(1) = 1.0$ and $\varepsilon = 0.08$. The roots are $m_1 = -0.09$ and $m_2 = 0.89$. Figure 2.5 shows the degeneration of the solution $y(k)$ and the formation of a boundary layer at $k = 0$.

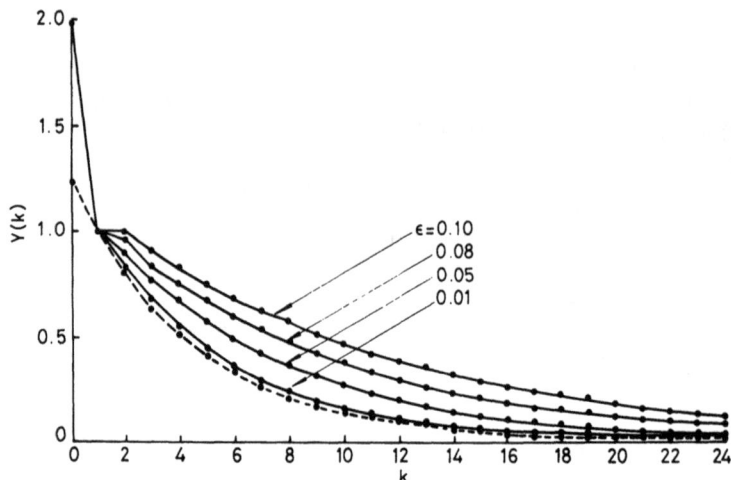

Fig. 2.5   *Degeneration of solution of Example 2.5*
————— exact solution $(\varepsilon \neq 0)$
– – – – – degenerate solution $(\varepsilon = 0)$

For the type of problem given by (2.29b), where the small parameter appears at the left end, we can consider it as a BVP only and not as an IVP as is easily seen from the stability considerations.

(ii)   *Boundary-value problem (BVP)*
Consider (2.29a) as a BVP with the given conditions $y(0)$, and $y(N)$. Then its solution is

$$y(k) = \frac{[y(0)m_2^N - y(N)]m_1^k + [y(N) - y(0)m_1^N]m_2^k}{m_2^N - m_1^N} \tag{2.36}$$

where m_1 and m_2 are given by (2.32) and (2.33). The approximate solution is given by

$$y^{(0)}(k) = y(N)\,(-a)^{K-N} \tag{2.37}$$

which is the same as that obtained by solving the degenerate problem (2.35) using the condition $y^{(0)}(N) = y(N)$. This means that, even in the case of a BVP for the type where the small parameter occurs at the right end, the condition $y(0)$ is lost in the process of degeneration and the boundary layer is associated with $k = 0$.

Consider the BVP for the type (2.29b) with $y(0)$ and $y(N)$. This structure is identical to that in differential equations. Its solution is given by (2.36) where m_1 and m_2 are given by

$$m_1 = -\frac{a}{2\varepsilon} + \frac{a}{2\varepsilon}(1 - 4\varepsilon/a^2)^{0\cdot 5}$$

$$= -1/a + 0(\varepsilon) \tag{2.38}$$

$$m_2 = -\frac{a}{2\varepsilon} - \frac{a}{2\varepsilon}(1 - 4\varepsilon/a^2)^{0\cdot 5}$$

$$= -a/\varepsilon + 1/a + 0(\varepsilon) \tag{2.39}$$

The approximate solution is

$$y^{(0)}(k) = y(0)\,(-1/a)^k \tag{2.40}$$

Next consider (2.29b) with $\varepsilon = 0$. Then the resulting degenerate problem is

$$ay^{(0)}(k + 1) + y^{(0)}(k) = 0 \tag{2.41}$$

and at most can satisfy any one of the given boundary conditions. If (2.41) is solved by using $y^{(0)}(0) = y(0)$, the solution is the same as (2.40). This shows that the end condition $y(N)$ is lost in the process of degeneration and the boundary layer exists at $k = N$. From the roots (2.38) and (2.39) it is seen once again that the problem possesses a two-time-scale property.

Example 2.6
Consider a BVP

$$\varepsilon y(k + 2) - 0\cdot 9\,y(k + 1) + y(k) = 0 \tag{2.42}$$

with $\varepsilon = 0\cdot 08$, $y(0) = 1\cdot 0$ and $y(10) = 10\cdot 0$.

The roots m_1 and m_2 are $1\cdot 25$ and $10\cdot 0$. Figure 2.6 illustrates the degeneration of the solution as ε tends to zero, the formation of the boundary layer and the consequent loss of the end condition $y(N)$.

2.2.2   Linear control systems

In state-variable representation, the general form for linear, shift-invariant, singularly perturbed discrete systems is given by (Phillips, 1980a; Naidu and Rao, 1985a):

$$x(k + 1) = A_{11}x(k) + \varepsilon^{1-j}A_{12}z(k) + B_1u(k) \tag{2.43a}$$

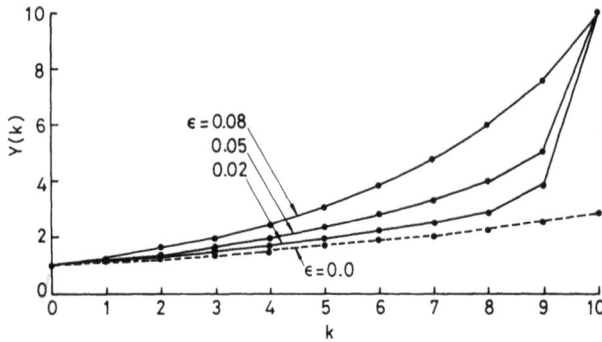

Fig. 2.6 *Degeneration of solution of Example 2.6*
—————— exact solution ($\varepsilon \neq 0$)
– – – – – degenerate solution ($\varepsilon = 0$)

$$\varepsilon^{2i}z(k+1) = \varepsilon^{j}A_{21}x(k) + \varepsilon A_{22}z(k) + \varepsilon^{j}B_{2}u(k) \qquad (2.43b)$$

$$0 \leqslant i \leqslant 1; \quad 0 \leqslant j \leqslant 1$$

The three limiting cases of (2.43) result in

(i) the C-model ($i = 0; j = 0$),

$$x(k+1) = A_{11}x(k) + \varepsilon A_{12}z(k) + B_{1}u(k) \qquad (2.44a)$$

$$z(k+1) = A_{21}x(k) + \varepsilon A_{22}z(k) + B_{2}u(k) \qquad (2.44b)$$

where the small parameter ε appears in the *column* of the system matrix,

(ii) the R-model ($i = 0; j = 1$),

$$x(k+1) = A_{11}x(k) + A_{12}z(k) + B_{1}u(k) \qquad (2.45a)$$

$$z(k+1) = \varepsilon A_{21}x(k) + \varepsilon A_{22}z(k) + \varepsilon B_{2}u(k) \qquad (2.45b)$$

where the small parameter ε appears in the *row* of the system matrix, and

(iii) the D-model ($i = 1; j = 1$),

$$x(k+1) = A_{11}x(k) + A_{12}z(k) + B_{1}u(k) \qquad (2.46a)$$

$$\varepsilon z(k+1) = A_{21}x(k) + A_{22}z(k) + B_{2}u(k) \qquad (2.46b)$$

where the small parameter ε is positioned in an identical fashion to that in a *differential* equation. Note that the replacement of $z(k)$ by $\varepsilon z(k)$ in the R-model (2.45) transforms into the C-model (2.44). In the above models, $x(k)$ and $z(k)$ are m- and n-dimensional state vectors, $u(k)$ is an r-dimensional control vector, and As and Bs are matrices of appropriate dimensions.

For our present discussion, it is enough if we consider the system (2.44) as an IVP with $x(k = 0) = x(0)$ and $z(k = 0) = z(0)$. The suppression of ε in system (2.44) results in the degenerate system

$$x^{(0)}(k+1) = A_{11}x^{(0)}(k) + B_1u(k), \quad x^{(0)}(k=0) = x(0) \tag{2.47a}$$

$$z^{(0)}(k+1) = A_{21}x^{(0)}(k) + B_2u(k), \quad z^{(0)}(k=0) \neq z(0) \tag{2.47b}$$

Here (2.47a) is a difference equation in $x^{(0)}(k)$ of order m whereas (2.47b) is an algebraic equation. It means that, once $x^{(0)}(k)$ is solved from (2.47a), $z^{(0)}(k)$ is automatically fixed by (2.47b), and hence $z^{(0)}(0)$ is not, in general, equal to $z(0)$. Thus the suppression of the small parameter ε in (2.44) leads to a low-order system (2.47) with a consequent loss of the initial condition $z(0)$. Hence (2.44) is in the singularly perturbed form.

The original system (2.44) and the degenerate system (2.47) are shown in Fig. 2.7, which illustrates the result of degeneration. Here we assume that $u(k)$ is independent of ε; otherwise $u(k)$ becomes $u^{(0)}(k)$ when ε is made equal to zero.

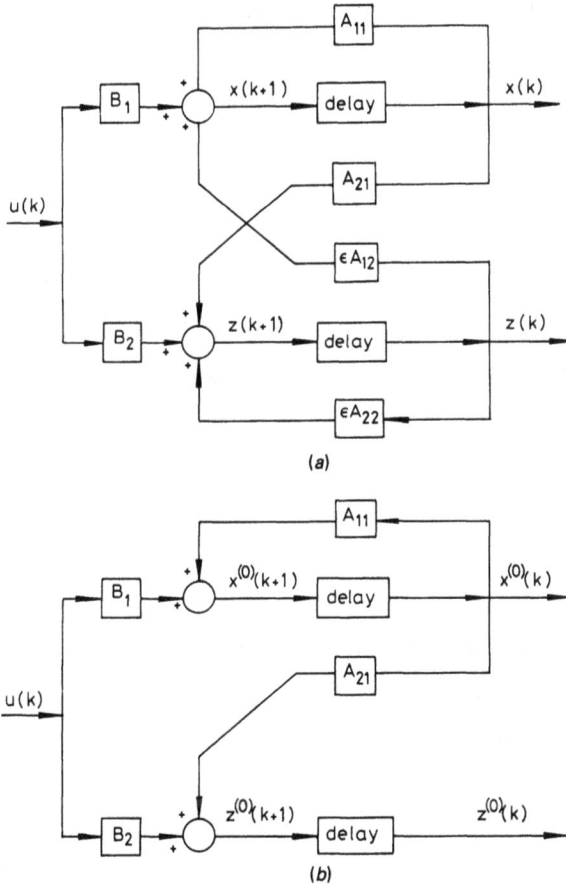

(a)

(b)

Fig. 2.7 *Singularly perturbed discrete control system*
a Original system
b Degenerate system

2.3   Singular perturbation method for continuous systems

In the previous Sections we have seen that in a singularly perturbed system the degeneration or the process of suppressing a small parameter relieves the stiffness, reduces the order and affects some of the given auxiliary conditions of the system. The main theme of the singular perturbation method is to retain the advantage of order reduction and to recover the lost conditions. We shall describe the method for IVP based on the original works of Vasileva (1963), because of its utmost importance in the theory of singular perturbations.

2.3.1   *Vasileva's method for initial value problems*

Consider a singularly perturbed nonlinear IVP

$$\frac{dx}{dt} = f(x,z,\varepsilon,t), \quad x(t=0) = x(0) \tag{2.48a}$$

$$\varepsilon \frac{dz}{dt} = g(x,z,\varepsilon,t), \quad z(t=0) = z(0) \tag{2.48b}$$

where x and z are m- and n-dimensional state vectors. In order to obtain a clear insight into the various aspects of the singular perturbation method as given by Vasileva (1963), we formulate (2.48) without a control vector u. The control vector u is included and dealt with in detail later.

By setting the small parameter $\varepsilon = 0$ in (2.48) the degenerate (low-order) problem is given by

$$\frac{dx^{(0)}}{dt} = f(x^{(0)}, z^{(0)}, 0, t) \tag{2.49a}$$

$$0 = g(x^{(0)}, z^{(0)}, 0, t) \tag{2.49b}$$

Solving (2.49b) for one of the several solutions,

$$z^{(0)}(t) = p(x^{(0)}, t) \tag{2.50}$$

Using (2.50) in (2.49),

$$\frac{dx^{(0)}}{dt} = f^0(x^{(0)}, t), \quad x^{(0)}(t=0) = x(0) \tag{2.51}$$

From (2.50), it is evident that $z^{(0)}(0)$ is not in general equal to $z(0)$.

The two main aspects of singular perturbation theory are the degeneration and asymptotic expansion. In degeneration, our interest is to find the conditions under which the full problem (2.48) tends to the degenerate problem (2.51). A theorem due to Tikhonov (1952) concerning the degeneration is given below after stating some assumptions (Wasow, 1965).

Assumption 2.1: The functions f and g are continuous in an open region of the space variables x, z, t.

Assumption 2.2: The equation

$$\frac{dz(\tau)}{d\tau} = g[x^{(0)}, z(\tau), 0, t] \tag{2.52}$$

in which $x^{(0)}$ and t are considered as fixed parameters, is called the boundary-layer equation of system (2.48).

Assumption 2.3: The root $z^{(0)} = p(x^{(0)}, t)$ of the equation $g(x^{(0)}, z^{(0)}, 0, t) = 0$ is called an isolated root in a domain D of the set of variables x and t, if there exists an $\varepsilon > 0$ such that the equation $g(x^{(0)}, z^{(0)}, 0, t) = 0$ has no solution other than $p(x^{(0)}, t)$ for $|z^{(0)} - p(x^{(0)}, t)| < \varepsilon$.

Assumption 2.4: The isolated root $z^{(0)} = p(x^{(0)}, t)$ is called stable in D if, for all points $(x, t) \in D$, the points $z^{(0)} = p(x^{(0)}, t)$ are asymptotically stable equilibrium points, in the sense of Lyapunov, of the auxiliary equation (2.52), as $\tau \to \infty$. This means that the Jacobian matrix g_z has all eigenvalues with negative real parts.

Assumption 2.5: The region of influence R of an isolated stable root $z^{(0)} = p(x^{(0)}, t)$ is the set of points $[x(0), z(0), t_0]$ such that the solution of the auxiliary equation (2.52) satisfying the initial condition $z(\tau = 0) = z(0)$ tends to the value $p(x(0), t_0)$, as $\tau \to \infty$.
   The following theorem is due to Tikhonov (1952).

Theorem 2.1: Let the assumptions 2.1 to 2.5 be satisfied. Then the solutions $x(t, \varepsilon)$ and $z(t, \varepsilon)$ of the full problem (2.48) are related to the solutions $x^{(0)}(t)$ and $z^{(0)}(t)$ of the degenerate problem (2.49) as

$$\underset{\varepsilon \to 0}{\text{limit}}\,[x(t, \varepsilon)] = x^{(0)}(t), \quad 0 \leqslant t \leqslant T \tag{2.53a}$$

$$\underset{\varepsilon \to 0}{\text{limit}}\,[z(t, \varepsilon)] = z^{(0)}(t), \quad 0 < t \leqslant T \tag{2.53b}$$

Here T is any number such that $z^{(0)} = p(x^{(0)}, t)$ is an isolated stable root of $g(x^{(0)}, z^{(0)}, t) = 0$ for $0 \leqslant t \leqslant T$. In other words, the convergence is uniform in $0 \leqslant t \leqslant T$ for $x(t, \varepsilon)$, and in any interval $0 < t_1 \leqslant t \leqslant T$ for $z(t, \varepsilon)$.
   Tikhonov's theorem is only the first step in the evaluation of approximate

solutions of singularly perturbed IVP. The second step is about the asymptotic expansions, and this has been analysed in a series of papers by Vasileva (1963). We begin by representing the solutions in the form of a series in powers of ε as

$$x(t,\varepsilon) = \sum_{i=0}^{\infty} x^{(i)}(t)\varepsilon^i; \quad z(t,\varepsilon) = \sum_{i=0}^{\infty} z^{(i)}(t)\varepsilon^i \tag{2.54}$$

and determine the coefficients $x^{(i)}(t)$ and $z^{(i)}(t)$ by means of formal substitution of (2.54) in (2.48) and comparison of coefficients of equal powers of ε. Then the following recursive equations are obtained:

$$\frac{dx^{(0)}}{dt} = f(x^{(0)}, z^{(0)}, 0, t) \tag{2.55a}$$

$$0 = g(x^{(0)}, z^{(0)}, 0, t) \tag{2.55b}$$

$$\frac{dx^{(1)}}{dt} = f_x x^{(1)} + f_z z^{(1)} + p_0$$

$$= f^1(x^{(0)}, z^{(0)}, x^{(1)}, z^{(1)}, t) \tag{2.56a}$$

$$\frac{dz^{(0)}}{dt} = g_x x^{(1)} + g_z z^{(1)} + q_0$$

$$= g^1(x^{(0)}, z^{(0)}, x^{(1)}, z^{(1)}, t) \tag{2.56b}$$

where p_0 and q_0 are polynomials consisting of $x^{(0)}, z^{(0)}, t$. We shall use the notation f^1 and g^1 to indicate all the coefficients on the right-hand side. Note that the zeroth order equation (2.55) is the same as that of the degenerate problem (2.49). It means that the series solutions (2.54) correspond to the outside region of the boundary layer at $t = 0$. Hence (2.54) is called the 'outer' series expansion.

The solution of (2.55) is obtained by using $x^{(0)}(t = 0) = x(0)$. On the other hand, the solution of (2.56) poses a problem, since the initial condition $x^{(i)}(t = 0)$ is not known. Once $x^{(i)}(t)$ is solved, $z^{(i)}(t)$ is automatically obtained from (2.56b). The outer series (2.54) cannot be expected to describe the solution of the IVP inside the boundary layer; but that is precisely where the initial value of $z^{(i)}(t)$ is to be found.

We have to relate the series (2.54) to the solutions of (2.48) in the boundary layer. For values of t that are small of order $0(\varepsilon)$, the solution to our perturbation problem can be found by means of a stretching transformation

$$t = \varepsilon t' \tag{2.57}$$

where t' is called the stretched variable. The stretched form of IVP (2.48) becomes

$$\frac{d\tilde{x}(t')}{dt'} = \varepsilon \tilde{f}[\tilde{x}(t'), \tilde{z}(t'), \varepsilon, \varepsilon t'] \tag{2.58a}$$

$$\frac{d\tilde{z}(t')}{dt'} = \tilde{g}[\tilde{x}(t'), \tilde{z}(t'), \varepsilon, \varepsilon t'] \tag{2.58b}$$

This has convergent expansions of the form

$$\tilde{x}(t', \varepsilon) = \sum_{i=0}^{\infty} \tilde{x}^{(i)}(t')\varepsilon^i; \quad \tilde{z}(t', \varepsilon) = \sum_{i=0}^{\infty} \tilde{z}^{(i)}(t')\varepsilon^i \tag{2.59}$$

Substitution of (2.59) in (2.58) and comparison of coefficients of ε result in

$$\frac{d\tilde{x}^{(0)}}{dt'} = 0 \tag{2.60a}$$

$$\frac{d\tilde{z}^{(0)}}{dt'} = \tilde{g}(\tilde{x}^{(0)}, \tilde{z}^{(0)}, 0) \tag{2.60b}$$

$$\frac{d\tilde{x}^{(1)}}{dt'} = \tilde{f}^1(\tilde{x}^{(0)}, \tilde{z}^{(0)}, 0) \tag{2.61a}$$

$$\frac{d\tilde{z}^{(1)}}{dt'} = \tilde{g}^1(\tilde{x}^{(0)}, \tilde{z}^{(0)}, \tilde{x}^{(1)}, \tilde{z}^{(1)}, t') \tag{2.61b}$$

$$\frac{d\tilde{x}^{(2)}}{dt'} = \tilde{f}^2(\tilde{x}^{(0)}, \tilde{z}^{(0)}, \tilde{x}^{(1)}, \tilde{z}^{(1)}, t') \tag{2.62a}$$

$$\frac{d\tilde{z}^{(2)}}{dt'} = \tilde{g}^2(\tilde{x}^{(0)}, \tilde{z}^{(0)}, \tilde{x}^{(1)}, \tilde{z}^{(1)}, \tilde{x}^{(2)}, \tilde{z}^{(2)}, t') \tag{2.62b}$$

The stretched problem (2.58) has initial conditions

$$\tilde{x}(t' = 0) = x(t = 0); \quad \tilde{z}(t' = 0) = z(t = 0) \tag{2.63}$$

This means that

$$\tilde{x}^{(0)}(t' = 0) = x(0); \quad \tilde{z}^{(0)}(t' = 0) = z(0) \tag{2.64a}$$

and $\quad \tilde{x}^{(i)}(t' = 0) = 0; \quad \tilde{z}^{(i)}(t' = 0) = 0, i > 0 \tag{2.64b}$

The series (2.59), which describes the behaviour of solutions inside the boundary layer, is called the 'inner' series. Still, we have not resolved the problem of determining the initial values $x^{(i)}(t = 0)$, $i > 0$, of the outer series. In order to relate the two types of expansions (2.54) and (2.59), we change the series (2.54) into a series involving t' and ε. Thus

$$x(t,\varepsilon) = x^{(0)}(t) + \varepsilon x^{(1)}(t) + \ldots$$

We expand the coefficients $x^{(0)}(t)$, $x^{(1)}(t)$, ..., around $t = 0$ as

$$x(t,\varepsilon) = [x^{(0)}(t=0) + t\dot{x}^{(0)}(t=0) + \ldots]$$
$$+ \varepsilon[x^{(1)}(t=0) + t\dot{x}^{(1)}(t=0) + \ldots]$$
$$+ \ldots \tag{2.65}$$

Now using $t' = t/\varepsilon$

$$\underset{\sim}{x}(t'\varepsilon,\varepsilon) = [x^{(0)}(t=0) + \varepsilon t'\dot{x}^{(0)}(t=0) + \ldots]$$
$$+ \varepsilon[x^{(1)}(t=0) + \varepsilon t'\dot{x}^{(1)}(t=0) + \ldots]$$
$$+ \ldots \tag{2.66}$$

Rearranging and summing according to powers of ε,

$$\underset{\sim}{x}(t',\varepsilon) = x^{(0)}(t=0) + \varepsilon[x^{(1)}(t=0) + t'\dot{x}^{(0)}(t=0)] + \ldots$$
$$= \underset{\sim}{x}^{(0)}(t') + \varepsilon\underset{\sim}{x}^{(1)}(t') + \ldots \tag{2.67a}$$

Similarly,

$$\underset{\sim}{z}(t',\varepsilon) = \underset{\sim}{z}^{(0)}(t') + \varepsilon\underset{\sim}{z}^{(1)}(t') + \ldots \tag{2.67b}$$

where

$$\underset{\sim}{x}^{(0)}(t') = x^{(0)}(t=0) \tag{2.68a}$$
$$\underset{\sim}{x}^{(1)}(t') = x^{(1)}(t=0) + t'\dot{x}^{(0)}(t=0) \tag{2.68b}$$

$$\ldots$$

$$\underset{\sim}{z}^{(0)}(t') = z^{(0)}(t=0) \tag{2.69a}$$
$$\underset{\sim}{z}^{(1)}(t') = z^{(1)}(t=0) + t'\dot{z}^{(0)}(t=0) \tag{2.69b}$$

$$\ldots$$

Note that the functions $\underset{\sim}{x}^{(0)}(t')$, $au\text{-}x^{(1)}(t')$, $\underset{\sim}{z}^{(0)}(t')$, $\underset{\sim}{z}^{(1)}(t')$, are polynomials in the stretched variables t'.

From (2.67), the initial values are

$$\underset{\sim}{x}^{(0)}(t'=0) = x^{(0)}(t=0); \quad \underset{\sim}{z}^{(0)}(t'=0) = z^{(0)}(t=0) \tag{2.70a}$$
$$\underset{\sim}{x}^{(i)}(t'=0) = x^{(i)}(t=0); \quad \underset{\sim}{z}^{(i)}(t'=0) = z^{(i)}(t=0), \quad i>0 \tag{2.70b}$$

The series (2.67) as functions of ε and t' has been derived from the outer series (2.54) which is a function of ε and t. This means that the series (2.67) and the inner series (2.59) are the formal solutions of the same stretched problem (2.58), but the difference lies in the initial conditions as shown by

(2.70) and (2.64). Then the recursive equations for series (2.67) are identical to (2.60) to (2.62), and are given by

$$\frac{d\underline{x}^{(0)}}{dt'} = 0 \tag{2.71a}$$

$$\frac{d\underline{z}^{(0)}}{dt'} = g(\underline{x}^{(0)}, \underline{z}^{(0)}, 0) \tag{2.71b}$$

$$\frac{d\underline{x}^{(1)}}{dt'} = f^1(\underline{x}^{(0)}, \underline{z}^{(0)}, 0) \tag{2.72a}$$

$$\frac{d\underline{z}^{(1)}}{dt'} = g^1(\underline{x}^{(0)}, \underline{z}^{(0)}, \underline{x}^{(1)}, \underline{z}^{(1)}, t') \tag{2.72b}$$

$$\frac{d\underline{x}^{(2)}}{dt'} = f^2(\underline{x}^{(0)}, \underline{z}^{(0)}, \underline{x}^{(1)}, \underline{z}^{(1)}, t') \tag{2.73a}$$

$$\frac{d\underline{z}^{(2)}}{dt'} = g^2(\underline{x}^{(0)}, \underline{z}^{(0)}, \underline{x}^{(1)}, \underline{z}^{(1)}, \underline{x}^{(2)}, \underline{z}^{(2)}, t') \tag{2.73b}$$

Thus the series (2.67) may also be described as the series solution of the stretched problem (2.58), and has the same initial values as the series solution of the unstretched problem (2.48). For small and moderate values of t' the functions $\underline{x}^{(i)}(t')$, $\underline{z}^{(i)}(t')$ differ considerably from $\bar{x}^{(i)}(t')$, $\bar{z}^{(i)}(t')$.

The initial values (2.70b) are chosen in such a way that the difference between $\bar{x}^{(i)}(t')$, $\bar{z}^{(i)}(t')$ and $\underline{x}^{(i)}(t')$, $\underline{z}^{(i)}(t')$ are exponentially small for large values of t'. That is

$$[\bar{x}^{(i)}(t') - \underline{x}^{(i)}(t')]_{t'=\infty} = 0 \tag{2.74a}$$

$$\underline{x}^{(i)}(t' = 0) = x^{(i)}(t = 0) = \int_0^\infty \left(\frac{d\bar{x}^{(i)}}{dt'} - \frac{d\underline{x}^{(i)}}{dt'} \right) dt', \quad i > 0 \tag{2.74b}$$

Once $x^{(i)}(t = 0)$ is known, the initial condition $\underline{z}^{(i)}(t' = 0) = z^{(i)}(t = 0)$ is automatically fixed from (2.56b).

The series (2.67) is called the *intermediate* series, since its function is only to link the outer series (2.54) and the inner series (2.59).

The total series solution to IVP (2.48) is given by the following theorem (Vasileva, 1963; Wasow, 1965).

Theorem 2.2: Let the assumptions 2.1 to 2.5 be satisfied. Then there exists

$\varepsilon_0 > 0, 0 \leqslant \varepsilon \leqslant \varepsilon_0$ and $R(t,\varepsilon)$ and $S(t,\varepsilon)$ uniformly bounded in the interval considered, such that

$$x(t,\varepsilon) = \sum_{i=0}^{j} [x^{(i)}(t) + \bar{x}^{(i)}(t') - \underline{x}^{(i)}(t')]\varepsilon^i + R(t,\varepsilon)\varepsilon^{j+1} \qquad (2.75a)$$

$$z(t,\varepsilon) = \sum_{i=0}^{j} [z^{(i)}(t) + \bar{z}^{(i)}(t') - \underline{z}^{(i)}(t')]\varepsilon^i + S(t,\varepsilon)\varepsilon^{j+1} \qquad (2.75b)$$

where

$x^{(i)}(t)$ and $z^{(i)}(t)$ = *outer* series functions

$\bar{x}^{(i)}(t')$ and $\bar{z}^{(i)}(t')$ = *inner* series functions

$\underline{x}^{(i)}(t')$ and $\underline{z}^{(i)}(t')$ = *intermediate* series functions.

In practice, one does not determine the series solutions in the same sequence that has been described earlier; i.e. one does not compute first all the coefficients of the outer series, then the terms of the inner series and finally those of the intermediate series. In actual working there is a considerable interplay in the evaluation of the various functions. The various steps involved in the process are given below using an example.

Example 2.7
Consider a nonlinear IVP (Wasow, 1965)

$$\frac{dx}{dt} = z, \quad x(t=0) = 1 \cdot 0 \qquad (2.76a)$$

$$\varepsilon \frac{dz}{dt} = x^2 - z^2, \quad z(t=0) = 0 \qquad (2.76b)$$

Using the relation $t' = t/\varepsilon$ in (2.76), the stretched problem becomes

$$\frac{dx(t')}{dt'} = \varepsilon z(t') \qquad (2.77a)$$

$$\frac{dz(t')}{dt'} = x^2(t') - z^2(t') \qquad (2.77b)$$

The sequence of steps in the method is given below up to first-order approximation.

Step 1: Substituting the outer series (2.54) in (2.76) and collecting coefficients of ε^0 on either side of the equations, we get the zeroth-order equations as

$$\frac{dx^{(0)}}{dt} = z^{(0)} \tag{2.78a}$$

$$0 = x^{(0)^2} - z^{(0)^2} \tag{2.78b}$$

Solving (2.78) with $x^{(0)}(t = 0) = x(0) = 1$,

$$x^{(0)}(t) = \exp(t); \quad z^{(0)}(t) = \exp(t) \tag{2.79}$$

Step 2: Using the inner series (2.59) in (2.77) and collecting coefficients of ε^0, we have

$$\frac{d\bar{x}^{(0)}}{dt'} = 0 \tag{2.80a}$$

$$\frac{d\bar{z}^{(0)}}{dt'} = \bar{x}^{(0)^2} - \bar{z}^{(0)^2} \tag{2.80b}$$

Solving the above equations with $\bar{x}^{(0)}(t' = 0) = x(0) = 1$ and $\bar{z}^{(0)}(t' = 0) = z(0) = 0$, we get

$$\bar{x}^{(0)}(t') = 1; \quad \bar{z}^{(0)}(t') = \tanh t' \tag{2.81}$$

Step 3: Inserting the intermediate series (2.67) in (2.77) and collecting coefficients of ε^0,

$$\frac{dx^{(0)}}{dt'} = 0 \tag{2.82a}$$

$$\frac{dz^{(0)}}{dt'} = x^{(0)^2} - z^{(0)^2} \tag{2.82b}$$

Using $x^{(0)}(t' = 0) = x(0) = 1$ and $z^{(0)}(t' = 0) = z^{(0)}(0) = 1$, we have

$$x^{(0)}(t') = 1; \quad z^{(0)}(t') = 1 \tag{2.83}$$

Step 4: By the process of substitution and comparison, the inner and intermediate series equations corresponding to a first-order approximation are obtained as

$$\frac{d\bar{x}^{(1)}}{dt'} = \bar{z}^{(0)} \tag{2.84a}$$

$$\frac{\mathrm{d}\tilde{z}^{(1)}}{\mathrm{d}t'} = 2\tilde{x}^{(0)}\tilde{x}^{(1)} - 2\tilde{z}^{(0)}\tilde{z}^{(1)} \qquad (2.84b)$$

and

$$\frac{\mathrm{d}\underset{\sim}{x}^{(1)}}{\mathrm{d}t'} = \underset{\sim}{z}^{(0)} \qquad (2.85a)$$

$$\frac{\mathrm{d}\underset{\sim}{z}^{(1)}}{\mathrm{d}t'} = 2\underset{\sim}{x}^{(0)}\underset{\sim}{x}^{(1)} - 2\underset{\sim}{z}^{(0)}\underset{\sim}{z}^{(1)} \qquad (2.85b)$$

where

$$\tilde{x}^{(1)}(t' = 0) = \tilde{z}^{(1)}(t' = 0) = 0; \quad \underset{\sim}{x}^{(1)}(t' = 0) = x^{(1)}(0);$$

$$\underset{\sim}{z}^{(1)}(t' = 0) = z^{(1)}(0).$$

Step 5: The initial condition $x^{(1)}(0)$ is evaluated using

$$\underset{\sim}{x}^{(1)}(0) = x^{(1)}(t' = 0) = \int_0^\infty \left[\frac{\mathrm{d}\tilde{x}^{(1)}}{\mathrm{d}t'} - \frac{\mathrm{d}\underset{\sim}{x}^{(1)}}{\mathrm{d}t'} \right] \mathrm{d}t' \qquad (2.86)$$

$$= \int_0^\infty [\tilde{z}^{(0)}(t') - \underset{\sim}{z}^{(0)}(t')]\mathrm{d}t'$$

$$= \int_0^\infty (\tanh t' - 1)\mathrm{d}t'$$

$$= -\log_e^2$$

Step 6: Using the value of $x^{(1)}(0)$ obtained in the above step, solve the first-order approximations of the outer series from

$$\frac{\mathrm{d}x^{(1)}}{\mathrm{d}t} = z^{(1)} \qquad (2.87a)$$

$$\frac{\mathrm{d}z^{(0)}}{\mathrm{d}t} = 2x^{(0)}x^{(1)} - 2z^{(0)}z^{(1)} \qquad (2.87b)$$

Solving the above we get

$$x^{(1)}(t) = 0\cdot5 + (x^{(1)}(0) - 0\cdot5)\exp(t) \qquad (2.88a)$$
$$z^{(1)}(t) = (x^{(1)}(0) - 0\cdot5)\exp(t) \qquad (2.88b)$$

where

$$x^{(1)}(0) = -\log_e^2$$

Note that once $x^{(1)}(0)$ is known, $z^{(1)}(0)$ is automatically fixed to be as $z^{(1)}(0) = -0{\cdot}5 - \log_e^2$.

Step 7: Having known $x^{(1)}(0)$ and $z^{(1)}(0)$, the equations of step 4 are now solved to be

$$\bar{x}^{(1)}(t') = \log_e \cosh t'; \quad \underline{x}^{(1)}(t') = -\log_e^2 + t' \tag{2.89a}$$

$$\underline{z}^{(1)}(t') = -0{\cdot}5 - \log_e^2 + t' \tag{2.89b}$$

The function $\bar{z}^{(1)}(t')$ cannot be solved explicitly, but has to be obtained numerically.

Step 8: The total series solution up to first-order approximation is given by

$$x(t,\varepsilon) = \sum_{i=0}^{1} [x^{(i)}(t) + \bar{x}^{(i)}(t') - \underline{x}^{(i)}(t')]\varepsilon^i \tag{2.90a}$$

$$z(t,\varepsilon) = \sum_{i=0}^{1} [z^{(i)}(t) + \bar{z}^{(i)}(t') - \underline{z}^{(i)}(t')]\varepsilon^i \tag{2.90b}$$

It is to be borne in mind that the intermediate series functions are either solved from the corresponding equations like (2.82) and (2.85) or formulated as polynomials in the stretched variable t' as given by (2.68) and (2.69). This is a very distinguishing feature of Vasileva's method in singular perturbation theory.

Although the method is described for IVP without inputs, it is equally applicable to problems with input functions. The following example illustrates Vasileva's method for a linear IVP with an exponential input function (Naidu, 1977).

Example 2.8
Consider a linear IVP with input as

$$\frac{dx}{dt} = -x + z, \quad x(t = 0) = x(0) \tag{2.91a}$$

$$\varepsilon \frac{dz}{dt} = -x - z + u, \quad z(t = 0) = z(0) \tag{2.91b}$$

where $u(t) = \exp(-t)$.

Following Vasileva's method given above, the general expressions for the outer, inner and intermediate series solutions up to first-order approximation are obtained and given below.

The exponential function $u(t) = \exp(-t)$ is represented as

$$
\left.
\begin{aligned}
u^{(0)}(t) &= \exp(-t); & u^{(1)}(t) &= 0 \\
\bar{u}^{(0)}(t') &= 1; & \bar{u}^{(1)}(t') &= -t' \\
\underline{u}^{(0)}(t') &= 1; & \underline{u}^{(1)}(t') &= -t'
\end{aligned}
\right\}
$$
(2.92)

The outer series solutions are

$$
\left.
\begin{aligned}
x^{(0)}(t) &= [x(0) - 1]\exp(-2t) + \exp(-t) \\
x^{(1)}(t) &= \{x^{(1)}(0) + 2[1 - x(0)]t\}\exp(-2t) \\
z^{(0)}(t) &= [-x(0) + 1]\exp(-2t) \\
z^{(1)}(t) &= \{z^{(1)}(0) - 2[1 - x(0)]t\}\exp(-2t)
\end{aligned}
\right\}
$$
(2.93)

The inner series solutions are

$$
\left.
\begin{aligned}
\bar{x}^{(0)}(t') &= x(0) \\
\bar{x}^{(1)}(t') &= x(0) + z(0) - 1 + 2[1 - x(0)]t' + [1 - x(0) - z(0)]\exp(-t') \\
\bar{z}^{(0)}(t') &= 1 - x(0) + [x(0) + z(0) - 1]\exp(-t') \\
\bar{z}^{(1)}(t') &= -3x(0) - z(0) + 3 + 2[x(0) - 1]t' \\
&\quad + \{3x(0) + z(0) - 3 + [x(0) + z(0) - 1]t'\}\exp(-t')
\end{aligned}
\right\}
$$
(2.94)

The intermediate series solutions are

$$
\left.
\begin{aligned}
\underline{x}^{(0)}(t') &= x(0) \\
\underline{x}^{(1)}(t') &= x^{(1)}(0) - [2x(0) - 1]t' \\
\underline{z}^{(0)}(t') &= 1 - x(0) \\
\underline{z}^{(1)}(t') &= z^{(1)}(0) + 2[x(0) - 1]t'
\end{aligned}
\right\}
$$
(2.95)

The initial values are

$$
x^{(1)}(0) = x(0) + z(0) - 1; \quad z^{(1)}(0) = -3x(0) - z(0) + 3
$$
(2.96)

The total series solution up to first-order approximation is given by (2.90).

Once again note the polynomial structure of the intermediate-series solutions (2.95).

For the system (2.91) with step input $u = 2\cdot0$; $x(0) = 2\cdot0$; $z(0) = -10\cdot0$ and $\varepsilon = 0\cdot2$, the series solutions for zeroth, first, and second-order approximations are obtained and compared with the high-order (exact) solutions as shown in Fig. 2.8. It is clearly seen that the increased order of approximation improves the series solutions.

The singular perturbation method for IVP due to Vasileva has the following special features:

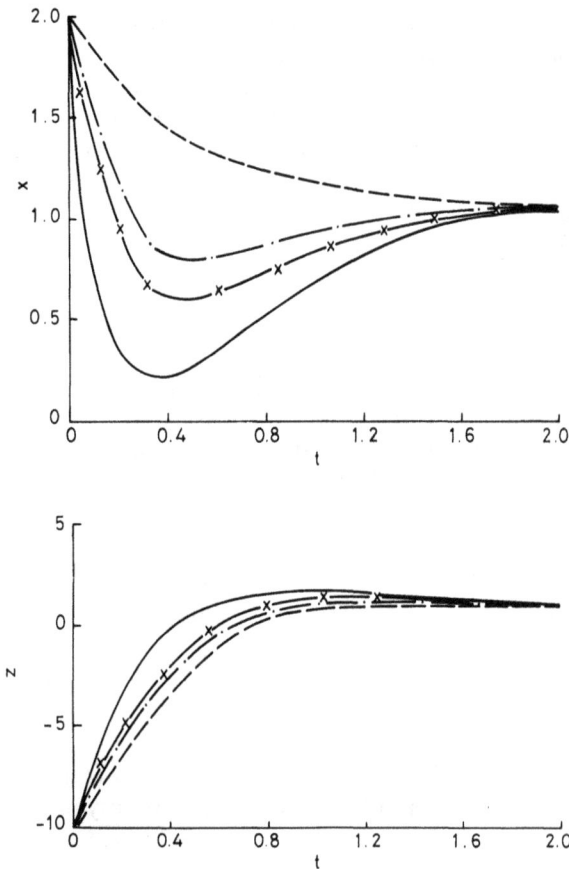

Fig. 2.8 *Exact and approximate solutions of Example 2.8*
——————— exact solution
— — — — — zeroth-order solution
——●—— first-order solution
——×—— second-order solution

(i) The approximate solution consists of the outer series solution plus the difference between the inner and intermediate series solutions.
(ii) The outer series solution is close to the full (exact) solution everywhere outside the boundary layer, and it also loses some of the initial conditions given for the full problem.
(iii) The difference between the inner and intermediate series solutions recovers the lost initial conditions, improves the solution inside the boundary layer, and has negligible contribution outside the boundary layer.
(iv) The intermediate series solutions are polynomials in the stretched vari-

able t' and are solved either from the corresponding series equations or formulated as polynomials in t'.

(v) Owing to the above features, it is observed that the inner series solution contains the exponentially decaying functions and the polynomial structure of the intermediate series solutions, so that the difference between inner and intermediate series solutions tends to be negligible as the stretched variable tends to infinity.

2.3.2   Boundary layer method

In Vasileva's singular perturbation method, we have seen that the approximate solution is composed of the outer series, plus the difference between the inner and the intermediate series solutions. The function of the inner and intermediate series solutions is to recover the lost initial conditions and improve the solution inside the boundary layer. A natural question arises whether a single series can substitute for the difference between the inner and intermediate series. The answer is the boundary-layer method where the single series is called boundary-layer correction (O'Malley, 1971; Hoppenstead, 1971). We now describe the boundary-layer method.

Let us reconsider the IVP given by (2.48). We shall seek a solution of the form

$$x(t,\varepsilon) = x_0(t,\varepsilon) + x_c(t',\varepsilon) \tag{2.97a}$$

$$z(t,\varepsilon) = z_0(t,\varepsilon) + z_c(t',\varepsilon) \tag{2.97b}$$

where $x_0(t,\varepsilon)$ and $z_0(t,\varepsilon)$, the outer solutions and $x_c(t',\varepsilon)$ and $z_c(t',\varepsilon)$, the boundary layer correction (BLC) solutions, all have asymptotic series expansions as

$$x_0(t,\varepsilon) = \sum_{i=0}^{\infty} x^{(i)}(t)\varepsilon^i; \quad z_0(t,\varepsilon) = \sum_{i=0}^{\infty} z^{(i)}(t)\varepsilon^i \tag{2.98a}$$

$$x_c(t,\varepsilon) = \sum_{i=0}^{\infty} x_c^{(i)}(t')\varepsilon^i; \quad z_c(t',\varepsilon) = \sum_{i=0}^{\infty} z_c^{(i)}(t')\varepsilon^i \tag{2.98b}$$

The outer solutions $x_0(t,\varepsilon)$ and $z_0(t,\varepsilon)$ are close to the exact solutions $x(t,\varepsilon)$ and $z(t,\varepsilon)$, respectively, for values away from $t=0$ or outside the boundary layer, while the boundary-layer corrections $x_c(t',\varepsilon)$ and $z_c(t',\varepsilon)$ are significant only near $t=0$ or inside the boundary layer. Thus, we ask that all the terms in the expansions for $x_c(t',\varepsilon)$ and $z_c(t',\varepsilon)$ tend to zero as t' tends to infinity.

The boundary layer method consists of first obtaining the outer series solution and next the BLC series solution. Substituting (2.97) in (2.48),

$$\frac{dx_0(t,\varepsilon)}{dt} + \frac{1}{\varepsilon}\frac{dx_c(t',\varepsilon)}{dt'} = f[x_0(t,\varepsilon) + x_c(t',\varepsilon), z_0(t,\varepsilon) + z_c(t',\varepsilon),\varepsilon,t]$$

$$\tag{2.99a}$$

$$\varepsilon \left[\frac{dz_0(t,\varepsilon)}{dt} + \frac{1}{\varepsilon} \frac{dz_c(t',\varepsilon)}{dt'} \right] = g[x_0(t,\varepsilon) + x_c(t',\varepsilon), \ z_0(t,\varepsilon) + z_c(t',\varepsilon), \varepsilon, t]$$

$$(2.99b)$$

The outer solution $x_0(t,\varepsilon)$ and $z_0(t,\varepsilon)$ must satisfy the original IVP (2.48). That is

$$\frac{dx_0(t,\varepsilon)}{dt} = f[x_0(t,\varepsilon), z_0(t,\varepsilon), \varepsilon, t] \qquad (2.100a)$$

$$\varepsilon \frac{dz_0(t,\varepsilon)}{dt} = g[x_0(t,\varepsilon), z_0(t,\varepsilon), \varepsilon, t] \qquad (2.100b)$$

Using (2.100) in (2.99), we get for the BLC solution

$$\frac{dx_c(t',\varepsilon)}{dt'} = \varepsilon f[x_0(t'\varepsilon,\varepsilon) + x_c(t',\varepsilon), \ z_0(t'\varepsilon,\varepsilon) + z_c(t',\varepsilon), t'\varepsilon, \varepsilon]$$

$$- \varepsilon f[x_0(t'\varepsilon,\varepsilon), z_0(t'\varepsilon,\varepsilon), t'\varepsilon, \varepsilon] \qquad (2.101a)$$

$$\frac{dz_c(t',\varepsilon)}{dt'} = g[x_0(t'\varepsilon,\varepsilon) + x_c(t',\varepsilon), \ z_0(t'\varepsilon,\varepsilon) + z_c(t',\varepsilon), t'\varepsilon, \varepsilon]$$

$$- g[x_0(t'\varepsilon,\varepsilon), z_0(t'\varepsilon,\varepsilon), t'\varepsilon, \varepsilon] \qquad (2.101b)$$

Now we have separated (2.100) for the outer solution and (2.101) for the BLC solution. The outer series expansions (2.98a) are substituted in (2.100), and by collecting coefficients of like powers of ε, we get the same equations (2.55) and (2.56) given in Vasileva's method. For the BLC solution, we expand (2.101) using (2.98b) and equate the coefficients. For zeroth-order approximation ($\varepsilon = 0$), we have the nonlinear problem

$$\frac{dx_c^{(0)}(t')}{dt'} = 0 \qquad (2.102a)$$

$$\frac{dz_c^{(0)}(t')}{dt'} = g[x^{(0)}(0) + x_c^{(0)}(t'), \ z^{(0)}(0) + z_c^{(0)}(t'), 0, 0]$$

$$- g[x^{(0)}(0), z^{(0)}(0), 0, 0] \qquad (2.102b)$$

For first-order approximation

$$\frac{dx_c^{(1)}(t')}{dt'} = f[x^{(0)}(0) + x_c^{(0)}(t'), \ z^{(0)}(0) + z_c^{(0)}(t'), 0, 0]$$

$$- f[x^{(0)}(0), z^{(0)}(0), 0, 0] \qquad (2.103a)$$

$$\frac{dz_c^{(1)}(t')}{dt'} = g_x[x^{(0)}(0) + x_c^{(0)}(t'), \ z^{(0)}(0) + z_c^{(0)}(t'),0,0]$$

(2.103b)

$$\times x_c^{(1)}(t') + g_z[x^{(0)}(0) + x_c^{(0)}(t'), \ z^{(0)}(0)$$

$$+ z_c^{(0)}(t'),0,0]z_c^{(1)}(t') + q_0$$

where q_0 is a polynomial in $x_c^{(0)}(t')$, $z_c^{(0)}(t')$.

Note that the functions $x_0(t'\varepsilon,\varepsilon)$ and $z_0(t'\varepsilon,\varepsilon)$ in (2.101) have the expansions around $t = 0$ as

$$\left.\begin{aligned}
x_0(t'\varepsilon,\varepsilon) &= [x^{(0)}(0) + t\dot{x}^{(0)}(0) + \ldots]_{t=t'\varepsilon} \\
&\quad + \varepsilon[x^{(1)}(0) + t\dot{x}^{(1)}(0) + \ldots]_{t=t'\varepsilon} \\
&\quad + \ldots \\
&= x^{(0)}(0) + \varepsilon[x^{(1)}(0) + t'\dot{x}^{(0)}(0)] + \ldots \\
&= \underline{x}(t',\varepsilon)
\end{aligned}\right\}$$

(2.104a)

Similarly,

$$\begin{aligned}
z_0(t'\varepsilon,\varepsilon) &= z^{(0)}(0) + \varepsilon[z^{(1)}(0) + t'\dot{z}^{(0)}(0)] + \ldots \\
&= \underline{z}(t',\varepsilon)
\end{aligned}$$

(2.104b)

Theorem 2.3: Under Assumptions 2.1–2.5 the total series solution is given by

$$x(t,\varepsilon) = \sum_{i=0}^{j} [x^{(i)}(t) + x_c^{(i)}(t')]\varepsilon^i + R(t,\varepsilon)\varepsilon^{j+1}$$

(2.105a)

$$z(t,\varepsilon) = \sum_{i=0}^{j} [z^{(i)}(t) + z_c^{(i)}(t')]\varepsilon^i + S(t,\varepsilon)\varepsilon^{j+1}$$

(2.105b)

where $R(t,\varepsilon)$ and $S(t,\varepsilon)$ are uniformly bounded in some interval (O'Malley, 1974a).

Our next concern is to allocate the initial conditions for solving the outer series equations and BLC series equations.

From (2.97) and (2.98),

$$x(0) = x^{(i)}(0) + x_c^{(i)}(0)$$

(2.106a)

$$z(0) = z^{(i)}(0) + z_c^{(i)}(0)$$

(2.106b)

Then

$$x_c^{(0)}(0) = x(0) - x^{(0)}(0) = 0; \quad z_c^{(0)}(0) = z(0) - z^{(0)}(0)$$

(2.107a)

$$x_c^{(i)}(0) = -x^{(i)}(0); \quad z_c^{(i)}(0) = -z^{(i)}(0), \quad i > 0$$

(2.107b)

It is also required that

$$\lim_{t' \to \infty} \left[x_c(t') \right] = 0; \quad \lim_{t' \to \infty} \left[z_c(t') \right] = 0 \tag{2.108}$$

Combining (2.107) and (2.108), we get

$$x^{(i)}(0) = -x_c^{(i)}(0) = - \int_0^\infty \left[\frac{dx_c^{(i)}(t')}{dt'} \right] dt', \quad i > 0 \tag{2.109}$$

This infinite integral exists since the integrand is dominated by functions of boundary-layer type which decay exponentially. The boundary-layer corrections represent the 'fast' mode, while the outer solutions represent the 'slow' mode of the problem.

Now let us show that the BLC is actually given by the difference between the inner and intermediate solutions, i.e.

$$x_c(t',\varepsilon) = \tilde{x}(t',\varepsilon) - \underline{x}(t',\varepsilon) \tag{2.110a}$$

$$z_c(t',\varepsilon) = \tilde{z}(t',\varepsilon) - \underline{z}(t',\varepsilon) \tag{2.110b}$$

For the IVP of (2.48), the equation for the inner solution is

$$\frac{d\tilde{x}(t',\varepsilon)}{dt'} = \varepsilon \tilde{f}[\tilde{x}(t',\varepsilon), \tilde{z}(t',\varepsilon), t'\varepsilon, \varepsilon] \tag{2.111a}$$

$$\frac{d\tilde{z}(t',\varepsilon)}{dt'} = \tilde{g}[\tilde{x}(t',\varepsilon), \tilde{z}(t',\varepsilon), t'\varepsilon, \varepsilon] \tag{2.111b}$$

and the equation for the intermediate solution is

$$\frac{d\underline{x}(t',\varepsilon)}{dt'} = \varepsilon fl[\underline{x}(t',\varepsilon), \underline{z}(t',\varepsilon), t'\varepsilon, \varepsilon] \tag{2.112a}$$

$$\frac{d\underline{z}(t',\varepsilon)}{dt'} = g[\underline{x}(t',\varepsilon), \underline{z}(t',\varepsilon), t', \varepsilon, \varepsilon] \tag{2.112b}$$

From (2.111) and (2.112),

$$\frac{d\tilde{x}(t',\varepsilon)}{dt'} - \frac{d\underline{x}(t',\varepsilon)}{dt'} = \varepsilon \tilde{f}[\tilde{x}(t',\varepsilon), \tilde{z}(t',\varepsilon), t'\varepsilon, \varepsilon] \\ - \varepsilon \tilde{f}[\underline{x}(t',\varepsilon), z(t',\varepsilon), t'\varepsilon, \varepsilon] \tag{2.113a}$$

$$\frac{d\tilde{z}(t',\varepsilon)}{dt'} - \frac{d\underline{z}(t',\varepsilon)}{dt'} = \tilde{g}[\tilde{x}(t',\varepsilon), \tilde{z}(t',\varepsilon), t'\varepsilon, \varepsilon] \\ - g[\underline{x}(t',\varepsilon), z(t',\varepsilon), t'\varepsilon, \varepsilon] \tag{2.113b}$$

Using (2.110) in (2.113),

$$\frac{dx_c(t',\varepsilon)}{dt'} = \varepsilon \bar{f}[x_c(t',\varepsilon) + x(t',\varepsilon),\ z_c(t',\varepsilon) + z(t',\varepsilon),\varepsilon t',\varepsilon]$$

$$- \varepsilon f[x(t',\varepsilon),z(t',\varepsilon),t'\varepsilon,\varepsilon] \tag{2.114a}$$

$$\frac{dz_c(t',\varepsilon)}{dt'} = \bar{g}[x_c(t',\varepsilon) + x(t',\varepsilon),\ z_c(t',\varepsilon) + z(t',\varepsilon),t'\varepsilon,\varepsilon]$$

$$- g[x(t',\varepsilon),z(t',\varepsilon),t'\varepsilon,\varepsilon] \tag{2.114b}$$

We know that the functions $x_0(t'\varepsilon,\varepsilon)$ and $z_0(t'\varepsilon,\varepsilon)$ appearing in the BLC (2.101) have the expansions around $t = 0$ as shown by (2.104). A close examination of (2.104) and (2.114) reveals that the $x_0(t'\varepsilon,\varepsilon)$ and $z_0(t'\varepsilon,\varepsilon)$ are equivalent to the intermediate functions $x(t',\varepsilon)$ and $z(t',\varepsilon)$. Thus (2.101) and (2.114) are equivalent.

Example 2.8
Let us reconsider the nonlinear IVP given in Example 2.7 and use the boundary-layer method. The outer series solutions being the same in the Vasileva's method and boundary-layer method, we shall be concerned with the boundary-layer corrections only. The BLC equation corresponding to (2.101) is

$$\frac{dx_c(t',\varepsilon)}{dt'} = \varepsilon[z_0(t',\varepsilon,\varepsilon) + z_c(t',\varepsilon)] - \varepsilon z_0(t',\varepsilon,\varepsilon) \tag{2.115a}$$

$$= \varepsilon z_c(t',\varepsilon)$$

$$\frac{dz_c(t',\varepsilon)}{dt'} = [x_0(t'\varepsilon,\varepsilon) + x_c(t',\varepsilon)]^2 - [z_0(t'\varepsilon,\varepsilon) + z_c(t',\varepsilon)]^2 \tag{2.115b}$$

$$- \{[x_0(t'\varepsilon,\varepsilon)]^2 - [z_0(t'\varepsilon,\varepsilon)]^2\}$$

$$= [x_c^2(t',\varepsilon)]^2 - [z_c^2(t',\varepsilon)]^2 + 2x_0(t'\varepsilon,\varepsilon)x_c(t',\varepsilon)$$

$$- 2z_0(t'\varepsilon,\varepsilon)z_c(t',\varepsilon)$$

Using (2.101) and (2.104) in (2.115), we get for the zeroth-order approximation.

$$\frac{dx_c^{(0)}(t')}{dt'} = 0 \tag{2.116a}$$

$$\frac{dz_c^{(0)}(t')}{dt'} = [x_c^{(0)}(t')]^2 - [z_c^{(0)}(t')]^2 + 2x^{(0)}(0)x_c^{(0)}(t') - 2z^{(0)}(0)z_c^{(0)}(t') \tag{2.116b}$$

For the first-order approximation

$$\frac{dx_c^{(1)}(t')}{dt'} = z_c^{(0)}(t') \tag{2.117a}$$

$$\frac{dz_c^{(1)}(t')}{dt'} = 2x_c^{(0)}(t')x_c^{(1)}(t') - 2z_c^{(0)}(t')z_c^{(1)}(t') \tag{2.117b}$$

$$+2\{x^{(0)}(0)x_c^{(1)}(t') + [x^{(1)}(0) + t'\dot{x}^{(0)}(0)]x_c^{(0)}(t')\}$$

$$-2\{z^{(0)}(0)z_c^{(1)}(t') + [z^{(1)}(0) + t'\dot{z}^{(0)}(0)]z_c^{(0)}(t')\}$$

Solving the above equations along with (2.107), we get

$$\left.\begin{array}{l} x_c^{(0)}(t') = 0; \quad z_c^{(0)}(t') = \tanh t' - 1 \\ x_c^{(1)}(t') = \log_e(2\cosh t') - t' \end{array}\right\} \tag{2.118}$$

The solution for $z_c^{(1)}(t')$ cannot be obtained explicitly. These are the same solutions obtained in (2.81) and (2.89).

It should be noted that the zeroth-order term $x_c^{(0)}(t')$ of the BLC is always zero owing to the appearance of the parameter ε on the RHS of the BLC equation (2.101a). As a result of working with similar problems, one can modify (2.97) as

$$x(t,\varepsilon) = x_0(t,\varepsilon) + \varepsilon x_c(t',\varepsilon) \tag{2.119a}$$

$$z(t,\varepsilon) = z_0(t,\varepsilon) + z_c(t',\varepsilon) \tag{2.119b}$$

so as to avoid the above situation of getting a zero value for $x_c^{(0)}(t')$ (O'Malley, 1971).

At this point, it is worthwhile to make the following important observations regarding Vasileva's method and the boundary-layer method:

(i) In Vasileva's method, the asymptotic solution is given by the outer series plus the difference between the inner and intermediate series solutions, whereas in the boundary-layer method, it is given by the outer series plus the boundary layer correction.

(ii) The asymptotic series solutions obtained by both the methods are identical since the difference between the inner and intermediate series solutions is equal to the boundary-layer correction and the outer series is the same in both the methods.

(iii) In working with a variety of control problems using both the methods (Naidu, 1977), it has been found that Vasileva's method has some advantages especially regarding computational aspects because:

(a) The structure of the equations describing the inner and intermediate series are identical (the difference being only with the associated auxiliary conditions) and can be written down easily by inspection of the corresponding outer series. This will facilitate the use of more or less identical computational routines for all the series. On the other hand, in the other method, the equations describing the boundary-layer corrections have to be formulated entirely in a different manner, and are quite unwieldy and difficult to handle for higher-order approximations.

(b) A very distinguishing feature of Vasileva's method is that the intermediate series solutions need not actually be solved from the corresponding equations, but can be formulated (or generated) as polynomials in the stretched variable.

Thus, although it appears that Vasileva's method requires three series, instead of the two series as in the boundary-layer method, in practice, only the two— outer and inner—series need to be solved, with less computational effort.

2.3.3 Boundary-value problems

Let us consider a nonlinear boundary-value problem (BVP) described by

$$\frac{dx}{dt} = f(x,z,\varepsilon,t) \qquad (2.120a)$$

$$\varepsilon \frac{dz}{dt} = g(x,z,\varepsilon,t) \qquad (2.120b)$$

where x and z are, respectively, $2m$- and $2n$-dimensional state vectors. The boundary conditions for (2.120) are

$$x_j(t = 0) = x_j(0), \quad x_{m+j}(t = T) = x_{m+j}(T), \quad j = 1 \text{ to } m \qquad (2.121a)$$

$$z_j(t = 0) = z_j(0), \quad x_{n+j}(t = T) = z_{n+j}(T), \quad j = 1 \text{ to } n \qquad (2.121b)$$

It is assumed that the function $f(x,z,\varepsilon,t)$ has continuous second partial derivatives and the function $g(x,z,\varepsilon,t)$ has continuous third partial derivates in a region R_1 with respect to x, z, t.

For $\varepsilon = 0$, the full problem (2.120) and (2.121) reduces to a degenerate problem

$$\frac{dx^{(0)}}{dt} = f(x^{(0)},z^{(0)},0,t) \qquad (2.122a)$$

$$0 = g(x^{(0)},z^{(0)},0,t) \qquad (2.122b)$$

satisfied by the boundary conditions

$$x_j^{(0)}(t = 0) = x_j(0); \quad x_{m+j}^{(0)}(t = T) = x_{m+j}(T), \quad j = 1 \text{ to } m \qquad (2.123)$$

As in the case of IVP, it is necessary to first obtain the conditions under which the full problem (2.120) along with (2.121) reduces to the degenerate problem (2.122) and (2.123). A theorem due to Tupchiev (1962) provides the answer and is given below.

Hypothesis 2.1: There exists a real nonsingular matrix $S(t)$ such that

$$S^{-1}(t)g_zS(t) = \begin{bmatrix} A(t) & 0 \\ 0 & D(t) \end{bmatrix} \quad (2.124)$$

where the matrix $S(t)$ is continuous with respect to t for $0 \leqslant t \leqslant T$. g_z is the partial derivative of g with respect to z evaluated along the solution $x^{(0)}(t)$ and $z^{(0)}(t)$ of the low-order problem (2.122) with (2.123), $A(t)$ is an $n \times n$ matrix each of whose eigenvalues has a negative real part for $0 \leqslant t \leqslant T$, and where $D(t)$ is an $n \times n$ matrix each of whose eigenvalues has a positive real part for $0 \leqslant t \leqslant T$.

Theorem 2.4: If Hypothesis 2.1 is true and if $x^{(0)}(t)$, $z^{(0)}(t)$ and the boundary data (2.121) lie in the region R_1, then $x(t,\varepsilon)$ and $z(t,\varepsilon)$ tend as ε tends to zero to $x^{(0)}(t)$ and $z^{(0)}(t)$ for all t, $0 < t < T$.

The next step is to obtain the asymptotic series expansions for the BVP of (2.120) and (2.121). We shall use Vasileva's method owing to its simplicity in obtaining the various controlling equations for the inner and intermediate series and the easy adaptability for computer simulation of the overall problem, especially in the case of nonlinear problems.

In Vasileva's method for BVP, we first note from (2.120)–(2.123) that the degeneration affects some of the boundary conditions and the boundary layers exist at both the initial and the final points. Accordingly it is necessary to introduce BLCs at the initial and final points. Each BLC consists of the difference between the inner and intermediate series solutions.

In this Section, we shall indicate only the important equations used in the method, reserving the detailed equations to a more appropriate situation of the two-point boundary value problems (TPBVP) arising in the open-loop optimal control systems in Chapter 4.

We seek the approximate solution for (2.120) and (2.121) as

$$x(t,\varepsilon) = x_o(t,\varepsilon) + x_i(t',\varepsilon) + x_f(t'',\varepsilon) \quad (2.125a)$$

$$z(t,\varepsilon) = z_o(t,\varepsilon) + z_i(t',\varepsilon) + z_f(t'',\varepsilon) \quad (2.125b)$$

where o refers to the outer solution and i and f refer to the BLC at the initial and final points, respectively, $t' = t/\varepsilon$ and $t'' = (t - T)/\varepsilon$ are the initial and final stretching transformations, respectively.

The problem corresponding to t' is

$$\frac{dx_i(t',\varepsilon)}{dt'} = hf_i[x_i(t',\varepsilon),z_i(t',\varepsilon),\varepsilon t',\varepsilon] \quad (2.126a)$$

$$\frac{dz_i(t',\varepsilon)}{dt'} = g_i[x_i(t',\varepsilon),z_i(t',\varepsilon),\varepsilon t',\varepsilon] \quad (2.126b)$$

and that corresponding to t'' is

$$\frac{dx_f(t'',\varepsilon)}{dt''} = \varepsilon f_f[x_f(t'',\varepsilon),z_f(t'',\varepsilon),\varepsilon t'' + T,\varepsilon] \qquad (2.127a)$$

$$\frac{dz_f(t'',\varepsilon)}{dt''} = g_f[x_f(t'',\varepsilon),z_f(t'',\varepsilon),\varepsilon t'' + T,\varepsilon] \qquad (2.127b)$$

It is to be noted that each of the outer and correction series solutions in (2.125) have series expansions as

$$\left.\begin{array}{l} x_o(t,\varepsilon) = \displaystyle\sum_{j=0}^{\infty} x^{(j)}(t)\varepsilon^j \\[4mm] x_i(t',\varepsilon) = \displaystyle\sum_{j=0}^{\infty} [\tilde{x}^{(j)}(t') - \underline{x}^{(j)}(t')]\varepsilon^j \\[4mm] x_f(t'',\varepsilon) = \displaystyle\sum_{j=0}^{\infty} [\tilde{x}^{(j)}(t'') - \underline{x}^{(j)}(t'')]\varepsilon^j \end{array}\right\} \qquad (2.128)$$

Similarly for the z variable.

By formal substitution and collection of coefficients, we get sets of recursive equations for the outer, inner and intermediate series for initial and final BLC. These equations will have the boundary conditions for the initial BLC as

$$\left.\begin{array}{ll} \tilde{x}_i^{(0)}(t'=0) = x(0); & \tilde{z}_i^{(0)}(t'=0) = z(0) \\[2mm] \tilde{x}_i^{(j)}(t'=0) = 0; & \tilde{z}_i^{(j)}(t'=0) = 0, \quad j>0 \\[2mm] \underline{x}_i^{(0)}(t'=0) = x(0); & \underline{z}_i^{(0)}(t'=0) = z^{(0)}(0) \\[2mm] \underline{x}_i^{(j)}(t'=0) = x^{(j)}(0); & \underline{z}_i^{(j)}(t'=0) = z^{(j)}(0), \quad j>0 \end{array}\right\} \qquad (2.129)$$

The values $x^{(j)}(0)$, $j>0$, are determined by imposing the conditions

$$[\tilde{x}^{(j)}(t') - \underline{x}^{(j)}(t')]_{t'=\infty} = 0 \qquad (2.130)$$

The other values $z^{(j)}(0)$ are automatically fixed once $x^{(j)}(0)$ are known.

Similarly for the final BLC

$$\left.\begin{array}{ll} \tilde{x}_f^{(0)}(t''=0) = x(T); & \tilde{z}_f^{(0)}(t''=0) = z(T) \\[2mm] \tilde{x}_f^{(j)}(t''=0) = 0; & \tilde{z}_f^{(j)}(t''=0) = 0, \quad j>0 \\[2mm] \underline{x}_f^{(0)}(t''=0) = x(T); & \underline{z}_f^{(0)}(t''=0) = z^{(0)}(T) \\[2mm] \underline{x}_f^{(j)}(t''=0) = x^{(j)}(T); & \underline{z}_f^{(j)}(t''=0) = z^{(j)}(T), \quad j>0 \end{array}\right\} \qquad (2.131)$$

The values $x^{(j)}(T)$, $j > 0$, are determined by imposing the condition

$$[\tilde{x}_i^{(j)}(t'') - \underline{x}_i^{(j)}(t'')]_{t''=-\infty} = 0 \tag{2.132}$$

and $z^{(j)}(T)$ is dictated by $x^{(j)}(T)$.

Let us note that the intermediate series functions $\underline{x}^{(j)}(t')$, and $\tilde{x}^{(j)}(t'')$ are polynomials in t' and t'', respectively, and this will simplify the overall computation by not solving for the same functions from the corresponding recursive equations.

2.3.4 Structural properties

We shall now discuss the structural properties—stability, controllability, and observability—of singularly perturbed continuous systems (Kokotovic and Haddad, 1975a; Chow and Kokotovic, 1976; Sannuti, 1977).

Consider the linear singularly perturbed system (2.18) with output vector as

$$\dot{x} = A_{11}x + A_{12}z + B_1u, \quad x(t=0) = x(0) \tag{2.133a}$$

$$\varepsilon\dot{z} = A_{21}x + A_{22}z + B_2u, \quad z(t=0) = z(0) \tag{2.133b}$$

$$y = C_1x + C_2z \tag{2.133c}$$

where y is a p-dimensional output vector and C_1 and C_2 are of appropriate dimensions. The degenerate ($\varepsilon = 0$) problem is

$$\dot{x}^{(0)} = A_{11}x^{(0)} + A_{12}z^{(0)} + B_1u^{(0)}, \quad x^{(0)}(t=0) = x(0) \tag{2.134a}$$

$$0 = A_{21}x^{(0)} + A_{22}z^{(0)} + B_2u^{(0)} \tag{2.134b}$$

$$y^{(0)} = C_1x^{(0)} + C_2z^{(0)} \tag{2.134c}$$

Rewriting (2.134) as

$$\dot{x}^{(0)} = A_0x^{(0)} + B_0u^{(0)} \tag{2.135a}$$

$$z^{(0)} = -A_{22}^{-1}\{A_{21}x^{(0)} + B_2u^{(0)}\} \tag{2.135b}$$

$$y^{(0)} = C_0x^{(0)} + D_0u^{(0)} \tag{2.135c}$$

where

$$A_0 = A_{11} - A_{12}A_{22}^{-1}A_{21}; \quad B_0 = B_1 - A_{12}A_{22}^{-1}B_2$$

$$C_0 = C_1 - C_2A_{22}^{-1}A_{21}; \quad D_0 = -C_2A_{22}^{-1}B_2$$

Note that $z^{(0)}(t = 0)$ is, in general, not equal to $z(0)$. The degenerate system (2.135) is also called the 'slow' sub system, where the 'fast' component due to the small parameter ε is eliminated. In order to find the fast component, one way is to stretch the system (2.133) and make $\varepsilon = 0$ assuming the slow variables as constant. That is,

$$\frac{dx_c^{(0)}}{dt'} = 0, \quad x_c^{(0)}(t' = 0) = 0 \tag{2.136a}$$

$$\frac{dz_c^{(0)}}{dt'} = A_{21}x_c^{(0)}(t') + A_{22}z_c^{(0)}(t') + B_2u_c^{(0)}(t'), \, z_c^{(0)}(t' = 0) = z(0) - z^{(0)}(0) \tag{2.136b}$$

$$y_c^{(0)}(t') = C_1x_c^{(0)}(t') + C_2z_c^{(0)}(t') \tag{2.136c}$$

or

$$\frac{dz_c^{(0)}}{dt'} = A_{22}z_c^{(0)}(t') + B_2u_c^{(0)}(t'), \quad z_c^{(0)}(t' = 0) = z(0) - z^{(0)}(0) \tag{2.137a}$$

$$y_c^{(0)}(t') = C_2z_c^{(0)}(t') \tag{2.137b}$$

and

$$x(t,\varepsilon) = x^{(0)}(t) + 0(\varepsilon) \tag{2.138a}$$

$$z(t,\varepsilon) = z^{(0)}(t) + z_c^{(0)}(t') + 0(\varepsilon) \tag{2.138b}$$

$$y(t,\varepsilon) = y^{(0)}(t) + y_c^{(0)}(t') + 0(\varepsilon) \tag{2.138c}$$

It is seen from (2.136)–(2.138) that $z_c^{(0)}(t')$ is the boundary-layer correction for the zeroth-order approximation and that (2.137a) is called the 'fast' subsystem. Thus the subsystems (2.135) and (2.137) operating on different time scales t and t', are called the 'slow' and 'fast' subsystems. The following theorems provide conditions for the controllability, stability and observability of the singularly perturbed system (2.133).

Theorem 2.5: If A_{22} is nonsingular and if the subsystems (2.135) and (2.137) are controllable, i.e.

$$\text{rank}\,[B_0 \vdots A_0B_0 \vdots \ldots \vdots A_0^{m-1}B_0] = m \tag{2.139a}$$

$$\text{rank}\,[B_2 \vdots A_{22}B_2 \vdots \ldots \vdots A_{22}^{n-1}B_0] = n \tag{2.139b}$$

then there exists $\varepsilon_0 > 0$ such that the singularly perturbed system (2.133) is controllable for all $0 < \varepsilon \leq \varepsilon_0$ (Kokotovic and Haddad, 1975a; Chow, 1977; Sannuti, 1977).

Theorem 2.6: If the real parts of all the eigenvalues of A_{22} are negative and the subsystem (2.135) is asymptotically stable, then there exists a $\varepsilon^* > 0$ such that the system (2.133) is asymptotically stable for all $0 < \varepsilon \leq \varepsilon^*$ (Wilde and Kokotovic, 1972; Klimuschev and Krasovskii, 1961).

Theorem 2.7: If subsystems (2.135) and (2.137) are observable, i.e.

$$\text{rank}\,[C_0 \vdots A_0 C_0 \vdots \dots \vdots A_0^{n-1} C_0] = m \tag{2.14a}$$

$$\text{rank}\,[C_2 \vdots A_{22} C_2 \vdots \dots \vdots A_{22}^{n-1} C_2] = n \tag{2.140b}$$

then there exists a $\varepsilon_1 > 0$ such that system (2.133) is observable for all $0 < \varepsilon \leqslant \varepsilon_1$ (Porter, 1977).

The above three theorems are important because they show that the structural properties (controllability, observability and stability) of the high-order system (2.133) for sufficiently small ε can be deduced from the corresponding properties of the two auxiliary, lower-order subsystem (2.135) and (2.137).

2.4 Singular perturbation method for discrete systems

In this Section, we present the singular perturbation method for discrete systems and bring out various features. It is based on the works of Rajagopalan and Naidu (1980a), Naidu and Rao (1981, 1982) and Rao and Naidu (1981). We discuss the initial and boundary-value problems separately.

2.4.1 Initial-value problems*

Consider the IVP (2.44) with $x(k=0) = x(0)$ and $z(k=0) = z(0)$. In order to get a clear insight into the method, consider the simplified second order form of (2.44); i.e.

$$x(k+1) = a_{11} x(k) + \varepsilon a_{12} z(k), \quad x(k=0) = x(0) \tag{2.141a}$$

$$z(k+1) = a_{21} x(k) + \varepsilon a_{22} z(k), \quad z(k=0) = z(0) \tag{2.141b}$$

The approximate solution of (2.141), obtained by ignoring terms with coefficients of ε and higher powers of ε, is given by

$$x(k) = a_{11}^k x(0) + \varepsilon^{k+1}\left(-\frac{a_{12}}{a_{11}} d_1 d_2^k\right) \tag{2.142a}$$

$$z(k) = a_{21} a_{11}^{k-1} x(0) + \varepsilon^k (d_1 d_2^k) \tag{2.142b}$$

where

$$d_1 = z(0) - \frac{a_{21}}{a_{11}} x(0); \quad d_2 = \frac{a_{11}a_{22} - a_{12}a_{21}}{a_{22}}$$

The degenerate form of (2.141) is

$$x^{(0)}(k+1) = a_{11} x^{(0)}(k) \tag{2.143a}$$

$$z^{(0)}(k+1) = a_{21} x^{(0)}(k) \tag{2.143b}$$

---

* Reprinted with permission from NAIDU, D. S. and RAO, A. K. (1981): 'Singular perturbation method for initial value problems with inputs in discrete control systems, *Int. J. Control*, 33, pp. 953–965.

Solving the above with $x^{(0)}(k=0) = x(0)$,

$$x^{(0)}(k) = a_{11}^k x(0) \qquad (2.144a)$$

$$z^{(0)}(k) = a_{21}a_{11}^{k-1}x(0) \qquad (2.144b)$$

A close examination of (2.142) and (2.144) reveals that:

(i) As ε tends to zero, the solution of $z(k)$ fails to uniformly converge to the degenerate solution $z^{(0)}(k)$ at $k = 0$, leading to the formation of a boundary layer at $k = 0$.

(ii) Once $x^{(0)}(k)$ is solved for, the solution $z^{(0)}(k)$ is automatically fixed and $z^{(0)}(0)$ is not, in general, equal to $z(0)$.

(iii) The initial condition $z(0)$, that is lost in the process of degeneration, can be recovered in the form of a boundary-layer correction (BLC) series.

The structure (2.142) suggests the transformations for the correction series to be as

$$v(k) = \frac{x(k)}{\varepsilon^{k+1}}; \quad w(k) = \frac{z(k)}{\varepsilon^k} \qquad (2.145)$$

The recovery of the lost initial condition $z(0)$ is demonstrated as follows.

Using (2.145) in (2.141), the transformed problem becomes

$$\varepsilon v(k+1) = a_{11}v(k) + a_{12}w(k) \qquad (2.146a)$$

$$w(k+1) = a_{21}v(k) + a_{22}w(k) \qquad (2.146b)$$

Note that (2.146) is still in the singularly perturbed form of type-D model (2.48). The degenerate form ($\varepsilon = 0$) of (2.146) is

$$w^{(0)}(k+1) = d_2 w^{(0)}(k) \qquad (2.147a)$$

$$v^{(0)}(k) = -\frac{a_{12}}{a_{11}} w^{(0)}(k) \qquad (2.147b)$$

Choosing $w^{(0)}(0) = z(0) - z^{(0)}(0)$ and substituting $z^{(0)}(0) = (a_{21}/a_{11})x(0)$ from (2.144b), the solution for (2.147) is

$$v^{(0)}(k) = -\frac{a_{21}}{a_{11}} d_1 d_2^k; \quad w^{(0)}(k) = d_1 d_2^k \qquad (2.148)$$

where d_1 and d_2 are as given in (2.142).

Note that the solutions (2.148) are the same terms associated with ε^{k+1} and ε^k in (2.142). The total series solution for the zeroth-order approximation is

$$x(k) = x^{(0)}(k) + \varepsilon^{k+1}v^{(0)}(k) \qquad (2.149a)$$

$$z(k) = z^{(0)}(k) + \varepsilon^k w^{(0)}(k) \qquad (2.149b)$$

which satisfy the given initial conditions. These ideas are now extended to the high-order problem (2.44) and for higher-order approximations.

We know that suppressing ε in (2.44) leads to the degenerate problem (2.47) which satisfies the given m initial conditions $x(0)$ and loses the remaining n initial conditions $z(0)$. To recover these conditions we need to use the transformations (2.145) and form the total series solution as

$$x(k,\varepsilon) = x_0(k,\varepsilon) + \varepsilon^{k+1}v(k,\varepsilon) \tag{2.150a}$$

$$z(k,\varepsilon) = z_0(k,\varepsilon) + \varepsilon^k w(k,\varepsilon) \tag{2.150b}$$

where $x_0(k,\varepsilon)$ and $z_0(k,\varepsilon)$ refer to the outer solution and $v(k,\varepsilon)$ and $w(k,\varepsilon)$ refer to the correction solution.

Substituting (2.150) in (2.44) and separating the outer and correction solutions, the equations for the outer solution are

$$x_0(k+1) = A_{11}x_0(k) + \varepsilon A_{12}z_0(k) + B_1 u(k) \tag{2.151a}$$

$$z_0(k+1) = A_{21}x_0(k) + \varepsilon A_{22}z_0(k) + B_2 u(k) \tag{2.151b}$$

and for the correction solution

$$\varepsilon v(k+1) = A_{11}v(k) + A_{12}w(k) \tag{2.152a}$$

$$w(k+1) \quad A_{21}v(k) + A_{22}w(k) \tag{2.152b}$$

Note that (2.152) is like the singularly perturbed D-model (2.46) and (2.151) is the same as the original problem (2.44), which is henceforth referred to as the outer problem.

Now, assume series solutions as

$$x(k,\varepsilon) = \sum_{j=0}^{\infty} x^{(j)}(k)\varepsilon^j; \quad z(k,\varepsilon) = \sum_{j=0}^{\infty} z^{(j)}(k)\varepsilon^j \tag{2.153a}$$

$$v(k,\varepsilon) = \sum_{j=0}^{\infty} v^{(j)}(k)\varepsilon^j; \quad w(k,\varepsilon) = \sum_{j=0}^{\infty} w^{(j)}(k)\varepsilon^j \tag{2.153b}$$

Note that in the above, we assumed $u(k)$ to be independent of the small parameter ε, for the sake of simplicity only.

Substituting (2.153a) in (2.44) and collecting coefficients of like powers of ε on either side, we get the outer series equations for the zeroth-order approximation as

$$x^{(0)}(k+1) = A_{11}x^{(0)}(k) + B_1 u(k) \tag{2.154a}$$

$$z^{(0)}(k+1) = A_{21}x^{(0)}(k) + B_2 u(k) \tag{2.154b}$$

and for the first-order approximation as

$$x^{(1)}(k+1) = A_{11}x^{(1)}(k) + A_{12}z^{(0)}(k) \tag{2.155a}$$

$$z^{(1)}(k+1) = A_{21}x^{(1)}(k) + A_{22}z^{(0)}(k) \tag{2.155b}$$

Similarly using (2.153b) in (2.152), we get the correction series equations for the zeroth-order approximation as

$$0 = A_{11}v^{(0)}(k) + A_{12}w^{(0)}(k) \tag{2.156a}$$

$$w^{(0)}(k+1) = A_{21}v^{(0)}(k) + A_{22}w^{(0)}(k) \tag{2.156b}$$

and for the first-order approximation as

$$v^{(0)}(k+1) = A_{11}v^{(1)}(k) + A_{12}w^{(1)}(k) \tag{2.157a}$$

$$w^{(1)}(k+1) = A_{21}v^{(1)}(k) + A_{22}w^{(1)}(k) \tag{2.157b}$$

The equations for higher-order approximations can be obtained in a similar way.

We observe that (2.154) and (2.155) are of order m while (2.156) and (2.157) are of order n.

Under the condition that A_{22} is nonsingular, the total series solution, composed of the outer and correction series solutions, is given by

$$x(k,\varepsilon) = \sum_{j=0}^{\infty} [x^{(j)}(k) + \varepsilon^{k+1}v^{(j)}(k)]\varepsilon^{j} \tag{2.158a}$$

$$z(k,\varepsilon) = \sum_{j=0}^{\infty} [z^{(j)}(k) + \varepsilon^{k}w^{(j)}(k)]\varepsilon^{j} \tag{2.158b}$$

The coefficients ε^{k+1} and ε^{k} in (2.158) indicate that the correction series terms $v^{(j)}(k)$ and $w^{(j)}(k)$ need to be solved for few values of k near $k = 0$ depending on the order of approximation.

An important aspect of singular perturbation analysis is the determination of the initial conditions required to solve the outer and correction series equations (2.154)–(2.157). These are obtained from (2.158) using the fact that the outer and correction series solutions together must satisfy all the given initial conditions. That is

$$
\begin{aligned}
x^{(0)}(0) &= x(0); & w^{(0)}(0) &= z(0) - z^{(0)}(0) \\
x^{(1)}(0) &= -v^{(0)}(0); & w^{(1)}(0) &= -z^{(1)}(0)
\end{aligned}
\tag{2.159}
$$

Example 2.9*

Consider a fifth-order problem (Phillips, 1980a),

$$
\begin{bmatrix} x_1(k+1) \\ x_2(k+1) \\ z_1(k+1) \\ z_2(k+1) \\ z_3(k+1) \end{bmatrix} = \begin{bmatrix} 0{\cdot}9014 & 0{\cdot}1179 & 0{\cdot}2100\varepsilon & 0{\cdot}0668\varepsilon & 0{\cdot}0842\varepsilon \\ -0{\cdot}0196 & 0{\cdot}8743 & 0{\cdot}0000\varepsilon & 0{\cdot}1000\varepsilon & 0{\cdot}1174\varepsilon \\ -0{\cdot}0071 & 0{\cdot}7342 & 0{\cdot}8070\varepsilon & 0{\cdot}0520\varepsilon & 0{\cdot}0843\varepsilon \\ -0{\cdot}7500 & -0{\cdot}0557 & -0{\cdot}1280\varepsilon & 0{\cdot}7743\varepsilon & -0{\cdot}0563\varepsilon \\ -0{\cdot}3060 & -0{\cdot}0169 & -0{\cdot}0440\varepsilon & 0{\cdot}5711\varepsilon & 0{\cdot}0529\varepsilon \end{bmatrix} \begin{bmatrix} x_1(k) \\ x_2(k) \\ z_1(k) \\ z_2(k) \\ z_3(k) \end{bmatrix}
$$

(2.160)

with initial conditions

$$x_1(0) = 1{\cdot}0; \quad z_1(0) = 0{\cdot}5; \quad z_3(0) = 0{\cdot}6$$

$$x_2(0) = -0{\cdot}8; \quad z_2(0) = 0{\cdot}2; \quad \varepsilon = 0{\cdot}25$$

The eigenvalues of the above problem are

$$p_{1,2} = 0{\cdot}8777 \pm j0{\cdot}1054; \quad p_3 = 0{\cdot}0179; \quad p_{4,5} = 0{\cdot}2055 \pm j0{\cdot}0236$$

The eigenvalues of the degenerate problem ($\varepsilon = 0$) are

$$p_{1,2} = 0{\cdot}8879 \pm j0{\cdot}0461$$

Using the method described above, the solutions are obtained for degenerate, zeroth- and first-order approximations and are shown in Fig. 2.9. The exact and series solutions coincide for a few values of k near $k = 0$ owing to the fact that the exact solution of (2.160) when evaluated in terms of the given initial conditions will contain terms of ε, ε^2, ..., depending on the value of k and accordingly the exact and first-, second-, etc., order approximations will have the same values up to that value of k.

2.4.2 Boundary-value problems

If the C-model (2.44) is considered as a BVP with, say, $x(k = N) = x(N)$ and $z(k = 0) = z(0)$, a similar analysis as for IVP in the previous Section reveals that the initial condition $z(0)$ is lost in the process of degeneration, and hence a boundary layer is formed at $k = 0$. This means that the various outer and correction series equations and the total series equations are the same as given for IVP. But the boundary conditions needed to solve the corresponding outer and correction series equations are different. That is

* Reprinted with permission from NAIDU, D. S. and RAO, A. K. (1985): 'Application of singular perturbation method to a steam power system', *Electric Power Systems Research*, **8**, pp. 219–226.

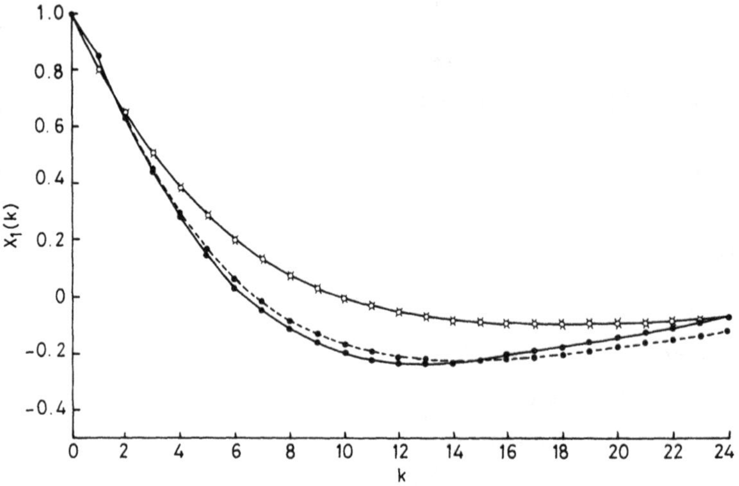

Fig. 2.9a and b *Exact and approximate solutions of $x_1(k)$ and $x_2(k)$ of Example 2.9*

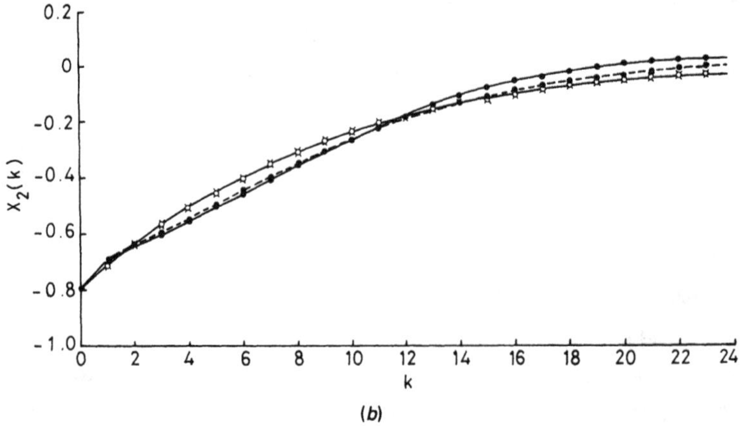

●———●———● exact solution
●---●---● first-order solution
×———×———× zeroth-order solution
○———○———○ degenerate solution
¤———¤———¤ solution common to degenerate and zeroth order

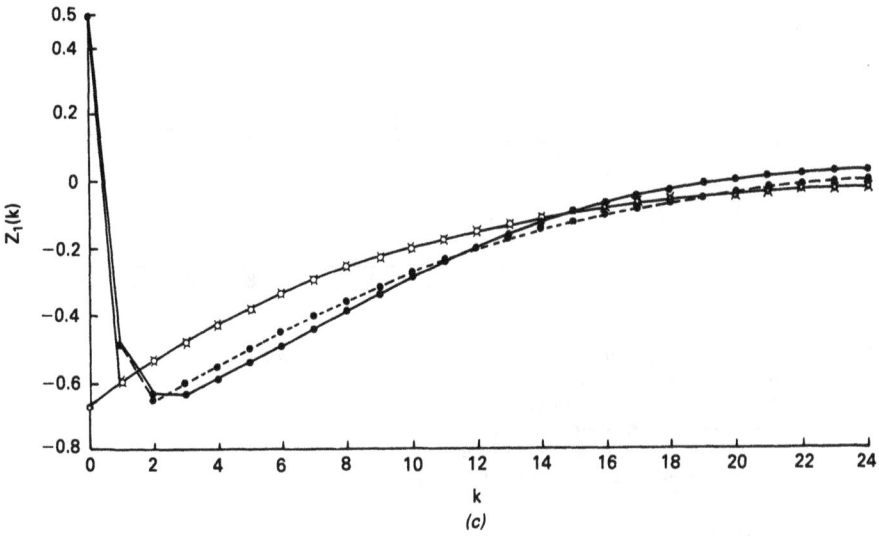

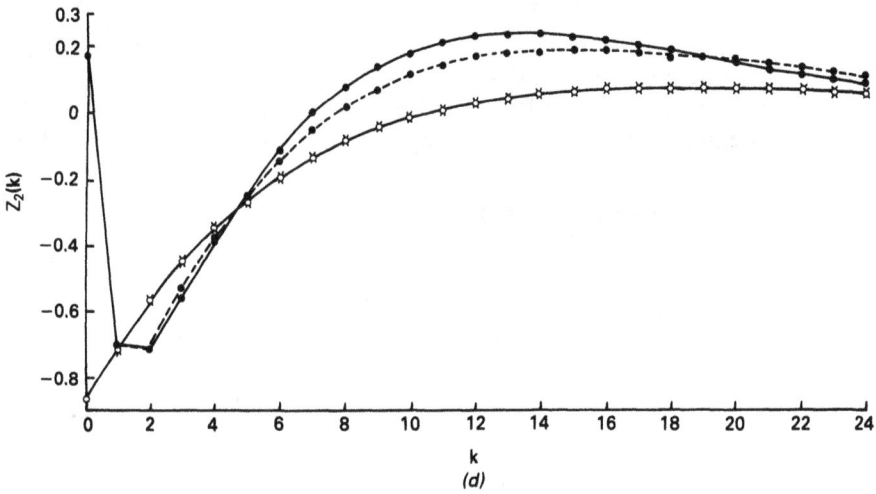

Fig. 2.9c, and d   *Exact and approximate solutions of $z_1(k)$ and $z_2(k)$ of Example 2.9*

●———●———● exact solution
●– – –●– – –● first-order solution
×–––×–––× zeroth-order solution
○———○———○ degenerate solution
¤———¤———¤ solution common to degenerate and zeroth order

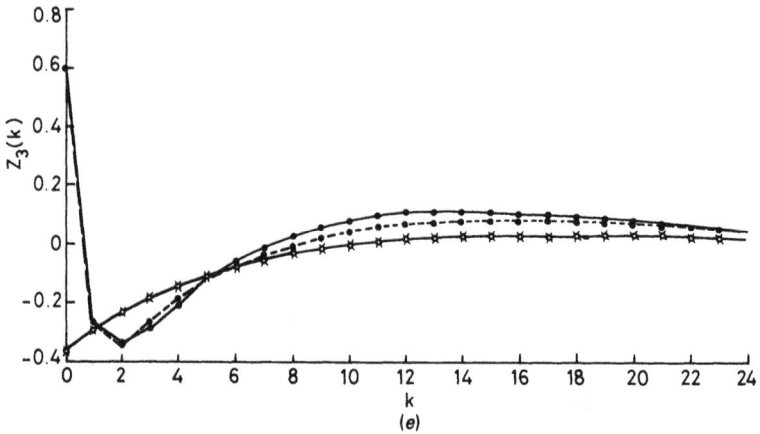

Fig. 2.9e *Exact and approximate solutions of $z_3(k)$ of Example 2.9*

●———●———● exact solution
●- - -●- - -● first-order solution
×———×———× zeroth-order solution
○———○———○ degenerate solution
¤———¤———¤ solution common to degenerate and zeroth order

$$x^{(0)}(N) = x(N); \quad w^{(0)}(0) = z(0) - z^{(0)}(0)$$

$$x^{(1)}(N) = 0; \qquad w^{(1)}(0) = -z^{(1)}(0) \tag{2.161}$$

The R-model (2.45), being only a transformation of the C-model, needs no separate treatment either as IVP or BVP.

The important case for BVP is the D-model (2.46). We shall consider its analysis in detail. For the sake of simplicity, consider the second-order problem of (2.46). That is

$$x(k+1) = a_{11}x(k) + a_{12}z(k) \tag{2.162a}$$

$$\varepsilon z(k+1) = a_{21}x(k) + a_{22}z(k) \tag{2.162b}$$

with boundary conditions $x(k=0) = x(0)$ and $z(k=N) = z(N)$.

The zeroth-order approximation for the solution of (2.162) is

$$x(k) = x(0)d_2^k + \varepsilon^{N-k+1}\left[\frac{a_{12}}{a_{22}}\left(\frac{a_{21}}{a_{22}}d_2^N x(0) + z(N)\right)a_{22}^{k-N}\right] \tag{2.163a}$$

$$z(k) = -\frac{a_{21}}{a_{22}}x(0)d_2^k + \varepsilon^{N-k}\left[\left(\frac{a_{21}}{a_{22}}d_2^N x(0) + z(N)\right)a_{22}^{k-N}\right] \tag{2.163b}$$

The degenerate problem of (2.162) is

$$x^{(0)}(k+1) = d_2 x^{(0)}(k) \tag{2.164a}$$

$$z^{(0)}(k) = -\frac{a_{21}}{a_{22}}x^{(0)}(k) \tag{2.164b}$$

A close examination of (2.162)–(2.164) reveals that

(i) If the degenerate problem (2.164) is solved with $x^{(0)}(k = 0) = x(0)$, the resulting solutions are the same as the first terms (i.e. not involving ε^{N-k+1} and ε^{N-k}) in (2.163).

(ii) As ε tends to zero, the solution $z(k)$ fails to converge uniformly to the degenerate solution $z^{(0)}(k)$ at $k = N$. In other words, $z(N)$ is lost in the process of degeneration. This leads to the formation of a boundary layer at $k = N$.

(iii) The structure of (2.163) indicates that the transformations required for the correction series are

$$v(k) = \frac{x(k)}{\varepsilon^{N-k+1}}; \quad w(k) = \frac{z(k)}{\varepsilon^{N-K}} \tag{2.165}$$

With this background, we are now ready to consider the high-order BVP (2.46). The total solution is sought in terms of the outer solution and correction solution, i.e.

$$x(k,\varepsilon) = x_0(k,\varepsilon) + \varepsilon^{N-k+1}v(k,\varepsilon) \tag{2.166a}$$

$$z(k,\varepsilon) = z_0(k,\varepsilon) + \varepsilon^{N-k}w(k,\varepsilon) \tag{2.166b}$$

The equation for the outer solution is the same as (2.46), whereas the equation for the correction solution, given by substituting (2.165) in (2.46), is

$$v(k+1) = \varepsilon A_{11}v(k) + A_{12}w(k) \tag{2.167a}$$

$$w(k+1) = \varepsilon A_{21}v(k) + A_{22}w(k) \tag{2.167b}$$

The series solutions are assumed to be the same as (2.153). Substitution of (2.153a) in (2.46) and collection of coefficients results in outer series equations for the zeroth-order approximation, as

$$x^{(0)}(k+1) = A_{11}x^{(0)}(k) + A_{12}z^{(0)}(k) + B_1u(k) \tag{2.168a}$$

$$0 = A_{21}x^{(0)}(k) + A_{22}z^{(0)}(k) + B_2u(k) \tag{2.168b}$$

or

$$x^{(0)}(k+1) = (A_{11} - A_{12}A_{22}^{-1}A_{21})x^{(0)}(k) + (B_1 - A_{12}A_{22}^{-1}B_2)u(k)$$

$$z^{(0)}(k) = -A_{22}^{-1}A_{21}x^{(0)}(k) - A_{22}^{-1}B_2u(k)$$

and for the first-order approximation as

$$x^{(1)}(k+1) = A_{11}x^{(1)}(k) + A_{12}z^{(1)}(k) \tag{2.169a}$$

$$z^{(0)}(k+1) = A_{21}x^{(1)}(k) + A_{22}z^{(1)}(k) \tag{2.169b}$$

Similarly, inserting (2.153b) in (2.167) results in correction series equations for the zeroth-order approximation as

$$v^{(0)}(k+1) = A_{12}w^{(0)}(k) \tag{2.170a}$$

$$w^{(0)}(k+1) = A_{22}w^{(0)}(k) \tag{2.170b}$$

and for the first-order approximation as

$$v^{(1)}(k+1) = A_{12}w^{(1)}(k) + A_{11}v^{(0)}(k) \tag{2.171a}$$

$$w^{(1)}(k+1) = A_{22}w^{(1)}(k) + A_{21}v^{(0)}(k) \tag{2.171b}$$

Similar equations can be obtained for higher-order approximations.

The total series solution, consisting of outer and correction series solutions, is given by

$$x(k,\varepsilon) = \sum_{j=0}^{\infty} [x^{(j)}(k) + \varepsilon^{N-k+1}v^{(j)}(k)]\varepsilon^j \tag{2.172a}$$

$$z(k,\varepsilon) = \sum_{j=0}^{\infty} [z^{(j)}(k) + \varepsilon^{N-k}w^{(j)}(k)]\varepsilon^j \tag{2.172b}$$

Here we assume that A_{22} is nonsingular.

The boundary conditions required for solving the series equations (2.168)–(2.171) are obtained from (2.172) and are given as

$$\left. \begin{array}{ll} x^{(0)}(0) = x(0); & w^{(0)}(N) = z(N) - z^{(0)}(N) \\ x^{(1)}(0) = 0; & w^{(1)}(N) = -z^{(1)}(N) \end{array} \right\} \tag{2.173}$$

Example 2.10*
Consider a third-order BVP as

$$\begin{bmatrix} x(k+1) \\ \varepsilon z_1(k+1) \\ \varepsilon z_2(k+1) \end{bmatrix} = \begin{bmatrix} 0 & 1 & 0 \\ 0 & 0 & 1 \\ 0{\cdot}6 & -0{\cdot}89 & 1{\cdot}83 \end{bmatrix} \begin{bmatrix} x(k) \\ z_1(k) \\ z_2(k) \end{bmatrix} + \begin{bmatrix} 0 \\ 0 \\ 1 \end{bmatrix} u(k) \tag{2.174}$$

with boundary conditions

$$x(0) = 2{\cdot}5; \quad z_1(8) = 8{\cdot}0; \quad z_2(8) = 1{\cdot}0$$

and $\varepsilon = 0{\cdot}1$ and $u(k)$ is a unit step function.

The eigenvalues of the problem are evaluated as

$$p_1 = 10{\cdot}0; \quad p_2 = 7{\cdot}5; \quad p_3 = 0{\cdot}8$$

and the eigenvalue of the degenerate model as $p = 0{\cdot}6742$ which show the singular perturbation character.

Using the method described above, the series equations up to the second approximation are obtained as shown in Fig. 2.10, which clearly illustrates

---

* Reprinted with permission from RAO, A. K. and NAIDU, D. S. (1981): 'Singularly perturbed boundary value problems in discrete systems', *Int. J. Control*, **34**, pp. 1163–1173.

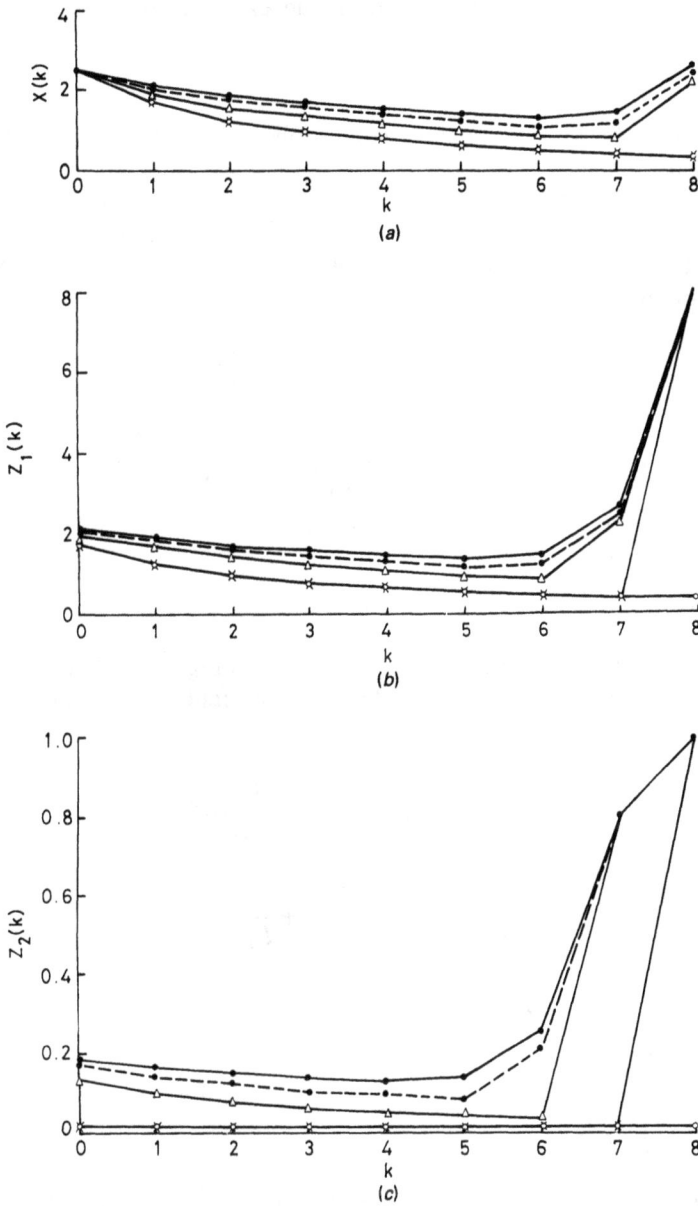

Fig. 2.10a, b and c *Exact and approximate solutions of Example 2.10*

●———●———● exact solution
●– – –●– – –● second-order solution
△———△———△ first-order solution
×– – –×– – –× zeroth-order solution
○———○———○ degenerate solution
¤———¤———¤ solution common to degenerate and zeroth order

that the approximate solutions approach the exact solution with increased order of approximation.

Example 2.11*
As a final example, we given an interesting application of the singular perturbation method.

Here, the transient behaviour of a two-winding transformer subjected to a step input is considered. The governing differential equations are cast in the singularly perturbed form by relating the coefficient of coupling with a small parameter.

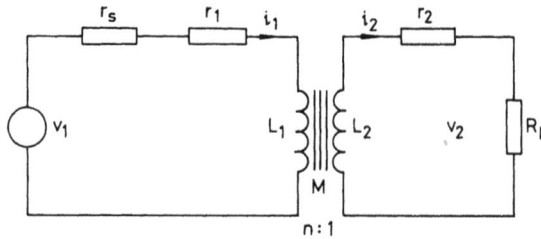

Fig. 2.11 *Two-winding transformer of Example 2.11*

Consider the two-winding transformer shown in Fig. 2.11. The state equations for the output voltage v_2 and the input current i_1 are easily obtained as

$$(1 - k^2)\frac{dv_2}{dt} = -\frac{R_2}{L_2}v_2 - \frac{kR_1R_2}{(L_1L_2)^{1/2}}i_1 + \frac{kR_L}{(L_1L_2)^{1/2}}v_1 \qquad (2.175a)$$

$$(1 - k^2)\frac{di_1}{dt} = -\frac{kR_2}{R_L(L_1L_2)^{1/2}}v_2 - \frac{R_1}{L_1}i_1 + \frac{1}{L_1}v_1 \qquad (2.175b)$$

where

$$R_1 = r_1 + r_s$$
$$R_2 = r_2 + R_L$$
$$k = M/(L_1L_2)^{1/2}$$

Since the coefficient of coupling k is nearly equal to unity, the quantity $1 - k^2$ is very small. Hence, a small parameter $\varepsilon = 1 - k^2$ is introduced. This means that the two derivatives in (2.175) are being multiplied by the

---

* Reprinted with permission from NAIDU, D. S. and SEN, S. (1982): 'Singular perturbation method for the transient analysis of a transformer', *Electric Power Systems Research*, **5**, pp. 307–313.

small parameter ε hence they are not in the singularly perturbed form. In order to avoid this, the following state variables are defined as

$$x = v_2 - R_L(L_1/L_2)^{1/2}i_1 \qquad\qquad\qquad (2.176a)$$

$$z = i_1 \qquad\qquad\qquad (2.176b)$$

Using (2.176) in (2.175), the singularly perturbed problem for a step input of $v_1 = U$ becomes

$$\frac{dx}{dt} = A_1x + A_2z + B_1U \quad x(t = 0) = x(0) \qquad\qquad (2.177a)$$

$$\varepsilon\frac{dz}{dt} = (A_3 + \varepsilon A_4)x + (A_5 + \varepsilon A_6)z + B_2U \quad z(t = 0) = z(0) \qquad (2.177b)$$

where

$$A_1 = -R_2/2L_2$$
$$A_2 = -R_L(R_2L_1 - R_1L_2)/2L_2(L_1L_2)^{1/2}$$
$$A_3 = -R_2/R_L(L_1L_2)^{1/2}$$
$$A_4 = R_2/2R_L(L_1L_2)^{1/2}$$
$$A_5 = -(R_2/L_2 + R_1/L_1), \quad A_6 = R_2/2L_2$$
$$B_1 = -R_L/2(L_1/L_2)^{1/2}, \quad B_2 = 1/L_1$$

$v_1 = U$ is the step input, and k is approximated by $1 - \varepsilon/2$.

Using the singular perturbation method explained in Section 2.3 the various results for the zeroth-order and first-order approximations are summarised below:

(i) *Zeroth-order approximation*
The outer series coefficients are:

$$x^{(0)}(t) = C_3 + [x(0) - C_3]\exp(C_1t)$$
$$z^{(0)}(t) = C_5 + C_4 - \exp(C_1t)$$

The inner series coefficients are

$$\bar{x}^{(0)}(t') = x(0)$$
$$\bar{z}^{(0)}(t') = D_1 + [z(0) - D_1]\exp(A_5t')$$

The intermediate series coefficients are

$$\underline{x}^{(0)}(t') = x(0)$$
$$\underline{z}^{(0)}(t') = z^{(0)}(0)$$

(ii) *first-order approximation*

The initial condition $x^{(1)}(0)$ is given by

$$x^{(1)}(0) = [z^{(0)}(0) - z(0)]A_2/A_5$$

The outer-series coefficients are

$$x^{(1)}(t) = [C_7/C_1 + x^{(1)}(0) + C_6t] \exp(C_1 t) - C_7/C_1$$

$$z^{(1)}(t) = -(A_3/A_5)\{[C_7/C_1 + x^{(1)}(0) + C_6t] \exp(C_1 t) - C_7/C_1\}$$
$$+ (C_6/A_2) \exp(C_1 t) + C_7/A_2$$

The inner-series coefficients are given by

$$\tilde{x}^{(1)}(t') = -D_4 + D_3 t' + D_4 \exp(A_5 t')$$

$$\tilde{z}^{(1)}(t') = D_6 + D_5 t' - (D_6 - D_7 t') \exp(A_5 t')$$

The intermediate-series coefficients are

$$\underline{x}^{(1)}(t') = x^{(1)}(0) + D_3 t'$$

$$\underline{z}^{(1)}(t') = z^{(1)}(0) + C_1 C_4 t'$$

where

$$C_1 = A_1 - A_2 A_3/A_5;\ C_2 = B_1 - A_2 B_2/A_5;\ C_3 = -C_2 U/C_1$$

$$C_4 = -[x(0) - C_3]A_3/A_5;\ C_5 = -B_2 U/A_5 - A_3 C_3/A_5$$

$$C_6 = [C_1 C_4 - A_4 x(0) + A_4 C_3 - A_6 C_4]A_2/A_5;\ C_7 = -(A_4 C_3 + A_6 C_5)A_2/A_5$$

$$D_1 = -[A_3 x(0) + B_2 U]/A_5;\ D_2 = A_2 z(0) - A_2 D_1;$$

$$D_3 = B_1 U + A_2 D_1 + A_1 x(0)$$

$$D_4 = D_2/A_5;\ D_5 = -A_3 D_3/A_5;$$

$$D_6 = -[A_6 D_1 - A_3 D_4 + A_4 x(0)]/A_5 - A_3 D_3/A_5^2$$

$$D_7 = A_6 z(0) + A_3 D_4 - A_6 D_1$$

For details, see Naidu and Sen (1982).

2.5 Discussion and conclusion

In this Chapter, we presented the singular perturbation method along with its characteristic features in continuous and discrete control systems. At first, we started with the idea of singular perturbation in differential equations. The theory has a rich literature well based on mathematical foundations as evidenced by excellent survey articles, monographs and books (Vasileva, 1963; Van Dyke, 1964; Wasow, 1965; Cole, 1968; Butuzov *et al.*, 1970; Lagerstrom and Casten, 1972; Nayfeh, 1973, 1981, 1985; Vasileva and Butuzov, 1973,

1978; Eckhaus, 1973, 1979; O'Malley, 1968, 1974*a*; Nayfeh and Mook, 1979; Campbell, 1980, 1982; Eckhaus and de Jager, 1982; Chang and Howes, 1984; Tikhonov *et al.*, 1984; Smith, 1985).

There are two aspects of singular perturbation in differential equations; namely regular degeneration and asymptotic expansions for IVP (2.48). The regular degeneration requires the stability of boundary-layer equation (2.52). Tikhonov (1952) stated that if a certain solution of (2.52) is asymptotically stable as $t' \to \infty$, then (2.48) degenerates regularly for each t restricted in a compact interval. His assumption was partly inadequate, and was corrected by Hoppensteadt (1971) who also gave a proof of Tikhonov's theorem and extended the result to the case $T = \infty$ in (2.53). This theorem has recently been reconsidered by Nipp (1983, 1985, 1986).

Flatto and Levinson (1955), Levin (1956) and Levin and Levinson (1954) studied the regular degeneration of (2.48) under the conditional stability of (2.52). Other studies on the conditional stability were made by Klimushev and Krasovskii (1961), Coppel (1967) and Hoppensteadt (1971).

Several contributions have been made on the second aspect of construction of asymptotic expansions. From a practical point of view, the second aspect is more important in obtaining the approximate solutions. Several approaches were taken by Vasileva (1963), Kaplun (1967), Hoppensteadt (1971) and O'Malley (1974*a*). Vasileva's method consists of obtaining the asymptotic solution in terms of outer, inner and intermediate series expansions. Kaplun (1967) obtained solutions by the method of matched asymptotic expansions (or inner and outer expansions). The basic ideas involved were present in Prandtl's boundary-layer theory (Prandtl, 1905; Friedrichs, 1955; Van Dyke, 1964; Fraenkel, 1969*a,b,c*). In Hoppensteadt (1971) and O'Malley (1971), the solution to (2.48) was constructed using the boundary-layer method.

In singular perturbations, the boundary-value problems (BVP) are of major interest. Several approaches have been made (Vishik and Lyusternik, 1957; Harris, 1960, 1973; Tupchiev, 1962; O'Malley, 1968, 1974a; Hoppensteadt, 1971).

The idea of singular perturbations in differential equations was first introduced to control theory by Kokotovic and Sannuti (1968), and there has been spectacular progress in this direction (Kokotovic and Perkins, 1972; Kokotovic *et al.*, 1976; Ardema, 1983*a*; Kokotovic, 1984, 1985; Saksena *et al.*, 1984). Particular problems considered are multivariable systems with state feedback (Porter, 1974, 1977, 1982; Suzuki and Muira, 1976; O'Reilly, 1979*a*; Grujic, 1979*a,b*); Bradshaw and Porter, 1979, 1981; Sannuti and Wasan, 1983; Kokotovic, 1984, 1985; Khorasani and Kokotovic, 1986), output feedback and observers (Balas, 1978; Chemouil and Wahdan, 1980; Fossard and Magni, 1980; O'Reilly, 1979*a,b*, 1980, 1983*a,b*, 1986; Javid, 1980, 1982; Khalil, 1981*b*; Saksena and Cruz, 1981*a*), eigenvalue and pole-placement problems (Chow, 1978*a*; Boglaev, 1979; Allemong and Kokotovic, 1980; Freiling, 1984), stability, sensitivity and controllability problems (Desoer and Shensa, 1970;

Shensa, 1971; Wilde and Kokotovic, 1972*a,b*; Siljak, 1972; Porter, 1974; Porter and Shenton, 1975a; Suzuki and Miura, 1976; Chow, 1977, 1978*a,b*; Desoer, 1977; Sannuti, 1977, 1978; Campbell, 1980, 1982; Sandell, 1979; Ioannou, 1981; Saksena and Kokotovic, 1981; Suzuki, 1981; Koda, 1982; Doraiswami, 1982; Khorasani and Pai, 1985*a,b*; Saberi and Khalil, 1985; Abed, 1986*a,b*; Steinberg, 1986) and other control problems (Moore, 1976; Hickin and Sinha, 1980; Fossard and Magni, 1980; Suzuki, 1981; Verghese *et al.*, 1981; Cobb, 1981; Fernando and Nicholson, 1982*a,b*, 1983*a,b*; Young and Kokotovic, 1982; Duc *et al.*, 1983; Eitelberg, 1982, 1983, 1985; Lastman *et al.*, 1984; Sobolev, 1984; Karpinskaya *et al.*, 1985).

An important problem in singular perturbations concerns the Van der Pol equation, periodic solutions, resonance, phase plane analysis and related topics (Friedrichs and Wasow, 1946; Flatto and Levinson, 1955; Halanay, 1966; Ackerberg and O'Malley, 1970; Ponzo and Wax, 1972*b*; Cohen, 1971, 1973; Weinstein and Smith, 1975; Guardabassi and Locatelli, 1975; O'Malley, 1976*b*; Grasman *et al.*, 1976; Wolfe, 1979; Koppel, 1979; Reiss, 1980; Guchenheimer, 1980*a,b*; Kassoy, 1982; Feilnichel, 1983*a*; Hoppensteadt, 1983; Keener, 1983; MacGillivray, 1983; Abed, 1985*a,b,c*; Hekimova and Bainov, 1985; Lange and Miura, 1985; Singh, 1986).

The fact that the theory of difference equations is in most respects akin to that of ordinary differential equations leads to the singular perturbation analysis of difference equations (Butuzov and Vasileva, 1970; Comstock and Hsiao, 1976; Hoppensteadt and Miranker, 1979; Reinehardt, 1979). Based on this, singular perturbation methods have been successfully developed for discrete control systems (Locatelli and Schiava, 1976; Phillips, 1980*a*, 1983; Rajagopalan and Naidu, 1980; Mahmoud, 1982*a,b*, 1986; Mahmoud and Singh, 1981*a*, 1984; Naidu and Rao, 1981, 1982, 1985*a,b*; Rao and Naidu, 1981; Atluri and Kao, 1981; Syrcos and Sannuti, 1983; Kando and Iwazumi, 1983*a,b*, 1984, 1985; Fernando and Nicholson, 1982*a,b*, 1983*a,b*; Litkouhi and Khalil, 1984, 1985; Tran and Sawan, 1983*a,b*, 1984*a,b,c*; Dragan, 1985).

An important source for difference equations is the numerical analysis of singular perturbation problems described by ordinary differential equations (Pearson, 1968; Bjurel *et al.*, 1970; Abrahamson *et al.*, 1974; Wiiloughby, 1974; Hemker and Miller, 1979; Miranker, 1980; Kreiss *et al.*, 1986).

At this stage it is necessary to bring out the various characteristic features of singular perturbations in differential and difference equations describing continuous and discrete control systems respectively.

(i) In a differential equation, the singularly perturbed structure is obtained only when the small parameter appears at the left end, i.e., multiplies the highest derivative. On the other hand, in a difference equation, the small parameter appearing either at left, right or both ends provides the singularly perturbed structure.

(ii) In a differential equation, the location of the boundary layer is decided by the sign of the coefficients of the differential equation, whereas in a difference equation the location of the boundary layer depends on the position (left, right or both ends) of the singular perturbation parameter.

(iii) In both cases, the degeneration process reduces the order of the system and loses some of the given auxiliary conditions.

(iv) The stretching transformations to the original continuous system remove the 'singularly perturbed character'. But the transformed discrete system still possesses the singular perturbation character, thus retaining the advantage of order reduction even in the computation of correction solutions.

(v) The approximate solution is given by the outer plus the difference between the inner and intermediate (or equivalent BLC) series solutions in the case of continuous problems, and it is given by the outer and correction series solutions for discrete problems.

(vi) A very important feature in discrete system is that the correction solution has multiplicative coefficients like ε^k *and* ε^{N-k} which are responsible for the fact that the correction solution needs to be evaluated only for a few values of k out of the given complete range. This is unlike the case of the continuous system, where the correction solution has to be evaluated for the given complete range of the time interval.

(vii) A peculiar character found in some problems of discrete systems is that the approximate solution and the exact solution are identical for a few values of k depending on the order of approximation. This is not found in continuous systems.

Chapter 3

Time scale analysis

The analysis and control of large-scale systems has always been a formidable task, not only due to its high order but also due to the fact that the majority of these systems possess interacting dynamic phenomena of widely different speeds giving rise to widely separated clusters of eigenvalues. In practice, we come across a number of systems having this character. For example, in a power-system model, voltage and frequency transients range from intervals of seconds, corresponding to generator voltage regulator, speed governor action and shaft energy storage, to several minutes corresponding to load-voltage regulator action, prime-mover fuel transfer times and thermal energy storage (Lapidus *et al.*, 1974).

The interaction of slow and fast phenomena in high-order systems makes them 'stiff', which demands expensive numerical routines. A stiff system has therefore a two-time-scale (slow and fast) property and need not, in general, be in the singularly perturbed structure with a small parameter multiplying the highest derivative or some of the state variables of the state equation. In other words, the singularly perturbed system is only one form of stiff or two-time-scale systems.

In this Chapter, we focus our attention on the time-scale analysis of continuous and discrete systems. Section 3.1 describes a block diagonalisation procedure to decouple a continuous system having the two-time property into a slow and a fast subsystem. The aspects of permutation and scaling are discussed in Section 3.2. Section 3.3 provides an interesting fact that singularly perturbed continuous systems can be viewed as two-time-scale systems. The feedback design for eigenvalue placement is the content of Section 3.4. Sections 3.5–3.7 discuss the above aspects of block diagonalisation and feedback design with reference to discrete systems. The last Section summarises the Chapter with a brief discussion.

3.1   Block diagonalisation of continuous systems

3.1.1   Problem and derivation

We consider a general two-time-scale stable system possessing two widely separated groups of eigenvalues, as

$$\dot{x} = A_{11}x + A_{12}z + B_1u \tag{3.1a}$$

$$\dot{z} = A_{21}x + A_{22}z + B_2u \tag{3.1b}$$

where x and z are m- and n-dimensional state vectors and u is an r-dimensional control vector, and the matrices A_{ij} and B_i are of appropriate dimensionality. We note that the association of a small parameter ε with $\dot{z}$ puts (3.1) in the singularly perturbed form. We assume that the m eigenvalues of the system (3.1) are small (or close to the origin) and the remaining n eigenvalues are large (or far from the origin), giving slow and fast responses respectively. In other words, the system (3.1) has m dominant modes and n non-dominant modes. We arrange the eigenspectrum $e(A)$ of system (3.1) in increasing order of absolute values as

$$e(A) = \{p_{s1}, \ldots, p_{sm}, p_{f1}, \ldots, p_{fn}\} \tag{3.2a}$$

$$e(A_s) = \{p_{s1}, \ldots, p_{sm}\} \tag{3.2b}$$

$$e(A_f) = \{p_{f1}, \ldots, f_n\} \tag{3.2c}$$

where

$$0 < |p_{s1}| < |p_{s2}| < \ldots < |p_{sm}| < |p_{f1}| < \ldots < |p_{fn}| \tag{3.2d}$$

The system (3.1) is said to possess a two-time-scale property, if the largest absolute eigenvalue of the slow eigenspectrum $e(A_s)$ is much smaller than the smallest absolute eigenvalue of the fast eigenspectrum $e(A_f)$ (Kokotovic, 1975; Phillips, 1980b; Mahmoud and Singh, 1981a). That is

$$\varepsilon = |p_{sm}|/|p_{f1}| \ll 1 \tag{3.3}$$

where the small parameter ε is a measure of separation of time scales. It should be noted that ε represents an intrinsic property of system (3.1) and does not have to appear explicitly in the system.

The basic idea of using time-scale analysis in obtaining low-order models is to decouple the slow and fast modes. This is achieved by using two-stage linear transformations (Chang, 1974). The first stage is to use

$$z_f = z + Lx \tag{3.4}$$

in (3.1) and choose the $(n \times m)$ matrix L such that

$$LA_{11} + A_{21} - LA_{12}L - A_{22}L = 0 \tag{3.5}$$

Then system (3.1) transforms into

$$\begin{bmatrix} \dot{x} \\ \dot{z}_f \end{bmatrix} = \begin{bmatrix} A_s & A_{12} \\ 0 & A_f \end{bmatrix} \begin{bmatrix} x \\ z_f \end{bmatrix} + \begin{bmatrix} B_1 \\ B_f \end{bmatrix} u \tag{3.6}$$

where

$$A_s = A_{11} - A_{12}L; \quad A_f = A_{22} + LA_{12}$$
$$B_f = B_2 + LB_1$$

The second stage is to apply the transformation

$$x_s = x - Mz_f \tag{3.7}$$

to (3.6) and choose the $(m \times n)$ matrix M such that

$$A_s M - MA_f + A_{12} = 0 \tag{3.8}$$

Then (3.6) transforms into

$$\begin{bmatrix} \dot{x}_s \\ \dot{z}_f \end{bmatrix} = \begin{bmatrix} A_s & 0 \\ 0 & A_f \end{bmatrix} \begin{bmatrix} x_s \\ z_f \end{bmatrix} + \begin{bmatrix} B_s \\ B_f \end{bmatrix} u \tag{3.9}$$

where $B_s = B_1 - MLB_1 - MB_2$.

The system (3.9) has the desired decoupled form where the slow and fast variables x_s and z_f can be solved independently of each other. Note that the transformations (3.4) and (3.7) relate the slow and fast variables x_s and z_f with the original variables x and z as

$$\begin{bmatrix} x_s \\ z_f \end{bmatrix} = \begin{bmatrix} I_1 - ML & -M \\ L & I_2 \end{bmatrix} \begin{bmatrix} x \\ z \end{bmatrix} \tag{3.10}$$

which interestingly has the inverse transformation

$$\begin{bmatrix} x \\ z \end{bmatrix} = \begin{bmatrix} I_1 & M \\ -L & I_2 - LM \end{bmatrix} \begin{bmatrix} x_s \\ z_f \end{bmatrix} \tag{3.11}$$

where I_1 and I_2 are unity matrices of appropriate dimensions. Note that (3.11) requires no matrix inversion.

3.1.2   Numerical aspects

We now consider the numerical solutions of the nonlinear equations (3.5) and (3.8) to obtain the values of L and M using an iterative procedure (Kokotovic, 1975; Avramovic et al., 1980). From (3.5),

$$L_{i+1} = A_{22}^{-1}(A_{21} + L_i A_{11} - L_i A_{12} L_i) \tag{3.12a}$$

$$L_0 = A_{22}^{-1} A_{21} \tag{3.12b}$$

We assume A_{22} to be nonsingular and ensure convergence of the iterative procedure. If we let

$$L = L_0 + D \tag{3.13}$$

then D is the solution of

$$DA_0 - (A_{22} + L_0 A_{12})D - DA_{12}D + L_0 A_0 = 0 \tag{3.14}$$

where

$$A_0 = A_{11} - A_{12}L_0$$

It has been shown that if the inequality (Kokotovic, 1975)

$$||A_{22}^{-1}|| \leq \frac{1}{3}(||A_0|| + ||A_{12}|| \, ||L_0||)^{-1} \tag{3.15}$$

is satisfied then a unique real root of (3.14) exists satisfying

$$0 \leq ||D|| \leq \frac{2||A_0|| \, ||L_0||}{(||A_0|| + ||A_{12}|| \, ||L_0||)} \tag{3.16}$$

From (3.5), (3.13) and (3.16), it can be readily seen that

$$||L_{i+1} - L|| \leq ||L_i - L|| \tag{3.17}$$

Similarly the iterative procedure to solve for M from (3.8) is

$$M_{i+1} = [(A_{11} - A_{12}L)M_i - M_i LA_{12}]A_{22}^{-1} + A_{12}A_{22}^{-1} \tag{3.18a}$$

with $\quad M_0 = A_{12}A_{22}^{-1} \tag{3.18b}$

It is found that (3.15) is a conservative condition, meaning that, for a majority of two-time-scale systems, the iterative algorithms (3.12) and (3.18) converge while the inequality (3.15) is satisfied without the factor $1/3$.

3.1.3 *Basic properties*

From (3.3) and (3.9), we can rewrite (3.3) as

$$\varepsilon = |\max\{e(A_s)\}|/|\min\{e(A_f)\}| \ll 1 \tag{3.19}$$

Using the norm properties of invertible matrices we have

$$||A_f^{-1}|| \ll ||A_s||^{-1} \tag{3.20}$$

From (3.12)

$$L_1 - L_0 = A_{22}^{-1}L_0 A_0 \tag{3.21a}$$

and hence $||L_1 - L_0|| \leq \varepsilon ||L_0|| \tag{3.21b}$

where $||A_{22}^{-1}|| \leq \varepsilon ||A_0||^{-1} \tag{3.21c}$

If the separation parameter ε is small and positive, we can write

$$||A_{22}^{-1}|| \ll ||A_0||^{-1} \tag{3.22}$$

Now $L = L_0 + 0(\varepsilon) = A_{22}^{-1}A_{21} + 0(\varepsilon)$ \hfill (3.23)

Also, $A_{22}^{-1}LA_{12} = 0(\varepsilon)$ \hfill (3.24a)

This gives $A_f = A_{22}[I_2 + A_{22}^{-1}LA_{12}]$

$$= A_{22}[I_2 + 0(\varepsilon)] \tag{3.24b}$$

or $||A_f^{-1}|| = [1 + 0(\varepsilon)] \, ||A_{22}^{-1}||$ \hfill (3.24c)

Similarly, $A_s = A_{11} - A_{12}L = A_{11} - A_{12}L_0 + 0(\varepsilon)$

$$= A_0 + 0(\varepsilon) \tag{3.25}$$

and

$$M = M_0 + 0(\varepsilon)$$

$$= A_{12}A_{22}^{-1} + 0(\varepsilon) \tag{3.26}$$

From the above analysis we conclude that a sufficient condition for separation of time scales is

$$||A_{22}^{-1}|| \leq [||A_0|| + ||A_{12}|| \, ||L_0||]^{-1} \tag{3.27}$$

From (3.24) and (3.25), we get

$$e(A_s) = e(A_0) [1 + 0(\varepsilon)] \tag{3.28a}$$

$$e(A_f) = e(A_{22}) [1 + 0(\varepsilon)] \tag{3.28b}$$

This shows that we get the first approximations of the slow and fast modes from the eigenvalues of A_0 and A_{22}, respectively.

Using the approximations of (3.24) and (3.25) in the unforced case of (3.9), we get

$$\dot{x}_{s0} = A_0 x_{s0} \tag{3.29a}$$

$$\dot{z}_{f0} = A_{22}z_{f0} \tag{3.29b}$$

where $A_0 = A_{11} - A_{12}A_{22}^{-1}A_{21}$.

Accordingly, the original states are given from (3.11) as

$$x = x_{s0} + M_0 z_{f0}$$

$$= x_{s0} + A_{12}A_{22}^{-1}z_{f0} + 0(\varepsilon) \tag{3.30a}$$

$$z = -L_0 x_{s0} + z_{f0} - L_0 M_0 z_{f0}$$

$$= -A_{22}^{-1}A_{21}x_{s0} + z_{f0} + 0(\varepsilon) \tag{3.30b}$$

We have seen how an $(m + n)$th order two-time-scale system (3.1) is decoupled into slow and fast subsystems of order m and n, respectively. To a first approximation, it is given by (3.29) involving the original submatrices A_{11}, A_{12}, A_{21} and A_{22}. For better approximations, we have to resort to an iterative procedure for the solution of L from (3.12), thereby getting A_s and A_f required for the decoupled system (3.9). From (3.6), we note that A_s and A_f are independent of M, which is required only for calculation of B_s in the case of a system with input.

An important assumption in the analysis is that the decoupled system (3.9) is completely controllable, which implies that the pairs A_s, B_s and A_f, B_f are controllable.

3.2 Permutation and scaling

In practice, a real physical two-time-scale model has its state variables arranged in an arbitrary order and the units of the state variables may be out of scale. Then the corresponding system matrix may not be in a form satisfying the inequalities (3.15) and (3.22). In such cases, the two-time-scale property can be exhibited by simple transformations like permutation and scaling.

Permutation is necessary to rearrange the given state variables such that the first m states of the transformed state vector correspond to the slow states and the next n states correspond to the n fast states. Scaling is required for readjustment of units and is used to reduce the norms of A_{22}^{-1}, A_0, A_{12} and L_0 as much as possible. In getting (3.20) from (3.19), the matrix eigenvalue and norm relationship

$$\max |e(A)| \leqslant \|A\| \tag{3.31}$$

where A is a square matrix, is used. If the matrix is ill-conditioned, the above relationship becomes

$$\max |e(A)| \ll \|A\| \tag{3.32}$$

Hence, if any one of the matrices A_{22}^{-1}, A_0, A_{12} and L_0 is ill-conditioned, the conditions (3.15) and (3.22) may be violated even though the system possesses inherently a two-time-scale property.

For a given system

$$\dot{y} = Ay \tag{3.33}$$

we use a scaling matrix

$$S = \text{diag}\,(D_m, D_n) \tag{3.34}$$

where D_m and D_n are diagonal matrices of dimensions m and n such that

$$y = Sy_s \tag{3.35}$$

Then the new state matrix is given by

$$y_s = SAS^{-1}y_s = \begin{bmatrix} \bar{A}_{11} & \bar{A}_{12} \\ \bar{A}_{21} & \bar{A}_{22} \end{bmatrix} y_s \qquad (3.36)$$

and obviously

$$e(\bar{A}_{22}) = e(A_{22}) \qquad (3.37a)$$

$$e(\bar{A}_0) = e(A_0) \qquad (3.37b)$$

Thus, scaling does not change the eigenvalues of the submatrices A_{22} and A_0.

At present there is no systematic procedure for scaling and no unique choice of permutation matrices. Most likely, their choices depend on the individual system matrix under consideration. However, a few guidelines are given below (Chow, 1982):

(i) If the A matrix is sparse, then the state variables can be arranged in ascending order of the magnitudes of the diagonal elements, because the diagonal elements correspond very roughly to the eigenvalues.

(ii) If the row norm of the off-diagonal elements of the ith row is large and the column norm of the off-diagonal elements of the ith column is proportionally small or vice versa, a scaling can be done.

(iii) Extending the idea in (ii), we are interested in the magnitudes of the products of the diagonally opposite elements; i.e. the value of $|a_{ij} \cdot a_{ji}|$, where a_{ij} is the (i,j)th element of A. Those elements whose products are comparatively large need to be scaled carefully, as scaling does not change their products.

(iv) In scaling, the objective is to decrease the condition numbers of the matrices A_{ij}. Ideally, the elements of the scaled matrix, $\bar{a}_{ij}$, should be scaled to $\bar{a}_{ij} = \sqrt{a_{ij} \cdot a_{ji}}$. However, owing to the large number of elements in A to be scaled, the scaling factors for some rows or columns may be conflicting. In these cases, scaling factors for large $|a_{ij} \cdot a_{ji}|$ values have to be used. Also, the scaling factors for the matrices A_{11} and A_{22} should be given first priority, while those of A_{12} and A_{21} should be adjusted such that their elements scaled by the scaling factors of A_{11} and A_{22} do not increase their norms considerably.

Example 3.1

In order to illustrate the method, let us consider a simple second-order system

$$\begin{bmatrix} \dot{x} \\ \dot{z} \end{bmatrix} = \begin{bmatrix} 0 & 1 \\ -0 \cdot 09 & -1 \end{bmatrix} \begin{bmatrix} x \\ z \end{bmatrix} \qquad (3.38)$$

which has the eigenvalues $p_1 = -0 \cdot 1$ and $p_2 = -0 \cdot 9$ and hence the separation ratio as $\varepsilon = p_1/p_2 = 1/9$.

Using the method given previously, we have

$$\begin{bmatrix} \dot{x}_s \\ \dot{z}_f \end{bmatrix} = \begin{bmatrix} A_s & 0 \\ 0 & A_f \end{bmatrix} \begin{bmatrix} x_s \\ z_f \end{bmatrix} \qquad (3.39)$$

where $A_s = -L;$ $A_f = -1 \cdot 0 + L$

and L is the iterative solution of

$$L_{k+1} = -[-0 \cdot 09 - L_k^2] \tag{3.40}$$

with $L_0 = 0 \cdot 09$

Using $L_0 = 0 \cdot 09,$ $L_1 = 0 \cdot 0981$

$$A_{s0} = -0 \cdot 09; \quad A_{f0} = -0 \cdot 91$$
$$A_{s1} = -0 \cdot 0981; \quad A_{f1} = -0 \cdot 9019 \tag{3.41}$$

From (3.34) we see clearly that the iterative solutions improve the eigenvalues to be closer to the exact eigenvalues. The inequality (3.15) becomes

$$1 \leqslant \frac{1}{3}(2 \times 0 \cdot 09)^{-1}$$

$$\leqslant \frac{1}{0 \cdot 54} \tag{3.42}$$

Example 3.2: Consider a fifth-order model of a steam power system given as (Chow and Kokotovic, 1976)*

$$\dot{y} = \begin{bmatrix} -2 & 0 & 0 & 0 & -4 \\ 4 \cdot 75 & -5 & 0 & 0 & 0 \\ 0 & 0 \cdot 167 & -0 \cdot 167 & 0 & 0 \\ 0 & 0 & 2 & -2 & 0 \\ 0 & 0 \cdot 025 & 0 \cdot 02333 & 0 \cdot 035 & -0 \cdot 1125 \end{bmatrix} y \tag{3.43}$$

We note that the norms of rows 1, 2 and 4 are larger than those of 3 and 5, and reindex the state variables using the permutation

$$P = (e_4, e_5, e_2, e_3, e_1)$$

where e_i is an elementary column vector whose ith entry is 1. With $y_p = Py$, the new system becomes

* Reprinted with permission from CHOW J.H. and KOKOTOVIC P.V. (1976): 'A decomposition of near optimum regulators for systems with slow and fast modes', *IEEE Trans. Aut. Control*, AC-21, pp. 701–705.

$$\dot{y}_p = \begin{bmatrix} -0\cdot1125 & 0\cdot02333 & 0\cdot035 & 0 & 0\cdot0225 \\ 0 & -0.167 & 0 & 0 & 0\cdot167 \\ 0 & 2 & -2 & 0 & 0 \\ -4 & 0 & 0 & -2 & 0 \\ 0 & 0 & 0 & 4\cdot75 & -5 \end{bmatrix} y_p$$

$$(3.44a)$$

For this system, condition (3.27) is not satisfied. However, using the scaling $y_s = Sy_p$, where

$$S = \text{diag} (4,1,1,2,1),$$

results in the system

$$\begin{bmatrix} \dot{x}_1 \\ \dot{x}_2 \\ \dot{z}_1 \\ \dot{z}_2 \\ \dot{z}_3 \end{bmatrix} = \begin{bmatrix} -0\cdot1125 & 0\cdot0932 & 0\cdot14 & 0 & 0\cdot1 \\ 0 & -0\cdot167 & 0 & 0 & 0\cdot167 \\ 0 & 2 & -2 & 0 & 0 \\ -2 & 0 & 0 & -2 & 0 \\ 0 & 0 & 0 & 2\cdot375 & -5 \end{bmatrix} \begin{bmatrix} x_1 \\ x_2 \\ z_1 \\ z_2 \\ z_3 \end{bmatrix}$$

$$(3.44b)$$

First let us note that system (3.44b) is in the form of (3.1) and

$$\|A_0\| = 0\cdot3032; \quad \|L_0\| \, \|A_{12}\| = 0\cdot2361; \quad \|A_{22}^{-1}\| = 0\cdot5609$$

Then the inequality (3.15), i.e.

$$0\cdot5609 \leqslant \frac{1}{3}(0\cdot3032 + 0\cdot2361)^{-1}$$

$$\leqslant \frac{1}{1\cdot6179} \qquad (3.45)$$

is satisfied and hence the system (3.44b) exhibits a two-time-scale property. The eigenvalues of system (3.44b) are

$$-0\cdot15 \pm j0\cdot15; \quad -1\cdot97 \pm j0\cdot28; \quad -5\cdot03$$

whereas the eigenvalues of the slow subsystem are

$$e(A_0) = -0\cdot166 \pm j0\cdot14$$

and those of the fast subsystem are

$$e(A_{22}) = -2; \quad -2; \quad -5.$$

Thus we see that the eigenvalues of the first approximate subsystems are close to those of the given system (3.44b).

3.3 Singularly perturbed continuous systems as two-time-scale systems

We are interested in the time-scale analysis of a singularly perturbed system given by

$$\dot{x} = A_{11}x + A_{12}z + B_1u, \quad x(t=0) = x(0) \tag{3.46a}$$

$$\varepsilon\dot{z} = A_{21}x + A_{22}z + B_2u, \quad z(t=0) = z(0) \tag{3.46b}$$

where x and z are m and n state vectors, u is an r vector, A_{22}^{-1} exists and $0 < \varepsilon \leq 1$.

Writing (3.46) in the form of (3.1) results in

$$\dot{x} = A_{11}x + A_{12}z + B_1u \tag{3.47a}$$

$$\dot{z} = \bar{A}_{21}x + \bar{A}_{22}z + \bar{B}_2u \tag{3.47b}$$

where $\bar{A}_{21} = A_{21}/\varepsilon, \quad \bar{A}_{22} = A_{22}/\varepsilon$

For (3.47) to possess the two-time-scale property, we want it to satisfy the condition (3.15), i.e.

$$||\bar{A}_{22}^{-1}|| \leq \frac{1}{3}(||A_0|| + ||A_{12}|| \, ||L_0||)^{-1} \tag{3.48}$$

where $A_0 = A_{11} - A_{12}L_0$

$$L_0 = \bar{A}_{22}^{-1}\bar{A}_{21} = A_{22}^{-1}A_{21}$$

The condition (3.48) can be rewritten as

$$\left\| \left(\frac{A_{22}}{\varepsilon}\right)^{-1} \right\| \leq 1/3 \, [||A_0|| + ||A_{12}|| \, ||A_{22}^{-1}A_{21}||]^{-1} \tag{3.49}$$

If

$$\varepsilon \leq 1/3 \, ||A_{22}^{-1}|| \, [||A_0|| + ||A_{12}|| \, ||A_{22}^{-1}A_{21}||] \tag{3.50}$$

then obviously inequality (3.49) is satisfied since $||(A_{22}/\varepsilon)^{-1}|| \leq ||A_0||^{-1}$. Thus a singularly perturbed system possesses a two-time-scale property (Kokotovic and Haddad, 1975a,b). In other words, the singularly perturbed form is only one kind of representation of the general two-time-scale system.

Let us now find the first approximate decoupled system for (3.47). That is

$$\begin{bmatrix} \dot{x}_{s0} \\ \dot{z}_{s0} \end{bmatrix} = \begin{bmatrix} A_{s0} & 0 \\ 0 & A_{f0} \end{bmatrix} \begin{bmatrix} x_{s0} \\ z_{s0} \end{bmatrix} + \begin{bmatrix} B_{s0} \\ B_{f0} \end{bmatrix} u \tag{3.51}$$

where $A_{s0} = A_{11} - A_{12}L_0;$ $A_{f0} = \tilde{A}_{22} + L_0A_{12}$

$\qquad B_{s0} = B_1 - M_0L_0B_1 - M_0\tilde{B}_2;$ $B_{f0} = \tilde{B}_2 + L_0B_1$

Using $L_0 = \tilde{A}_{22}^{-1}\tilde{A}_{21} = A_{22}^{-1}A_{21}$

$\qquad M_0 = A_{12}\tilde{A}_{22}^{-1} = \varepsilon A_{12}A_{22}^{-1}$

we have

$$
\left.
\begin{aligned}
A_{s0} &= A_{11} - A_{12}A_{22}^{-1}A_{21} = A_0 \\
\varepsilon A_{f0} &= A_{22} + \varepsilon A_{22}^{-1}A_{21}A_{12} = A_{22} + 0(\varepsilon) \\
B_{s0} &= B_1 - A_{12}A_{22}^{-1}B_2 - \varepsilon A_{12}A_{22}^{-1}A_{21}B_1 \\
&= B_0 + 0(\varepsilon) \\
\varepsilon B_{f0} &= B_2 + \varepsilon A_{22}^{-1}A_{21} = B_2 + 0(\varepsilon)
\end{aligned}
\right\}
\tag{3.52}
$$

Thus (3.51) becomes

$$
\begin{bmatrix} \dot{x}_{s0} \\ \dot{z}_{f0} \end{bmatrix} = \begin{bmatrix} A_0 & 0 \\ 0 & \tilde{A}_{22} \end{bmatrix} \begin{bmatrix} x_{s0} \\ z_{f0} \end{bmatrix} + \begin{bmatrix} B_0 \\ \tilde{B}_2 \end{bmatrix} u
\tag{3.53}
$$

where $\tilde{A}_{22} = A_{22}/\varepsilon;$ $\tilde{B}_2 = B_2/\varepsilon$.

We try to obtain the approximate decoupled system (3.53) in an alternative way from (3.46) using singular perturbation analysis. Making $\varepsilon = 0$ in (3.46) is equivalent to assuming that the fast transients are instantaneous. Then the resulting degenerate (or slow) system becomes

$$
\dot{x}^{(0)} = A_{11}x^{(0)} + A_{12}z^{(0)} + B_1u^{(0)}, \quad x^{(0)}(0) = x(0)
\tag{3.54a}
$$

$$
0 = A_{21}x^{(0)} + A_{22}z^{(0)} + B_2u^{(0)}, \quad z^{(0)}(0) \neq z(0)
\tag{3.54b}
$$

Rewriting (3.54) as

$$
\dot{x}^{(0)} = A_0x^{(0)} + B_0u^{(0)}
\tag{3.55a}
$$

$$
z^{(0)} = -A_{22}^{-1}(A_{21}x^{(0)} + B_2u^{(0)})
\tag{3.55b}
$$

Thus we have separated the slow subsystem (3.55) from the original system (3.46). Now in order to get the fast subsystem we note that the fast modes contribute to the system dynamical behaviour during a short initial period and assume that the slow variables $x^{(0)}$, $z^{(0)}$ and $u^{(0)}$ are constant during fast transients. That is, we have

$$
x^{(0)} = \text{constant}; \quad \dot{z}^{(0)} = 0
$$

Using these in (3.46b) and (3.54b), we get

$$
\varepsilon[\dot{z} - \dot{z}^{(0)}] = A_{22}(z - z^{(0)}) + B_2(u - u^{(0)})
\tag{3.56}
$$

Defining

$$z_{f0} = z - z^{(0)}; \quad u_{f0} = u - u^{(0)} \tag{3.57}$$

the fast subsystem becomes

$$\varepsilon \frac{dz_{f0}}{dt} = A_{22}z_{f0} + B_2 u_{f0}, \quad z_{f0}(0) = z(0) - z^{(0)}(0) \tag{3.58a}$$

or

$$\frac{dz_{f0}}{dt} = \frac{A_{22}}{\varepsilon} z_{f0} + \frac{B_2}{\varepsilon} u_{f0} \tag{3.58b}$$

We can also rewrite (3.58a) in terms of a stretching transformation $t = \varepsilon t'$ to yield

$$\frac{dz^{(0)}(t')}{dt'} = A_{22}z^{(0)}(t') + B_2 u^{(0)}(t'), \quad z^{(0)}(t' = 0) = z(0) - z^{(0)}(0) \tag{3.59}$$

Thus, the slow and fast subsystems are given by (3.55) and (3.58), respectively, which are identical to (3.53). The subsystems (3.55) and (3.59) show that they operate on different time scales, i.e. slow time t and fast time t'.

We recall that the approximate slow and fast subsystems (3.55) and (3.59) have been obtained in Chapter 2 as the degenerate and boundary-layer correction equations (2.51) and (2.102) corresponding to a zeroth-order approximation. This reveals the fact that the singular perturbation method provides the first approximation of time-scale analysis using block diagonalisation for dynamical continuous systems with slow and fast modes.

3.4 Feedback control design

We present a two-stage feedback control design for the two-time-scale system (3.1). The design procedure is based on the explicitly invertible block diagonalising transformations (3.10) and (3.11) (Phillips, 1980b). We assume that the original system (3.1) is completely controllable, which implies the same for the decoupled subsystem (3.9); i.e. the pairs A_s, B_s and A_f, B_f are controllable.

Consider the eigenvalue placement problem for (3.1) using a linear state feedback control

$$u = Fy, \quad y' = (x', z') \tag{3.60}$$

where F is the $r \times (m + n)$-gain matrix. The design procedure is implemented in two stages.

In the first stage, we use the transformation (3.10) to convert (3.1) into (3.9) and design an $(r \times m)$ feedback matrix F_1 to place the eigenvalues of

$(A_s + B_sF_1)$ at m desired locations. Let

$$u = u_1 + u_2 = [F_1 \quad 0]y_d + u_2 \tag{3.61}$$

The substitution of (3.61) into (3.9) yields

$$\dot{y}_d = \begin{bmatrix} \dot{x}_s \\ \dot{z}_f \end{bmatrix} = \begin{bmatrix} A_s + B_sF_1 & 0 \\ B_fF_1 & A_f \end{bmatrix} \begin{bmatrix} x_s \\ z_f \end{bmatrix} + \begin{bmatrix} B_s \\ B_f \end{bmatrix} u_2 \tag{3.62}$$

Since (3.62) is lower-triangular, the application of (3.10) is simplified by noting that $M = 0$. Define

$$T_r = \begin{bmatrix} I_1 & 0 \\ \bar{L} & I_2 \end{bmatrix}; \qquad T_r^{-1} = \begin{bmatrix} I_1 & 0 \\ -\bar{L} & I_2 \end{bmatrix} \tag{3.63}$$

where $\bar{L}$ is a real root of

$$\bar{L}(A_s + B_sF_1) - A_f\bar{L} + B_fF_1 = 0 \tag{3.64}$$

The transformation $q = T_r y_d$ to (3.62) yields

$$\dot{q} = \begin{bmatrix} A_s + B_sF_1 & 0 \\ 0 & A_f \end{bmatrix} q + \begin{bmatrix} B_s \\ B_f + \bar{L}B_s \end{bmatrix} u_2 \tag{3.65}$$

We note again that the pair $(A_f, B_f + \bar{L}B_s)$ is completely controllable.

In the second stage we design an $(r \times m)$ feedback matrix F_2 to place the eigenvalues of $A_f + (B_f + \bar{L}B_s)F_2$ at n desired locations. Substituting the feedback control

$$u_2 = [0 \quad F_2]q \tag{3.66a}$$

into (3.61) and using $y_d = T_d y$ and $q = T_r y_d$, we get

$$u = \{[F_1 \quad 0]T_d + [0 \quad F_2]T_rT_d\}y \tag{3.66b}$$

where $\quad T_d = \begin{bmatrix} I_1 - ML & -M \\ L & I_2 \end{bmatrix}$ from (3.10)

Expressing T_r and T_d in terms of L, M and $\bar{L}$, we obtain the final form of the feedback gain matrix F in (3.60) as

$$F = [F_2L + (F_1 - F_2\bar{L})(I - ML) - F_1M + F_2(I + \bar{L}M)] \tag{3.67}$$

Example 3.3*
Consider an eighth-order power system problem (Phillips, 1980b) whose system

---

* Reprinted with permission from PHILLIPS, R. G. (1980): 'A two-stage design of linear feedback controls', *IEEE Trans.*, **AC-25**, pp. 1220–1223.

matrices A and B are given as

$$A = \begin{bmatrix} -5 & 0 & 0 & 0 & 4{\cdot}75 & 0 & 0 & 0 \\ 0 & -2 & 0 & 0 & 0 & -2 & 0 & 0 \\ -0{\cdot}8 & -0{\cdot}11 & -3{\cdot}99 & -0{\cdot}93 & 0 & -0{\cdot}07 & 10 & -9{\cdot}1 \\ 0 & 0 & 1{\cdot}32 & -1{\cdot}39 & 0 & 0 & 0 & -0{\cdot}28 \\ 0 & 0 & 0 & 0 & -0{\cdot}2 & 0 & 0 & 0 \\ 0{\cdot}17 & 0 & 0 & 0 & 0 & -0{\cdot}17 & 0 & 0 \\ 0 & 0 & 0{\cdot}2 & 0 & 0 & 0 & -0{\cdot}5 & 0 \\ 0{\cdot}01 & 0{\cdot}01 & -0{\cdot}06 & 0{\cdot}12 & 0 & 0{\cdot}01 & 0 & -0{\cdot}11 \end{bmatrix}$$

$$(3.68a)$$

$$B' = \begin{bmatrix} 0 & 0 & 10 & 0 & 0 & 0 & 0 & 0 \\ 0 & 0 & 0 & 0 & 4 & 0 & 0 & 0 \end{bmatrix} \qquad (3.68b)$$

The open-loop eigenvalues are

$$\left. \begin{array}{ll} p_1 = -4{\cdot}35; & p_{5,6} = -0{\cdot}13 \pm j0{\cdot}21 \\ p_2 = -5{\cdot}0; & p_7 = -0{\cdot}17 \\ p_3 = -2{\cdot}0; & p_8 = -0{\cdot}2 \\ p_4 = -1{\cdot}39 \end{array} \right\} \qquad (3.69)$$

The minimum open-loop separation ratio is $|p_5|/|p_4| = 0{\cdot}19$, and this shows that the model has four slow and four fast variables. After two iterations L is approximated by

$$L_1 = \begin{bmatrix} 0 & 0 & 0 & 0 \\ 0{\cdot}034 & 0 & 0 & 0 \\ -0{\cdot}001 & -0{\cdot}001 & 0{\cdot}048 & -0{\cdot}001 \\ 0{\cdot}002 & 0{\cdot}007 & 0{\cdot}014 & 0{\cdot}092 \end{bmatrix} \qquad (3.70a)$$

and M by

$$M_1 = \begin{bmatrix} 0{\cdot}989 & 0 & 0 & 0 \\ -0{\cdot}137 & 1{\cdot}091 & 0 & 0 \\ -0{\cdot}006 & -0{\cdot}012 & 2{\cdot}359 & -1{\cdot}966 \\ 0{\cdot}006 & 0{\cdot}032 & 2{\cdot}48 & -1{\cdot}625 \end{bmatrix} \qquad (3.70b)$$

resulting in

$$A_{s1} = \begin{bmatrix} 5 & 0 & 0 & 0 \\ 0{\cdot}068 & -2 & 0 & 0 \\ 0{\cdot}056 & -0{\cdot}033 & -4{\cdot}355 & -0{\cdot}011 \\ 0 & 0{\cdot}002 & 1{\cdot}323 & -1{\cdot}363 \end{bmatrix} \qquad (3.71a)$$

$$B'_{s1} = \begin{bmatrix} -3.958 & 0.548 & 0.025 & -0.024 \\ 0 & 0 & 9.123 & -0.981 \end{bmatrix} \qquad (3.71b)$$

For desired eigenvalues

$$p_{1,2}(\text{des}) = -8 \pm j2; \quad p_3(\text{des}) = -6; \quad p_4(\text{des}) = -4 \qquad (3.72)$$

a feedback gain matrix F_1 is obtained as

$$F_1 = \begin{bmatrix} -0.615 & -20.865 & 0 & 0 \\ 0.009 & -0.016 & -0.777 & -1.843 \end{bmatrix} \qquad (3.73)$$

To design a feedback matrix F_2 for the pair $(A_f, B_f + \bar{L}B_s)$, we first find $\bar{L}_1$ as

$$\bar{L}_1 = \begin{bmatrix} -0.02 & 6.178 & 0 & 0 \\ 0.007 & -0.033 & 0 & 0 \\ -0.001 & -0.002 & 0.057 & 0.132 \\ 0 & -0.002 & 0.015 & 0.035 \end{bmatrix} \qquad (3.74)$$

resulting in

$$A_f = \begin{bmatrix} -0.2 & 0 & 0 & 0 \\ 0.164 & -0.167 & 0 & 0 \\ -0.003 & -0.006 & -0.015 & -0.439 \\ 0.008 & 0.023 & 0.137 & -0.262 \end{bmatrix} \qquad (3.75)$$

$$[B_f - \bar{L}B_s]' = \begin{bmatrix} 0.533 & 0.045 & -0.001 & 0.002 \\ 0 & 0 & 0.088 & 0.031 \end{bmatrix} \qquad (3.76)$$

For desired eigenvalues of

$$p_{5,6}(\text{des}) = -2 \pm j1; \quad p_7(\text{des}) = -1; \quad p_8(\text{des}) = -0.5 \qquad (3.77)$$

a feedback gain matrix F_2 is obtained as

$$F_2 = \begin{bmatrix} -2.615 & -49.15 & 0 & 0 \\ -0.234 & -1.142 & -2.266 & -39.916 \end{bmatrix} \qquad (3.78)$$

The actual closed-loop eigenvalues using the composite feedback system (3.66) are

$$p_{1,2}(\text{cl}) = -7 \cdot 99 \pm j1 \cdot 99 \qquad p_{5,6}(\text{cl}) = -2 \pm j1$$

$$p_3(\text{cl}) = -6 \cdot 09 \qquad p_7(\text{cl}) = -1 \cdot 03$$
$$\left. \right\} \qquad (3.79)$$
$$p_4(\text{cl}) = -3 \cdot 93 \qquad p_8(\text{cl}) = -0 \cdot 5$$

These are very close to their desired locations.

3.5   Block diagonalisation of discrete systems

In Sections 3.1–3.4 we presented the time-scale analysis to reduced-order modelling and feedback control design of continuous systems. In the remaining Sections of the Chapter we try to give a similar treatment of a time-scale approach to discrete systems. We shall observe some similarities and differences between the continuous and discrete systems.

3.5.1   *Problem formulation and derivation*

Consider a linear discrete control system

$$\begin{bmatrix} x(k+1) \\ z(k+1) \end{bmatrix} = \begin{bmatrix} A_{11} & A_{12} \\ A_{21} & A_{22} \end{bmatrix} \begin{bmatrix} x(k) \\ z(k) \end{bmatrix} + \begin{bmatrix} B_1 \\ B_2 \end{bmatrix} u(k) \qquad (3.80)$$

with initial conditions $x(k=0) = x(0)$ and $z(k=0) = z(0)$ where $x(k)$ and $z(k)$ are m- and n-dimensional state vectors, $u(k)$ is an r-dimensional control vector and A_{ij} and B_j are matrices of consistent dimensionality. We assume that the system (3.80) is asymptotically stable and its eigenspectrum consists of a cluster of m large eigenvalues and a cluster of n small eigenvalues. Let the eigenvalues of system (3.80) be arranged as

$$1 > |p_{s1}| > \ldots > |p_{sm}| > |p_{f1}| > \ldots > |p_{fn}| \qquad (3.81)$$

If the condition

$$\varepsilon = |p_{f1}|/|p_{sm}| \ll 1 \qquad (3.82)$$

is satisfied, then the system (3.80) possesses a two-time-scale (slow and fast) property (Mahmoud and Singh, 1981a; Mahmoud, 1982a,b). In other words, the stable discrete system (3.80) is said to exhibit a two-time-scale behaviour if the largest absolute eigenvalue of the fast eigenspectrum is much smaller than (has wider separation with) the smallest absolute eigenvalue of the slow eigenspectrum.

In the literature one usually finds 'stiffness' being stamped on a system in which the ratio of eigenvalues is given by condition (3.3) for continuous systems and condition (3.82) for discrete systems (Miranker, 1980). This ratio appears explicitly as the small parameter ε in the case of singularly perturbed systems.

It is evident from conditions (3.2) and (3.3) for continuous systems that the 'small' eigenvalues produce 'slow' modes and 'large' eigenvalues produce 'fast' modes. Similarly from conditions (3.81) and (3.82) we see that, in discrete systems, the 'slow' modes are generated by 'large' eigenvalues and the 'fast' modes are generated by 'small' eigenvalues.

In order to block-diagonalise the system (3.80) or separate it into slow and fast subsystems, we use a two-stage linear transformation similar to (3.10). In the first step, we remove the block A_{21} to make (3.80) upper-triangular by using the transformation

$$z_f(k) = z(k) + Lx(k) \qquad (3.83)$$

where the $(n \times m)$ matrix L is a real root of

$$LA_{11} + A_{21} - LA_{12}L - A_{22}L = 0 \qquad (3.84a)$$

or $\quad A_{22}L + LA_{12}L - LA_{11} - A_{21} = 0 \qquad (3.84b)$

Using the transformation (3.83) in system (3.80), we get

$$\begin{bmatrix} x(k+1) \\ z_f(k+1) \end{bmatrix} = \begin{bmatrix} A_s & A_{12} \\ 0 & A_f \end{bmatrix} \begin{bmatrix} x(k) \\ z_f(k) \end{bmatrix} + \begin{bmatrix} B_1 \\ B_f \end{bmatrix} u(k) \qquad (3.85)$$

where

$$A_s = A_{11} - A_{12}L; \quad A_f = A_{22} + LA_{12}$$
$$B_f = B_2 + LB_1$$

In the second stage, we apply the transformation

$$x_s(k) = x(k) - Mz_f(k) \qquad (3.86)$$

to (3.85) and choose the $(m \times n)$ matrix M such that

$$A_s M - MA_f + A_{12} = 0 \qquad (3.87)$$

Then (3.85) reduces to

$$\begin{bmatrix} x_s(k+1) \\ z_f(k+1) \end{bmatrix} = \begin{bmatrix} A_s & 0 \\ 0 & A_f \end{bmatrix} \begin{bmatrix} x_s(k) \\ z_f(k) \end{bmatrix} + \begin{bmatrix} B_s \\ B_f \end{bmatrix} u(k) \qquad (3.88)$$

where

$$B_s = B_1 - MLB_1 - MB_2$$

Now system (3.88) is in the desired decoupled form. Note that the slow and fast variables $x_s(k)$ and $z_f(k)$ are related to the original state variables $z(k)$ and $z(k)$ by means of the transformations (3.83) and (3.86), which are combined to form

$$\begin{bmatrix} x_s(k) \\ z_f(k) \end{bmatrix} = \begin{bmatrix} I_1 - ML & -M \\ L & I_2 \end{bmatrix} \begin{bmatrix} x(k) \\ z(k) \end{bmatrix} \tag{3.89}$$

and this has an inverse transformation

$$\begin{bmatrix} x(k) \\ z(k) \end{bmatrix} = \begin{bmatrix} I_1 & M \\ -L & I_2 - LM \end{bmatrix} \begin{bmatrix} x_s(k) \\ z_f(k) \end{bmatrix} \tag{3.90}$$

Note that (3.90) needs no matrix inversion.

3.5.2 Basic properties and numerical aspects

The condition (3.82) can be rewritten as

$$\varepsilon = \max |e(A_f)|/\min |e(A_s)| \ll 1 \tag{3.91}$$

where

$$e(A_f) = \{p_{f1}, \ldots, p_{fn}\}$$
$$e(A_s) = \{p_{s1}, \ldots, p_{sm}\}$$

Using the norm properties of invertible matrices the time separation property can be expressed as

$$||A_f|| \ll 1/||A_s^{-1}|| \quad \text{or} \quad ||A_s^{-1}||^{-1} \tag{3.92}$$

The iterative solution of (3.84) is given as (Phillips, 1980a; Kando and Iwazumi, 1983a,b),

$$L_{i+1} = (A_{22}L_i + L_i A_{12}L_i - A_{21})A_{11}^{-1} \tag{3.93}$$

with initial value $L_0 = -A_{21}A_{11}^{-1}$.

Similarly for (3.87) we have

$$M_{i+1} = A_{11}^{-1}(A_{12}LM_i + M_i LA_{12} + M_i A_{22} - A_{12}) \tag{3.94}$$

with initial value $M_0 = -A_{11}^{-1}A_{12}$.

Using these approximate values of L_0 and M_0 in (3.88) we get

$$A_{s0} = A_{11} - A_{12}L_0 = A_{11} + A_{12}A_{21}A_{11}^{-1} = A_{11} + 0(\varepsilon)$$
$$A_{f0} = A_{22} + L_0 A_{12} = A_{22} - A_{21}A_{11}^{-1}A_{12}$$
$$B_{s0} = B_1 - M_0 L_0 B_1 - M_0 B_2 = B_1 + 0(\varepsilon)$$
$$B_{f0} = B_2 + L_0 B_1 = B_2 - A_{21}A_{11}^{-1}B_1$$

and

$$\begin{bmatrix} x_{s0}(k+1) \\ x_{f0}(k+1) \end{bmatrix} = \begin{bmatrix} A_{11} & 0 \\ 0 & A_{f0} \end{bmatrix} \begin{bmatrix} x_{s0}(k) \\ x_{f0}(k) \end{bmatrix} + \begin{bmatrix} B_1 \\ B_{f0} \end{bmatrix} u(k) \tag{3.95}$$

We shall show in the next Section that this decoupled system (3.95) corresponds to the first approximation in singular perturbation analysis. A different approach to the approximation is considered by Mahmoud and Singh (1981a).

We summarise below the basic structural properties of the subsystems (3.88) or (3.95).

(i) If the spectral radius of A_{11} has an absolute value less than 1 and the reduced subsystem (3.95) is asymptotically stable, then the discrete two-time-scale system (3.80) is asymptotically stable.

(ii) If

$$\text{rank}\,[B_1, A_{11}B_1, \ldots, A_{11}^{m-1}B_1] = m \qquad (3.96a)$$

$$\text{rank}\,[B_{f0}, A_{f0}B_{f0}, \ldots, A_{f0}^{n-1}B_{f0}] = n \qquad (3.96b)$$

then the two-time-scale system (3.80) is completely controllable. The above two properties show that the stability and controllability of the $(m + n)$-dimensional system (3.80) can be deduced from that of the m- and n-dimensional decoupled system (3.95).

As L and M matrix recursions approach equilibrium, since $e(A_s)$ and $e(A_f)$ are sufficiently far apart, controllability of our original system pair (A, B) guarantees controllability of the subsystem pairs (A_s, B_s), (A_f, B_f).

However, in real-time applications, computation-time requirements will make exact equilibrium values of L and M unreachable in general. For such cases a slightly modified condition for controllability is proposed by Phillips (1980a).

Example 3.4

We consider a digital flight-control system. The discrete model of an aircraft (longitudinal) control system is represented by (Naidu and Price, 1987).

$$
\begin{bmatrix} x_1(k+1) \\ x_2(k+1) \\ x_3(k+1) \\ z_1(k+1) \\ z_2(k+1) \end{bmatrix} =
\begin{bmatrix}
0\cdot9237 & -0\cdot3081 & 0\cdot0 & 0\cdot053 & -0\cdot0904 \\
0\cdot0397 & 0\cdot9955 & 0\cdot0 & -0\cdot1075 & 0\cdot5889 \\
0\cdot0871 & 1\cdot8995 & 1\cdot0 & -0\cdot6353 & 0\cdot3940 \\
-0\cdot0355 & 0\cdot0101 & 0\cdot0 & 0\cdot0078 & 0\cdot1374 \\
0\cdot0696 & -0\cdot0127 & 0\cdot0 & -0\cdot0971 & 0\cdot2874
\end{bmatrix}
\begin{bmatrix} x_1(k) \\ x_2(k) \\ x_3(k) \\ z_1(k) \\ z_2(k) \end{bmatrix}
$$

$$
+
\begin{bmatrix}
0\cdot0428 & -0\cdot0004 & -0\cdot154 \\
-0\cdot4846 & -0\cdot5154 & -0\cdot0022 \\
-0\cdot1615 & -0\cdot0675 & -0\cdot0053 \\
-0\cdot202 & -0\cdot2893 & 0\cdot0051 \\
-0\cdot8067 & -0\cdot8522 & -0\cdot0084
\end{bmatrix}
\begin{bmatrix} u_1(k) \\ u_2(k) \\ u_3(k) \end{bmatrix}
$$

where, x_1 is velocity, x_2 is pitch angle, x_3 is altitude, z_1 is angle of attack, z_2 is pitch angular velocity, u_1 is elevator deflection, u_2 is flap deflection, u_3 is throttle position. The states x_1 and x_3 are scaled down by a factor of 100 to facilitate easy representation.

The eigenvalues of the discrete model are 1·01, 0·9621 ± j0·1753, 0·2173, 0·0729, suggesting three slow states (x_1, x_2, x_3) and two fast states (z_1, z_2) with separation ratio as

$$|p_{s3}|/|p_{f1}| = 4.5$$

Applying the time-scale technique to the discrete model, the following results are obtained. The slow subsystem matrix is given by

$$A_s = \begin{bmatrix} 0\cdot9125 & -0\cdot3103 & 0\cdot0 \\ 0\cdot1070 & 1\cdot0117 & 0\cdot0 \\ 0\cdot1453 & 1\cdot9065 & 1\cdot0 \end{bmatrix}$$

with eigenvalues as

$$1\cdot0; \quad 0\cdot9621 \pm j0\cdot1754$$

The fast subsystem matrix becomes

$$A_f = \begin{bmatrix} 0\cdot0097 & 0\cdot1313 \\ -0\cdot0999 & 0\cdot2805 \end{bmatrix}$$

with eigenvalues as

$$0\cdot2173; \quad 0\cdot0728$$

The input matrices become

$$B_s = \begin{bmatrix} -0\cdot124 & -0\cdot1734 & -0\cdot1518 \\ -1\cdot041 & -1\cdot0856 & 0\cdot0054 \\ 1\cdot0533 & 1\cdot2192 & -0\cdot0213 \end{bmatrix}$$

$$B_f = \begin{bmatrix} -0\cdot1978 & -0\cdot2858 & 0\cdot0015 \\ -0\cdot7976 & -0\cdot8374 & 0\cdot0107 \end{bmatrix}$$

An examination of the above results indicates that the decoupled system is a close approximation to the original high-order system.

3.6 Singularly perturbed discrete systems as two-time-scale systems

We consider the time-scale analysis of a singularly perturbed discrete system

$$\begin{bmatrix} x(k+1) \\ z(k+1) \end{bmatrix} = \begin{bmatrix} A_{11} & \varepsilon A_{12} \\ A_{21} & \varepsilon A_{22} \end{bmatrix} \begin{bmatrix} x(k) \\ z(k) \end{bmatrix} + \begin{bmatrix} B_1 \\ B_2 \end{bmatrix} u(k) \qquad (3.97)$$

with initial conditions $x(k=0) = x(0)$ and $z(k=0) = z(0)$.

Substituting $\bar{A}_{12} = \varepsilon A_{12}$ and $\bar{A}_{22} = \varepsilon A_{22}$ in system (3.97) results in the form of system (3.80). Now using the transformation (3.89) in the resulting system, the decoupled system becomes

$$\begin{bmatrix} x_s(k+1) \\ z_f(k+1) \end{bmatrix} = \begin{bmatrix} A_s & 0 \\ 0 & A_f \end{bmatrix} \begin{bmatrix} x_s(k) \\ z_f(k) \end{bmatrix} + \begin{bmatrix} B_s \\ B_f \end{bmatrix} u(k) \qquad (3.98)$$

where $A_s = A_{11} - \varepsilon A_{12}L; \quad A_f = \varepsilon(A_{22} + LA_{12})$

$B_s = B_1 - MLB_1 - MB_2; \quad B_f = B_2 + LB_1$

and L and M are the solutions of

$$LA_{11} + A_{21} - \varepsilon A_{22}L - \varepsilon LA_{12}L = 0 \qquad (3.99a)$$

$$A_sM - MA_f + \varepsilon A_{12} = 0 \qquad (3.99b)$$

The condition for the two-time scale property of (3.97) is

$$||(A_f|| \ll ||A_s^{-1}||^{-1} \qquad (3.100a)$$

or $\varepsilon||(A_{22} + LA_{12})|| \ll ||(A_{11} + \varepsilon A_{12}L)^{-1}||^{-1} \qquad (3.100b)$

For ε sufficiently small, the approximate solutions are

$A_{s0} = A_{11}; \quad \varepsilon A_{f0} = A_{22} - A_{21}A_{11}^{-1}A_{12}$

$B_{s0} = B_{11}; \quad B_{f0} = B_2 - A_{21}A_{11}^{-1}B_1$

so that the approximate decoupled system is

$$\begin{bmatrix} x_{s0}(k+1) \\ z_{f0}(k+1) \end{bmatrix} = \begin{bmatrix} A_{11} & 0 \\ 0 & \varepsilon A_{f0} \end{bmatrix} \begin{bmatrix} x_s(k) \\ z_f(k) \end{bmatrix} + \begin{bmatrix} B_1 \\ B_{f0} \end{bmatrix} u(k) \qquad (3.101)$$

Now we try to obtain the system (3.101) in an alternative way from (3.97) using the singular perturbation method. The degenerate (slow) system is obtained by making $\varepsilon = 0$ in (3.97). That is

$$x^{(0)}(k+1) = A_{11}x^{(0)}(k) + B_1u^{(0)}(k), \quad x^{(0)}(k=0) = x(0) \qquad (3.102a)$$

$$z^{(0)}(k+1) = A_{21}x^{(0)}(k) + B_2u^{(0)}(k), \quad z^{(0)}(k=0) \neq z(0) \qquad (3.102b)$$

Here we note that $z(k)$ has lost its initial condition $z(0)$ in the process of degeneration. In order to recover this lost initial condition, we use boundary-layer corrections as discussed in Section 2.4. The transformations for corrections are given as

$$x_c(k) = x(k)/\varepsilon^{k+1}; \quad z_c(k) = z(k)/\varepsilon^k$$
$$u_c(k) = u(k)/\varepsilon^{k+1}$$

(3.103)

Using transformation (3.103) in system (3.97), the transformed system becomes

$$\varepsilon x_c(k+1) = A_{11}x_c(k) + A_{12}z_c(k) + B_1u_c(k)$$

(3.104a)

$$z_c(k+1) = A_{21}x_c(k) + A_{22}z_c(k) + B_2u_c(k)$$

(3.104b)

By suppressing ε in (3.104), we get

$$0 = A_{11}x_c^{(0)}(k) + A_{12}z_c^{(0)}(k) + B_1u_c^{(0)}(k)$$

(3.105a)

$$z_c^{(0)}(k+1) = A_{21}x_c^{(0)}(k) + A_{22}z_c^{(0)}(k) + B_2u_c^{(0)}(k)$$

(3.105b)

or

$$z_c^{(0)}(k+1) = (A_{22} - A_{21}A_{11}^{-1}A_{12})z_c^{(0)}(k) + (B_2 - A_{21}A_{11}^{-1}B_1)u_c^{(0)}(k)$$

(3.106a)

$$x_c^{(0)}(k) = -A_{11}^{-1}[A_{12}z_c^{(0)}(k) + B_1u_c^{(0)}(k)]$$

(3.106b)

Let us consider the complete solutions of $x(k)$ and $z(k)$ only and for the present ignoring $u(k)$ to simplify the analysis, we have

$$x(k) = x^{(0)}(k) + \varepsilon^{k+1}x_c^{(0)}(k)$$

(3.107a)

$$z(k) = z^{(0)}(k) + \varepsilon^k z_c(k)$$

(3.107b)

Our current interest is only in the degenerate (slow) subsystem (3.102a) and the correction (fast) subsystem (3.106). They are given (without inputs) as

$$x^{(0)}(k+1) = A_{11}x^{(0)}(k) \qquad x^{(0)}(k=0) = x(0)$$

(3.108a)

$$z_c^{(0)}(k+1) = A_{f0}x_c^{(0)}(k) \qquad z_c^{(0)}(k=0) = z(0) - z^{(0)}(0)$$

(3.108b)

Note from (3.107b) that the fast subsystem solution is given by

$$\varepsilon^k z_c^{(0)}(k) = (\varepsilon A_{f0})^k z_c^{(0)}(k)$$

(3.109)

From (3.101),

$$z_{f0}(k) = (\varepsilon A_{f0})^k z_{f0}(k)$$

(3.110)

By comparing (3.109) and (3.110) we see that the approximate degenerate and boundary-layer correction solutions are equivalent to the slow and fast subsystem solutions. Thus in discrete systems as in continuous systems we get the fact that the singular perturbation method and time-scale analysis give identical results for the first approximation.

Example 3.5
Let us try to use the above analysis for a fifth-order steam-power system given by

$$
\begin{bmatrix} x_1(k+1) \\ x_2(k+1) \\ z_1(k+1) \\ z_2(k+1) \\ z_3(k+1) \end{bmatrix} = \begin{bmatrix} 0{\cdot}9014 & 0{\cdot}1179 & 0{\cdot}0525 & 0{\cdot}0167 & 0{\cdot}021 \\ -0{\cdot}0196 & 0{\cdot}8743 & 0 & 0{\cdot}025 & 0{\cdot}0293 \\ -0{\cdot}0071 & 0{\cdot}7342 & 0{\cdot}2018 & 0{\cdot}013 & 0{\cdot}0211 \\ -0{\cdot}75 & -0{\cdot}0557 & -0{\cdot}032 & 0{\cdot}1936 & -0{\cdot}0141 \\ -0{\cdot}306 & -0{\cdot}017 & -0{\cdot}011 & 0{\cdot}1428 & 0{\cdot}0132 \end{bmatrix} \begin{bmatrix} x_1(k) \\ x_2(k) \\ z_1(k) \\ z_2(k) \\ z_3(k) \end{bmatrix}
$$

(3.111)

with eigenvalues

$$p_{1,2} = 0{\cdot}8777 \pm j0{\cdot}1054; \quad p_3 = 0{\cdot}0179$$

$$p_{4,5} = 0{\cdot}2055 \pm j0{\cdot}0236$$

The approximate slow subsystem is

$$x_{s0}(k+1) = A_{11}x_{s0}(k) \tag{3.112}$$

whose eigenvalues are given by $0{\cdot}8768 \pm j0{\cdot}1018$ and the approximate fast subsystem is

$$x_{f0}(k+1) = A_{f0}x_{f0}(k) \tag{3.113}$$

with $A_{f0} = A_{22} - A_{21}A_{11}^{-1}A_{12}$

whose eigenvalues are given by $0{\cdot}0172; 0{\cdot}2052 \pm j0{\cdot}0226$. Thus we see that the eigenvalues of the slow and fast subsystems are, to a first approximation, close to the eigenvalues of the given two-time-scale system.

3.7   Closed-loop eigenvalue placement

In this Section we shall use the time-scale analysis of singularly perturbed systems and obtain reduced-order modelling for eigenvalue placement (Phillips, 1980a). Given (3.97), the corresponding decoupled system is (3.98). Now place n_f fast modes using the fast subsystem

$$z_f(k+1) = A_f z_f(k) + B_f u_f(k) \tag{3.114}$$

$$u_f(k) = G_f x_f(k) = G_f[Lx(k) + z(k)] \tag{3.115}$$

It is important to look at this partial feedback in both the transformed system (3.98) and original system (3.97). In the transformed system (3.98),

$$
\begin{bmatrix} x_s(k+1) \\ z_f(k+1) \end{bmatrix} = \begin{bmatrix} A_s & B_s G_f \\ 0 & A_f + B_f G_f \end{bmatrix} \begin{bmatrix} x_s(k) \\ z_f(k) \end{bmatrix} + \begin{bmatrix} B_s \\ B_f \end{bmatrix} u(k) \tag{3.116}
$$

Notice that, if we were to re-blockdiagonalise this system, our block-diagonal terms would not change and the slow submatrix A_s will be preserved. In addition, the fast eigenvalues would be provided by the designed fast submatrix $(A_f + B_f G_f)$.

When the partial feedback (3.115) is fed back into the original system (3.97),

$$\begin{bmatrix} x(k+1) \\ z(k+1) \end{bmatrix} = \begin{bmatrix} A_{11} + B_1 G_f L & \varepsilon A_{12} + B_1 G_f \\ A_{21} + B_2 G_f L & \varepsilon A_{22} + B_2 G_f \end{bmatrix} \begin{bmatrix} x(k) \\ z(k) \end{bmatrix} + \begin{bmatrix} B_1 \\ B_2 \end{bmatrix} u(k) \quad (3.117)$$

We have designed G_f to place the n_f fast modes in n_f fast desired (stable) locations.

Now, if it is also desired to place the m_s slow eigenvalues, we select G_s such that

$$x_s(k+1) = A_s x_s(k) + B_s u_s(k) \tag{3.118}$$

$$u_s(k) = G_s x_s(k) = G_s(I_1 - ML)x(k) - G_s Mz(k) \tag{3.119}$$

The composite control is

$$\begin{aligned} u_c(k) &= u_f(k) + u_s(k) \\ &= G_f L x(k) + G_f z(k) + G_s(I_1 - ML)x(k) - G_s Mz(k) \\ &= [G_f L + G_s(I_1 - ML)]x(k) + (G_f - G_s M)z(k) \end{aligned} \tag{3.120}$$

and it places m_s slow eigenvalues of $(A_s + B_s G_s)$ and n_f fast eigenvalues of $(A_f + B_f G_f)$ at the desired locations. Note that the feedback design has been carried out without handling at any stage a high-order matrix.

Example 3.6

Consider (Phillips 1980*a*)

$$\begin{bmatrix} x_1(k+1) \\ x_2(k+1) \\ z_1(k+1) \\ z_2(k+1) \end{bmatrix} = \begin{bmatrix} 1 & 0.6 & 0.4 & 0.8 \\ -0.2 & 1.3 & 0 & 0.7 \\ 0 & 0.1 & 0.05 & 0.7 \\ 0.1 & 0.1 & 0 & 0.1 \end{bmatrix} \begin{bmatrix} x_1(k) \\ x_2(k) \\ z_1(k) \\ z_2(k) \end{bmatrix} + \begin{bmatrix} 1 \\ 1 \\ 1 \\ 1 \end{bmatrix} u(k) \quad (3.121)$$

The open-loop eigenvalues are

$$p_{1,2} = 1.16 \pm j0.37, \quad p_3 = 0.1; \quad p_4 = 0.0235$$

The desired closed-loop eigenvalues are at

$$p_1 = 0.98; \quad p_2 = 0.96; \quad p_3 = 0.2; \quad p_4 = 0.15$$

After some iterations for L and M, we have

$$G_s = [0\cdot077118 \quad -0\cdot30607]$$

$$u_s(k) = [0\cdot077118, -0\cdot30607, 0\cdot0060567, -0\cdot164035]\begin{bmatrix} x(k) \\ z(k) \end{bmatrix}$$

Incorporating fast feedback control,

$$G_f = [0\cdot64234 \quad -0\cdot42944]$$

$$u_f(k) = [-0\cdot048113, 0\cdot059422, 0\cdot62834, -0\cdot41281]\begin{bmatrix} x(k) \\ z(k) \end{bmatrix}$$

gives closed-loop eigenvalues of

$$p_1 = 0\cdot97994; \quad p_2 = 0\cdot96001; \quad p_3 = 0\cdot19972; \quad p_4 = 0\cdot15028$$

3.8   Three-time scale systems*

Let us consider a general three-time-scale system. Based on a block diagonal-isation procedure (Chang, 1974; Kokotovic, 1975), a decomposition technique is presented where the original system is decomposed into 'slow', 'fast 1', and 'fast 2' (faster than fast 1) subsystems. The central idea is to use three-stage linear transformations. A numerical algorithm is presented for the iterative solution of the resulting Riccati-type equations. A ninth-order example is provided to illustrate the technique (Naidu and Ravinder, 1985).

3.8.1   Statement of the problem

Consider a stable, linear time-invariant system

$$\dot{y} = Ay + Bu \tag{3.122}$$

where y is an n-dimensional state vector, u is an r-dimensional control vector and A and B are $n \times n$- and $n \times r$-dimensional matrices, respectively. Let the system (3.122) be rewritten as

$$\begin{bmatrix} \dot{x} \\ \dot{z}_1 \\ \dot{z}_2 \end{bmatrix} = \begin{bmatrix} A_{11} & A_{12} & A_{13} \\ A_{21} & A_{22} & A_{23} \\ A_{31} & A_{32} & A_{33} \end{bmatrix} \begin{bmatrix} x \\ z_1 \\ z_2 \end{bmatrix} + \begin{bmatrix} B_1 \\ B_2 \\ B_3 \end{bmatrix} u \tag{3.123}$$

where x, z_1 and z_2 are n_1, n_2 and n_3 vectors such that $n_1 + n_2 + n_3 = n$. We assume that the system (3.123) possesses three distinct groups of eigenvalues

---

* Reprinted with permission from NAIDU, D. S., and RAVINDER, R. (1985): 'On three-time scale analysis', Proc. 24th IEEE Conference on Decision and Control, Fort Lauderdale, FL, USA, Dec. 1985

such that n_1 eigenvalues are close to the origin, and n_2 and n_3 eigenvalues are far and farther from the origin. In other words, the system (3.123) has n_1 'slow' modes and n_2 and n_3 'fast' and 'faster' modes, and thereby possesses three-time-scale character. For system (3.123), let us arrange the eigenspectrum $p(A)$ in increasing order of absolute values as

$$p(A) = (p_1, \ldots, p_{n1}, p_{n1+1}, \ldots, p_{n1+n2}, p_{n1+n2+1}, \ldots, p_n) \qquad (3.124)$$

where

$$0 < |p_1| < |p_2| < \ldots < |p_{n1}| < |p_{n1+1}| < \ldots$$

$$\ldots < |p_{n1+n2}| < |p_{n1+n2+1}| < \ldots < |p_n| \qquad (3.125)$$

The system (3.123) is said to possess a three-time-scale property (Kokotovic, 1975) if

$$|p_{n1}| \ll |p_{n1+1}| \qquad (3.126a)$$

and

$$|p_{n1+n2}| \ll |p_{n1+n2+1}| \qquad (3.126b)$$

The main aim is to decouple (3.123) into three subsystems as

$$\begin{bmatrix} \dot{x}_s \\ \dot{z}_{f1} \\ \dot{z}_{f2} \end{bmatrix} = \begin{bmatrix} A_s & 0 & 0 \\ 0 & A_{f1} & 0 \\ 0 & 0 & A_{f2} \end{bmatrix} \begin{bmatrix} x_s \\ z_{f1} \\ z_{f2} \end{bmatrix} + \begin{bmatrix} G_1 \\ G_2 \\ G_3 \end{bmatrix} u \qquad (3.127)$$

where x_s, z_{f1}, and z_{f2} are n_1-, n_2- and n_3-dimensional state vectors. The three-time-scale property of system (3.127) can be expressed as

$$|\max p(A_s)| \ll |\min p(A_{f1})| \qquad (3.128a)$$

and

$$|\max p(A_{f1})| \ll |\min p(A_{f2})| \qquad (3.128b)$$

For any square matrix F and its norm $||F||$, we have

$$|\max p(F)| \leq ||F|| \qquad (3.129a)$$

and

$$|\min p(F)|^{-1} \leq ||F^{-1}|| \qquad (3.129b)$$

if F^{-1} exists. Then using (3.129), the three-time-scale property (3.128) becomes

$$||A_{f2}^{-1}|| \ll ||A_{f1}||^{-1} \qquad (3.130a)$$

and

$$||A_{f1}^{-1}|| \ll ||A_s||^{-1} \qquad (3.130b)$$

3.8.2   *Three-stage transformations*

We decompose system (3.123) into three subsystems (3.127) by using three-stage linear transformations. The *first stage* is to apply the change of variables

$$
\begin{bmatrix} x \\ z_1 \\ z_{f2} \end{bmatrix} = \begin{bmatrix} I_1 & 0 & 0 \\ 0 & I_2 & 0 \\ L_{31} & L_{32} & I_3 \end{bmatrix} \begin{bmatrix} x \\ z_1 \\ z_{f2} \end{bmatrix} \tag{3.131}
$$

to system (3.123) and choose the $(n_3 \times n_1)$ matrix L_{31} and $(n_3 \times n_2)$ matrix L_{32} such that

$$
L_{31}A_{11} - L_{31}A_{13}L_{31} + L_{32}A_{21} - L_{32}A_{23}L_{31} - A_{33}L_{31} + A_{31} = 0 \tag{3.132a}
$$

and

$$
L_{31}A_{12} - L_{31}A_{13}L_{32} + L_{32}A_{22} - L_{32}A_{23}L_{32} - A_{33}L_{32} + A_{32} = 0 \tag{3.132b}
$$

Then the system (3.123) reduces to

$$
\begin{bmatrix} \dot{x} \\ \dot{z}_1 \\ \dot{z}_{f2} \end{bmatrix} = \begin{bmatrix} \bar{A}_{11} & \bar{A}_{12} & \bar{A}_{13} \\ \bar{A}_{21} & \bar{A}_{22} & \bar{A}_{23} \\ 0 & 0 & A_{f2} \end{bmatrix} \begin{bmatrix} x \\ z_1 \\ z_{f2} \end{bmatrix} + \begin{bmatrix} B_1 \\ B_2 \\ G_3 \end{bmatrix} u \tag{3.133}
$$

where

$$
\left.
\begin{aligned}
& A_{f2} = A_{33} + L_{31}A_{13} + L_{32}A_{23} \\
& \bar{A}_{11} = A_{11} - A_{13}L_{31}; \quad \bar{A}_{12} = A_{12} - A_{13}L_{32} \\
& \bar{A}_{21} = A_{21} - A_{23}L_{31}; \quad \bar{A}_{23} = A_{23}; \quad \bar{A}_{13} = A_{13} \\
& G_3 = B_3 + L_{32}B_2 + L_{31}B_1
\end{aligned}
\right\} \tag{3.134}
$$

The *second stage* is to apply the transformation

$$
\begin{bmatrix} x \\ z_{f1} \\ z_{f2} \end{bmatrix} = \begin{bmatrix} I_1 & 0 & 0 \\ L_{21} & I_2 & L_{23} \\ 0 & 0 & I_3 \end{bmatrix} \begin{bmatrix} x \\ z_1 \\ z_{f2} \end{bmatrix} \tag{3.135}
$$

to the system (3.133) and choose the $(n_2 \times n_1)$ matrix L_{21} and $(n_2 \times n_3)$ matrix L_{23} such that

$$
L_{21}\bar{A}_{11} - L_{21}\bar{A}_{12}L_{21} - \bar{A}_{22}L_{21} + \bar{A}_{21} = 0 \tag{3.136a}
$$

and

$$
L_{21}\bar{A}_{13} - L_{21}\bar{A}_{12}L_{23} + L_{23}A_{f2} - \bar{A}_{22}L_{23} + \bar{A}_{23} = 0 \tag{3.136b}
$$

Then the system (3.133) is transformed to

$$
\begin{bmatrix} \dot{x} \\ \dot{z}_{f1} \\ \dot{z}_{f2} \end{bmatrix} = \begin{bmatrix} A_s & \bar{A}_{12} & \bar{A}_{13} \\ 0 & A_{f1} & 0 \\ 0 & 0 & A_{f2} \end{bmatrix} \begin{bmatrix} x \\ z_{f1} \\ z_{f2} \end{bmatrix} + \begin{bmatrix} B_1 \\ G_2 \\ G_3 \end{bmatrix} u \tag{3.137}
$$

where

$$
\left. \begin{aligned}
A_s &= \bar{A}_{11} - \bar{A}_{12}L_{21}; \quad A_{f1} = \bar{A}_{22} + L_{21}\bar{A}_{12} \\
\bar{A}_{12} &= \bar{A}_{12}; \quad \bar{A}_{13} = \bar{A}_{13} - \bar{A}_{12}L_{23} \\
G_2 &= B_2 + L_{21}B_1 + L_{23}G_3
\end{aligned} \right\} \tag{3.138}
$$

Finally, in the *third stage*, we use the transformation

$$
\begin{bmatrix} x \\ z_{f1} \\ z_{f2} \end{bmatrix} = \begin{bmatrix} I_1 & L_{12} & L_{13} \\ 0 & I_2 & 0 \\ 0 & 0 & I_3 \end{bmatrix} \begin{bmatrix} x \\ z_{f1} \\ z_{f2} \end{bmatrix} \tag{3.139}
$$

to system (3.137) and choose the $(n_1 \times n_2)$ matrix L_{12} and $(n_1 \times n_3)$ matrix L_{13} such that

$$
L_{12}A_{f1} + A_s L_{12} + \bar{A}_{12} = 0 \tag{3.140a}
$$

and

$$
L_{13}A_{f2} + A_s L_{13} + \bar{A}_{13} = 0 \tag{3.140b}
$$

The transformation (3.139) converts system (3.137) into

$$
\begin{bmatrix} \dot{x}_s \\ \dot{z}_{f1} \\ \dot{z}_{f2} \end{bmatrix} = \begin{bmatrix} A_s & 0 & 0 \\ 0 & A_{f1} & 0 \\ 0 & 0 & A_{f2} \end{bmatrix} \begin{bmatrix} x_s \\ z_{f1} \\ z_{f2} \end{bmatrix} + \begin{bmatrix} G_1 \\ G_2 \\ G_3 \end{bmatrix} u \tag{3.141}
$$

where

$$
G_1 = B_1 + L_{12}G_2 + L_{13}G_3
$$

The system (3.141) is now in the desired decoupled diagonal form. The distinguishing feature of the transformation

$$
\begin{bmatrix} x_s \\ z_{f1} \\ z_{f2} \end{bmatrix} = \begin{bmatrix} \bar{L}_{11} & \bar{L}_{12} & \bar{L}_{13} \\ \bar{L}_{21} & \bar{L}_{22} & \bar{L}_{23} \\ \bar{L}_{31} & \bar{L}_{32} & I_3 \end{bmatrix} \begin{bmatrix} x \\ z_1 \\ z_2 \end{bmatrix} \tag{3.142}
$$

is that the inverse transformation is obtained simply from

$$\begin{bmatrix} x \\ z_1 \\ z_1 \end{bmatrix} = \begin{bmatrix} \check{I}_1 & \check{L}_{12} & \check{L}_{13} \\ \check{L}_{21} & \check{L}_{22} & \check{L}_{23} \\ \check{L}_{31} & \check{L}_{32} & \check{L}_{33} \end{bmatrix} \begin{bmatrix} x_s \\ z_{f1} \\ z_{f2} \end{bmatrix} \qquad (3.143)$$

where

$$\left. \begin{aligned} \check{L}_{11} &= I_1 + L_{12}L_{21} + L_{12}L_{23}L_{31} \\ \check{L}_{12} &= L_{12} + L_{13}L_{23} + L_{12}L_{23}L_{32} \\ \check{L}_{13} &= L_{13} + L_{12}L_{23} \\ \check{L}_{21} &= L_{21} + L_{23}L_{31}; \quad \check{L}_{22} = I_2 + L_{23}L_{32}; \quad \check{L}_{23} = L_{23} \\ \check{L}_{31} &= L_{31}; \quad \check{L}_{32} = L_{32} \end{aligned} \right\} \qquad (3.144)$$

and

$$\left. \begin{aligned} \tilde{L}_{12} &= -L_{12}; \quad \tilde{L}_{13} = -L_{13} \\ \tilde{L}_{21} &= -L_{21}; \quad \tilde{L}_{22} = -I_2 + L_{21}L_{12} \\ \tilde{L}_{23} &= -L_{23} + L_{21}L_{13}; \quad \tilde{L}_{31} = -L_{31} + L_{32}L_{21} \\ \tilde{L}_{32} &= L_{32} + L_{31}L_{21} - L_{32}L_{21}L_{12} \\ \tilde{L}_{33} &= L_{31}L_{13} + L_{32}L_{23} - L_{32}L_{21}L_{13} \end{aligned} \right\} \qquad (3.145)$$

Let us note from (3.142)–(3.145) that the inverse transformation (3.143) for (3.142) does not require any matrix inversion.

3.8.3 Numerical algorithm

From the above analysis, we note that the evaluation of matrices A_s, A_{f1}, and A_{f2} and G_1, G_2 and G_3 requires the solutions for L_{11}, L_{12} etc., from the Riccati-type equations (3.132), (3.136) and (3.140). Because of the nature of these equations, we have to resort to numerical solution using an iterative procedure (Kokotovic, 1975). From (3.132), the iterative solution for L_{31} and L_{32} is given as

$$\begin{aligned} L_{31}(i+1) = A_{33}^{-1}[A_{31} &+ L_{31}(i)A_{11} + L_{32}(i)A_{21} - L_{31}(i)A_{13}L_{31}(i) \\ &- L_{32}(i)A_{23}L_{31}(i)] \end{aligned} \qquad (3.146a)$$

and

$$\begin{aligned} L_{32}(i+1) = A_{33}^{-1}[A_{32} &+ L_{31}(i)A_{12} + L_{32}(i)A_{22} - L_{31}(i)A_{13}L_{32}(i) \\ &- L_{32}(i)A_{23}L_{32}(i)] \end{aligned} \qquad (3.146b)$$

with starting values

$$L_{31}(0) = A_{33}^{-1}A_{31}; \quad L_{32}(0) = A_{33}^{-1}A_{32} \qquad (3.147)$$

Next we consider the iterative solution of (3.136) to obtain L_{21} and L_{23}. That is

$$L_{21}(i+1) = \bar{A}_{22}^{-1}[\bar{A}_{21} + L_{21}(i)\bar{A}_{11} - L_{21}(i)\bar{A}_{12}L_{21}(i)] \qquad (3.148a)$$

and

$$L_{23}(i+1) = \bar{A}_{22}^{-1}[A_{23} + L_{21}\bar{A}_{13} + L_{23}(i)A_{f2} - L_{21}\bar{A}_{12}L_{23}(i)] \qquad (3.148b)$$

with starting values

$$L_{21}(0) = A_{22}^{-1}\bar{A}_{21}; \quad L_{23}(0) = A_{22}^{-1}\bar{A}_{23} \qquad (3.149)$$

Finally the iterative solution of (3.140) to get L_{12} and L_{13} is given by

$$L_{12}(i+1) = [-L_{12}(i)L_{21}\bar{A}_{12} - A_s L_{12}(i) - \bar{A}_{12}]\bar{A}_{22}^{-1} \qquad (3.150a)$$

and

$$L_{13}(i+1) = [-L_{13}(i)L_{31}A_{13} - L_{13}(i)L_{32}A_{23} - A_s - \bar{A}_{13}]A_{33}^{-1} \qquad (3.150b)$$

with initial values

$$L_{12}(0) = -\bar{A}_{12}\bar{A}_{22}^{-1}; \quad L_{13}(0) = -\bar{A}_{13}A_{33}^{-1} \qquad (3.151)$$

Let us note that the above iterative solutions demand that the submatrices A_{22} and A_{33} are invertible.

Example 3.7
In order to demonstrate the above decomposition technique, let us consider a ninth-order example whose system matrix is (Jamshidi, 1983)

$$A = \begin{bmatrix}
-0\cdot311 & 0\cdot000 & 1\cdot028 & 0\cdot000 & 0\cdot000 & 2\cdot000 & 0\cdot100 & 6\cdot300 & -0\cdot400 \\
1\cdot010 & -0\cdot650 & -0\cdot084 & 0\cdot000 & 0\cdot830 & -1\cdot700 & 3\cdot000 & 2\cdot400 & 0\cdot200 \\
-0\cdot110 & 0\cdot000 & -1\cdot260 & 0\cdot000 & -6\cdot300 & 1\cdot040 & 0\cdot000 & 4\cdot100 & 0\cdot010 \\
0\cdot000 & 0\cdot100 & 0\cdot000 & -13\cdot000 & 0\cdot000 & 0\cdot480 & 0\cdot000 & 0\cdot000 & 0\cdot000 \\
0\cdot000 & -0\cdot050 & -0\cdot250 & 0\cdot000 & -17\cdot600 & -0\cdot300 & 0\cdot000 & 0\cdot000 & 0\cdot000 \\
0\cdot000 & 4\cdot500 & -3\cdot800 & -640\cdot000 & 375\cdot000 & -97\cdot000 & 1\cdot12 & 0\cdot000 & -0\cdot800 \\
0\cdot000 & -2\cdot90 & 2\cdot75 & 11\cdot1 & 0\cdot000 & 1\cdot01 & -143\cdot0 & 0\cdot500 & 0\cdot000 \\
2\cdot500 & -5\cdot7 & -4\cdot0 & -0\cdot000 & -0\cdot000 & -0\cdot910 & 0\cdot0 & -184\cdot0 & -4\cdot20 \\
1\cdot80 & -3\cdot1 & 3\cdot18 & 0\cdot000 & 1\cdot76 & -0\cdot70 & 1\cdot53 & 3\cdot2 & -207\cdot0
\end{bmatrix}$$

The eigenvalues of the above system matrix are found to be

$$-0\cdot3035; \quad -0\cdot7990; \quad -1\cdot569$$

$$-15\cdot3097; \quad -20\cdot3809$$

$$-92\cdot2556; \quad -142\cdot957; \quad -183\cdot181; \quad -207\cdot567$$

and the system possesses three 'slow' modes, two 'fast 1' modes and four 'fast 2' (faster than 1) modes. Using the iterative procedure explained in the previous Section, we summarise the following results. First we give the results corresponding to the *first approximation*, i.e. using the starting values only without any iterations, and then give the *final approximation* after a few iterations to keep the error within reasonable limits.

(a) *Slow subsystem:* At the first approximation, the slow subsystem matrix A_s is

$$A_s^1 = \begin{bmatrix} -0.4000 & 0.1733 & 1.906 \\ 1.044 & -0.8084 & -0.0379 \\ -0.0592 & -0.0945 & -1.4634 \end{bmatrix}$$

whose eigenvalues are

$$-0.3146; \quad -0.8028; \quad -1.5563$$

The final approximation after a few iterations gives the slow subsystem matrix as

$$A_s^5 = \begin{bmatrix} -0.3923 & 0.1689 & 1.1956 \\ 1.0516 & -0.8135 & -0.0380 \\ -0.0593 & -0.0945 & -1.4633 \end{bmatrix}$$

with eigenvalues

$$-0.3046; \quad -0.8018; \quad -1.5627$$

(b) *Fast 1 subsystem:* At the first approximation, the fast 1 subsystem A_{f1} is given by

$$A_{f1}^1 = \begin{bmatrix} -16.0741 & 1.7972 \\ 1.8423 & -18.589 \end{bmatrix}$$

with eigenvalues

$$-14.7861; \quad -20.1928$$

The final approximation gives

$$A_{f1}^5 = \begin{bmatrix} -16.7853 & 2.3492 \\ 2.582 & -18.9049 \end{bmatrix}$$

with eigenvalues

$$-15.3095; \quad -20.3904$$

(c) Fast 2 (faster) subsystem: With the starting values for the iteration (at the first approximation), the fast 2 (faster) subsystem matrix A_{f2} is given by

$$
A_{f2}^1 = \begin{bmatrix}
-97 \cdot 5000 & 1 \cdot 1200 & 0 \cdot 0000 & -0 \cdot 8000 \\
1 \cdot 0100 & -143 \cdot 0000 & 0 \cdot 5000 & 0 \cdot 0000 \\
-0 \cdot 9100 & 0 \cdot 0000 & -184 \cdot 0000 & -4 \cdot 2000 \\
-0 \cdot 7000 & 1 \cdot 5300 & 3 \cdot 2000 & -207 \cdot 0000
\end{bmatrix}
$$

whose eigenvalues are

$$-97 \cdot 4701; \quad -143 \cdot 0256; \quad -184 \cdot 5976; \quad -206 \cdot 4068$$

At the *final* approximation after a few iterations, we have

$$
A_{f2}^5 = \begin{bmatrix}
-92 \cdot 2791 & 0 \cdot 9384 & 0 \cdot 0769 & -0 \cdot 8119 \\
1 \cdot 0671 & -142 \cdot 9600 & 0 \cdot 4721 & 0 \cdot 0039 \\
-0 \cdot 9367 & 0 \cdot 0920 & -183 \cdot 7692 & -4 \cdot 1881 \\
-0 \cdot 6918 & 1 \cdot 5768 & 3 \cdot 2318 & -206 \cdot 9934
\end{bmatrix}
$$

whose eigenvalues are

$$-92 \cdot 2553; \quad -142 \cdot 9591; \quad -184 \cdot 3448; \quad -206 \cdot 4022$$

Let us note that the eigenvalues of the slow, fast 1 and fast 2 (faster) subsystems at the final approximation are closer to the exact values than those of the corresponding subsystems at the first approximation (Naidu and Ravinder, 1985).

3.9 Discussion and conclusion

In this Chapter we have discussed the time-scale analysis of continuous and discrete control systems. The central idea has been the application of a two-stage block diagonalisation to decouple a given two-time-scale system into two low-order slow and fast subsystems. The analysis involves an iterative solution of nonlinear algebraic equations. An important aspect of permutation and scaling has been discussed in Section 3.2. It is interesting to note from Section 3.3 that the singular perturbation method provides to a first approximation the same results as given by time-scale analysis. The feedback design giving the required eigenvalue placement has been provided in Section 3.4. The remaining Sections have been devoted to the time-scale analysis of discrete systems.

The separation of states into slow and fast variables is not an easy task, but demands physical insight and ingenuity on the part of the analyst (Kokotovic, 1981; Chow, 1982). In the absence of empirical estimates of the speed of state variables, physical parameters such as time constants, loop gains, masses and inertias are examined to label the state variables as 'slow' and 'fast'. Sometimes a permutation and/or scaling of states is required to order the states according to their speed of response (Kokotovic *et al.*, 1980). The separation of time scales has its foundation in some linear transformations and finally reduces to iterative solutions of algebraic Riccati equations (Chang, 1974; Chow, 1975; Kokotovic, 1975; Javid and Kokotovic, 1977; Anderson, 1982; O'Malley and Anderson, 1982; Avramovic et al., 1980; Chemouli and Wahdan, 1980; Phillips, 1980*a,b*, 1983; Magni and Fossard, 1982; Sigh, 1982; Young, 1982*b*, Kando and Iwazumi, 1983*c*; Peponides and Kokotovic, 1983; Siljak, 1983; Chow and Kokotovic, 1985; Khalil, 1984; Kokotovic, 1984, 1985). The iterative procedure is related to the simultaneous subspace iterations.

An alternative method for the separation of time scales is the quasi-steady-state iterations (Phillips, 1983; Kokotovic *et al.*, 1980; Winkelman *et al.*, 1980). Time-scale decoupling for linear time-varying systems is discussed by O'Malley and Anderson (1982).

A class of nonlinear systems separable into slow and fast modes is also outlined by Kokotovic *et al.* (1980) and multi-time-scale analysis is proposed by repeated application of two-time-scale decomposition (Winkelman *et al.*, 1980).

In model (Davison, 1966) and aggregation (Aoki, 1968; Duc et al., 1983) methods applied to two-time-scale systems, the slow subsystem is retained while neglecting the fast subsystem. A model approach to state decomposition is given by Litz and Roth (1981).

The time-scale analysis of discrete systems is of recent origin (Locatelli and Schiavoni, 1976). In the beginning, attempts to model discrete systems with slow and fast behaviour in a strictly singularly perturbed structure faced some stability problems (Comstock and Hsiao, 1976; Reinhardt, 1979). With the formulation of some other types of singularly perturbed structures, there has been considerable interest in the analysis of two-time-scale discrete systems (Phillips, 1980*a*; Rajagopalan and Naidu, 1980*a*; Naidu and Rao, 1981; Mahmoud and Singh, 1981*a*, 1984; Mahmoud, 1982*a,b,c*, 1986; Syrcos and Sannuti, 1983; Tran and Sawan, 1983*a,b*, 1984*a,b,c*; Kando and Iwazumi, 1983*a,b*, 1984, 1985; Litkouhi and Khalil, 1985; Mahmoud *et al.*, 1985*a*, Naidu and Rao, 1985*a*).

Another interesting aspect is the discretisation of continuous singularly perturbed systems which pose problems regarding the choice of discretising interval with respect to either slow or fast time scales (Locatelli and Schaivoni, 1976; Hemker and Miller, 1979; Miranker, 1981; Blankenship, 1981; Litkouhi and Khalil, 1984, 1985).

An important topic is the multi-time-scale systems having more than one distinct small parameter and giving rise to widely separated groups of eigenvalues (Vasileva, 1963; O'Malley, 1968, 1974*a*; Hoppensteadt, 1971; Jamshidi,

1972, 1974, 1976; Asatani, 1974; Chen and O'Malley, 1974*a*; Grujic, 1976*a*,*b*; Ozguner, 1979; Kando and Iwazumi, 1981, 1983*c*; Dragan and Halanay, 1982; Mahmoud *et al.*, 1982; Singh, 1982; Coderch, *et al.*, 1983*a*; Silva-Madriz, 1986; Silva-Madriz and Sastry, 1984, 1986).

Chapter 4

Open-loop optimal control of continuous systems

Optimisation is a very important aspect in system theory. The computational burden for demanding optimum performance of the system is quite heavy. In general, the 'curse' of dimensionality 'plagues' optimal design, as the computational complexity increases rapidly with the order of the system. For singularly perturbed systems, the situation is further aggravated by the 'scare' of stiffness. Thus the singular perturbation method which alleviates the difficulties of dimensionality and stiffness is most welcome to optimal control theory, especially in open-loop optimal control, where we have to face the two-point boundary-value problem (TPBVP).

In this Chapter, our main intention is to describe the singluar perturbation method in order to obtain asymptotic power-series expansions for the singularly perturbed TPBVP arising in the open-loop optimal control of linear and nonlinear continuous systems. Thus in Section 4.1 we consider a linear, time-invariant singularly perturbed system along with minimisation of a standard quadratic performance index as a free-end-point problem. Using the necessary conditions for optimisation, we arrive at the singularly perturbed TPBVP in terms of the state and co-state variables with appropriate boundary conditions.

In Section 4.2 we present the singular perturbation method by constructing asymptotic power-series expansions in terms of an outer solution based on the reduced (or degenerate) problem and the initial and final boundary-layer corrections (BLCs) based on the corresponding stretched problems. The main purpose of BLC is to recover the boundary conditions lost in the process of degeneration. The asymptotic solution for optimal control and cost is also given

The fixed-end-point linear optimal control problem is discussed in Section 4.3. In Section 4.4 we address an important class of nonlinear singularly perturbed system and obtain the asymptotic power-series expansions for the TPBVP. Section 4.5 concludes the Chapter with a discussion on related works.

4.1   Formulation of the linear problem

For a singularly perturbed, linear, time-invariant continuous system, we formulate the open-loop optimal control problem leading to a singularly perturbed two-point boundary-value problem (TPBVP). Consider

$$\frac{dx}{dt} = A_{11}x + A_{12}z + B_1u, \quad x(t=0) = x(0) \tag{4.1a}$$

$$\varepsilon\frac{dz}{dt} = A_{21}x + A_{22}z + B_2u, \quad z(t = 0) = z(0) \tag{4.1b}$$

and a quadratic performance index

$$J = \frac{1}{2}y'(T)Sy(T) + \frac{1}{2}\int_0^T (y'Dy + u'Ru)dt \tag{4.2}$$

where

$$Y = \begin{bmatrix} x \\ z \end{bmatrix}; \quad S = \begin{bmatrix} S_{11} & \varepsilon S_{12} \\ \varepsilon S_{12}' & \varepsilon S_{22} \end{bmatrix}; \quad D = \begin{bmatrix} D_{11} & D_{12} \\ D_{12}' & D_{22} \end{bmatrix}$$

x and z are m- and n-dimensional state vectors, u is an r-dimensional control vector, ε is the small positive parameter responsible for singular perturbation, A_{ij} and B_i are matrices of appropriate dimensionality. We assume that S and D are symmetric and positive-semidefinite matrices and R is a symmetric and positive-definite matrix.

It is well known that the free end-point problem defined by (4.1) and (4.2) has a unique optimal control which minimises J (Athans and Falb, 1966). We are interested, however, in obtaining an asymptotic solution of the problem as the small parameter ε tends to zero. To obtain necessary and sufficient conditions for the optimal control, we introduce the Hamiltonian

$$H(x,z,p,q,t,\varepsilon) = \frac{1}{2}(x'D_{11}x + 2x'D_{12}z + z'D_{22}z + u'Ru)$$

$$+ p'(A_{11}x + A_{12}z + B_1u) + q'(A_{21}x + A_{22}z + B_2u) \tag{4.3}$$

Using elementary calculus of variations (Athans and Falb, 1966), we have, along an optimal trajectory, $\partial H/\partial u = 0$ leading to

$$u = -R^{-1}(B_1'p + B_2'q) \tag{4.4}$$

where the co-state variables p and q satisfy equations

$$\frac{dp}{dt} = -\frac{\partial H}{\partial x} = -D_{11}x - D_{12}z - A_{11}'p - A_{21}'q \tag{4.5a}$$

$$\varepsilon \frac{dq}{dt} = - \frac{\partial H}{\partial z} = - D_{12}'x - D_{22}z - A_{12}'p - A_{22}'q \tag{4.5b}$$

on $0 \leq t \leq T$ and the terminal conditions

$$p(T) = S_{11}x(T) + \varepsilon S_{12}z(T) \tag{4.6a}$$

$$q(T) = S_{12}'x(T) + S_{22}z(T) \tag{4.6b}$$

Together with (4.4)–(4.6), we have the original state equations as

$$\frac{dx}{dt} = \frac{\partial H}{\partial p} = A_{11}x + A_{12}z + B_1u \tag{4.7a}$$

$$\varepsilon \frac{dz}{dt} = \frac{\partial H}{\partial q} = A_{21}x + A_{22}z + B_2u \tag{4.7b}$$

with initial conditions

$$x(t=0) = x(0); \quad z(t=0) = z(0) \tag{4.8}$$

Note that since $\partial^2 H/\partial u^2 = R$ is positive definite, the optimal control (4.4) will minimise J in (4.2).

Using (4.4) in (4.7) and with (4.5), we have

$$\frac{dx}{dt} = A_{11}x + A_{12}z - E_{11}p - E_{12}q \tag{4.9a}$$

$$\frac{dp}{dt} = - D_{11}x - D_{12}z - A_{11}'p - A_{21}'q \tag{4.9b}$$

$$\varepsilon \frac{dz}{dt} = A_{21}x + A_{22}z - E_{12}'p - E_{22}q \tag{4.9c}$$

$$\varepsilon \frac{dq}{dt} = - D_{12}'x - D_{22}z - A_{12}'p - A_{22}'q \tag{4.9d}$$

where

$$E_{11} = B_1R^{-1}B_1'; \quad E_{12} = B_1R^{-1}B_2'; \quad E_{22} = B_2R^{-1}B_2'$$

The $2(m+n)$th- order problem (4.9), together with the boundary conditions (4.6) and (4.8), is called a singularly perturbed TPBVP in the sense that as ε is made zero, the order of the problem drops from $2(m+n)$ to $2m$, thereby loosing the boundary conditions $z(0)$ and $q(T)$ (Kokotovic and Sannuti, 1968; Hadlock, 1973; O'Malley, 1972a,b,c; Sannuti, 1975; Ardema, 1983b). The reduced problem obtained by making $\varepsilon = 0$ is given by

$$\frac{dx^{(0)}}{dt} = A_{11}x^{(0)} + A_{12}z^{(0)} - E_{11}p^{(0)} - E_{12}q^{(0)} \tag{4.10a}$$

$$\frac{dp^{(0)}}{dt} = -D_{11}x^{(0)} - D_{12}z^{(0)} - A'_{11}p^{(0)} - A'_{21}q^{(0)} \tag{4.10b}$$

$$0 = A_{21}x^{(0)} + A_{22}z^{(0)} - E'_{12}p^{(0)} - E_{22}q^{(0)} \tag{4.10c}$$

$$0 = -D'_{12}x^{(0)} - D_{22}z^{(0)} - A'_{12}p^{(0)} - A'_{22}q^{(0)} \tag{4.10d}$$

with boundary conditions

$$x^{(0)}(t=0) = x(0); \quad p^{(0)}(T) = p(T) \tag{4.11a}$$

$$z^{(0)}(t=0) \ne z(0); \quad q^{(0)}(T) \ne q(T) \tag{4.11b}$$

Note that (4.10) can be solved for $z^{(0)}$ and $q^{(0)}$ as linear functions of $x^{(0)}$ and $p^{(0)}$ provided that the $2n \times 2n$ matrix

$$\begin{bmatrix} A_{22} & -E_{22} \\ -D_{22} & -A'_{22} \end{bmatrix}$$

in nonsingular. We shall obtain

$$z^{(0)} = \overline{K}_{11}x^{(0)} + \overline{K}_{12}p^{(0)} \tag{4.12a}$$

$$q^{(0)} = \overline{K}_{21}x^{(0)} + \overline{K}_{22}p^{(0)} \tag{4.12b}$$

Where K_{ij} can be explicitly obtained and (4.10) becomes

$$\frac{dx^{(0)}}{dt} = K_{11}x^{(0)} + K_{12}p^{(0)} \tag{4.13a}$$

$$\frac{dp^{(0)}}{dt} = K_{21}x^{(0)} + K_{22}p^{(0)} \tag{4.13b}$$

For example, $K_{11} = A_{11} + A_{12}\overline{K}_{11} - E_{12}\overline{K}_{21}$.

Here, we solve the reduced order TPBVP (4.13) along with the boundary conditions (4.11a). Solving for $x^{(0)}$ and $p^{(0)}$, $z^{(0)}$ and $q^{(0)}$ are automatically fixed from (4.12). Hence the boundary conditions $z^{(0)}$ and $q(T)$ are lost as shown by (4.11b) in the process of degeneration.

The reduced problem (4.13) can also be obtained by an alternative approach. First we neglect ε in state equations (4.1) and then obtain the necessary conditions for optimal control. This again leads to a TPBVP which is identical to (4.13). Thus the reduced problem obtained by setting $\varepsilon = 0$ in the necessary conditions of the full problem is the same as the problem obtained by setting $\varepsilon = 0$ in the state equations and then applying the necessary conditions. In other words, the sequence of the processes of degeneration and optimisation is interchangeable. The process of first degenerating and then optimising is attractive since it involves fewer algebraic manipulations.

4.2   Singular perturbation method

In the previous Section we have seen that the singularly perturbed, linear TPBVP described by (4.9), (4.6) and (4.8), under the action of degeneration, becomes the reduced order TPBVP described by (4.10) and (4.11). Consequent to order reduction, the reduced-order problem looses some of the auxiliary conditions corresponding to initial and final points. Thus we are led naturally to incorporate two boundary-layer corrections (BLCs) at the initial and final points.

Our main interest is to determine asymptotic expansions for the states x and z and co-states p and q of the full problem (4.9) in terms of the small positive parameter ε as ε tends to zero. Once the expansions for the co-states are obtained, we can determine an analogous expansion for the control function u from (4.4). The method for linear systems is based on the works of Hadlock (1973), O'Malley (1972a,b, 1975), O'Malley and Kung (1975) and Naidu (1977). Thus we seek asymptotic solutions of the form

$$x(t,\varepsilon) = x_o(t,\varepsilon) + \varepsilon x_i(t',\varepsilon) + \varepsilon x_f(t'',\varepsilon) \qquad (4.14a)$$

$$z(t,\varepsilon) = z_o(t,\varepsilon) + z_i(t',\varepsilon) + z_f(t'',\varepsilon) \qquad (4.14b)$$

$$p(t,\varepsilon) = p_o(t,\varepsilon) + \varepsilon p_i(t',\varepsilon) + \varepsilon p_f(t'',\varepsilon) \qquad (4.14c)$$

$$q(t,\varepsilon) = q_o(t,\varepsilon) + q_i(t',\varepsilon) + q_f(t'',\varepsilon) \qquad (4.14d)$$

where the suffix o refers to outer solution and i and f refer to boundary-layer corrections at the initial point $t = 0$ and final point $t = T$, respectively. The stretching transformations at the initial and final points are given by

$$t' = t/\varepsilon; \quad t'' = (T-t)/\varepsilon \qquad (4.15)$$

The 'outer solution' (or 'outer expansion') satisfies the system (4.9) and has asymptotic power-series expansions as ε tends to zero of the form

$$\{x_o(t,\varepsilon), \ z_o(t,\varepsilon), \ p_o(t,\varepsilon), \ q_o(t,\varepsilon)\}$$

$$\sim \sum_{j=0}^{\infty} \{x^{(j)}(t), \ z^{(j)}(t), \ p^{(j)}(t), \ q^{(j)}(t)\} \varepsilon^j \qquad (4.16)$$

The 'boundary-layer correction' at the initial point $t = 0$ has asymptotic power-series expansions as ε tends to zero, as

$$\{\varepsilon x_i(t',\varepsilon),\ z_i(t',\varepsilon),\ \varepsilon p_i(t',\varepsilon),\ q_i(t',\varepsilon)\}$$

$$\sim \sum_{j=0}^{\infty} \{\varepsilon x_i^{(j)}(t'),\ z_i^{(j)}(t'),\ \varepsilon p_i^{(j)}(t'),\ q_i^{(j)}(t')\}\varepsilon^j \qquad (4.17)$$

These terms tend to zero as the initial stretched variable t' tends to infinity.

The 'boundary-layer correction' at the final point $t = T$ has asymptotic power-series expansions as ε tends to zero, as

$$\{\varepsilon x_f(t'',\varepsilon),\ z_f(t'',\varepsilon),\ \varepsilon p_f(t'',\varepsilon),\ q_f(t'',\varepsilon)\}$$

$$\sim \sum_{j=0}^{\infty} \{\varepsilon x_f^{(j)}(t''),\ z_f^{(j)}(t''),\ \varepsilon p_f^{(j)}(t''),\ q_f^{(j)}(t'')\}\varepsilon^j \qquad (4.18)$$

These coefficients tend to zero as the final stretched co-ordinate t'' tends to infinity.

Note that an expansion like (4.14) converges asymptotically to the outer expansion (x_0, z_0, p_0, q_0) within $0 < t < T$ to all orders ε^j as ε tends to zero. Further, $x(t,\varepsilon) \to x^{(0)}(t)$ and $p(t,\varepsilon) \to p^{(0)}(t)$ uniformly on $0 \leq t \leq T$ as $\varepsilon \to 0$, while convergence of $z(t,\varepsilon)$ to $z^{(0)}(t)$ and of $q(t,\varepsilon)$ to $q^{(0)}(t)$ will break down at both ends $t = 0$ and $t = T$ (O'Malley, 1972a,b; 1975).

The reason behind attaching ε for corrections to x and p in (4.14) is that it will simplify the solutions of the correction equations and the evaluation of their boundary conditions.

An important assumption in using (4.14) is that the initial BLCs $x_i(t')$, $z_i(t')$, $p_i(t')$, $q_i(t')$ at $t = T$ and the final BLCs $x_f(t'')$, $z_f(t'')$, $p_f(t'')$, $q_f(t'')$ at $t = 0$ are asymptotically negligible. Hence the boundary conditions (4.6) and (4.8) are asymptotically equivalent to

$$x(0) = x^{(j)}(0) + \varepsilon x_i^{(j)}(0); \quad z(0) = z^{(j)}(0) + z_i^{(j)}(0) \qquad (4.19a)$$

$$p(T) = p^{(j)}(T) + \varepsilon p_f^{(j)}(0); \quad q(T) = q^{(j)}(T) + q_f^{(j)}(0) \qquad (4.19b)$$

Thus

$$x^{(0)}(0) = x(0); \quad z_i^{(0)}(0) = z(0) - z^{(0)}(0) \qquad (4.20a)$$

$$p^{(0)}(T) = p(T); \quad q_f^{(0)}(0) = q(T) - q^{(0)}(T) \qquad (4.20b)$$

$$x^{(j)}(0) = -x_i^{(j-1)}(0); \quad z_i^{(j)}(0) = -z^{(j)}(0), \quad j \geq 1 \qquad (4.20c)$$

$$p^{(j)}(T) = -p_f^{(j-1)}(0); \quad q_f^{(j)}(0) = -q^{(j)}(T), \quad j \geq 1 \qquad (4.20d)$$

Note that the boundary conditions $x^{(j)}(0)$ and $p^{(j)}(T)$ are given in terms of the next lower-order corrections.

4.2.1 Outer expansion

By substituting the outer expansion (4.16) into (4.9) and collecting coefficients of like powers of ε we get a set of recursive equations. For the zeroth-order (ε^0) approximation, we have the same equations (4.10) or (4.12). For first-order (ε^1) approximation, we have

$$\frac{dx^{(1)}}{dt} = A_{11}x^{(1)} + A_{12}z^{(1)} - E_{11}p^{(1)} - E_{12}q^{(1)} \tag{4.21a}$$

$$\frac{dp^{(1)}}{dt} = -D_{11}x^{(1)} - D_{12}z^{(1)} - A'_{11}p^{(1)} - A'_{21}q^{(1)} \tag{4.21b}$$

$$\frac{dz^{(0)}}{dt} = A_{21}x^{(1)} + A_{22}z^{(1)} - E'_{12}p^{(1)} - E_{22}q^{(1)} \tag{4.21c}$$

$$\frac{dq^{(0)}}{dt} = -D'_{12}x^{(1)} - D_{22}z^{(1)} - A'_{12}p^{(1)} - A'_{22}q^{(1)} \tag{4.21d}$$

The solution of the above differential equations in $x^{(1)}$ and $p^{(1)}$ of reduced order $2m$ requires the boundary conditions $x^{(1)}(0)$ and $p^{(1)}(T)$ given in (4.20). Similar equations can be written down for higher-order approximations.

4.2.2   Initial boundary-layer correction

Since the outer expansion (4.16) satisfies the original system (4.9) and the boundary-layer correction at $t = T$ is asymptotically negligible at $t = 0$, (4.14) and (4.9) imply that the initial BLC must satisfy the system of equations,

$$\frac{dx_i}{dt'} = \frac{dx}{dt} - \frac{dx_o}{dt} = \varepsilon A_{11}x_i + A_{12}z_i - \varepsilon E_{11}p_i - E_{12}q_i \tag{4.22a}$$

$$\frac{dp_i}{dt'} = \frac{dp}{dt} - \frac{dp_o}{dt} = \varepsilon D_{11}x_i - D_{12}z_i - \varepsilon A'_{11}p_i - A'_{21}q_i \tag{4.22b}$$

$$\frac{dz_i}{dt'} = \varepsilon\left(\frac{dz}{dt} - \frac{dz_o}{dt}\right) = \varepsilon A_{11}x_i + A_{22}z_i - \varepsilon E'_{12}p_i - E_{22}q_i \tag{4.22c}$$

$$\frac{dq_i}{dt'} = \varepsilon\left(\frac{dq}{dt} - \frac{dq_o}{dt}\right) = -\varepsilon D'_{12}x_i - D_{22}z_i - \varepsilon A'_{12}p_i - A'_{22}q_i \tag{4.22d}$$

Now substituting asymptotic power series expansions (4.17) for initial BLC into (4.22), we obtain a system of equations. In particular, the lowest-order terms will satisfy

$$\frac{dx_i^{(0)}}{dt'} = A_{12}z_i^{(0)} - E_{12}q_i^{(0)} \tag{4.23a}$$

$$\frac{dp_i^{(0)}}{dt'} = -D_{12}z_i^{(0)} - A'_{21}q_i^{(0)} \tag{4.23b}$$

$$\frac{dz_i^{(0)}}{dt'} = A_{22}z_i^{(0)} - E_{22}q_i^{(0)} \tag{4.23c}$$

$$\frac{dq_i^{(0)}}{dt'} = -D_{22}z_i^{(0)} - A_{22}'q_i^{(0)} \tag{4.23d}$$

This is a set of differential equations in $z_i^{(0)}(t')$ and $q_i^{(0)}(t')$ which is solved by using the condition $z_i^{(0)}(0) = z(0) - z^{(0)}(0)$ and using the fact that (4.13) should have exponentially decaying solutions only. Thus all the initial corrections $x_i^{(0)}(t')$, $p_i^{(0)}(t')$, $z_i^{(0)}(t')$, $q_i^{(0)}(t')$ are uniquely determined as decaying components. Further, (4.20) gives $x^{(1)}(0) = -x_i^{(0)}(0)$, which is required for solving the first-order outer equations (4.21).

For higher-order $(j \geq 1)$ approximations of initial BLC, we have

$$\frac{dx_i^{(j)}}{dt'} = A_{12}z_i^{(j)} - E_{12}q_i^{(j)} + M_x^{(j-1)}(t') \tag{4.24a}$$

$$\frac{dp_i^{(j)}}{dt'} = -D_{12}z_i^{(j)} - A_{21}'q_i^{(j)} + M_p^{(j-1)}(t') \tag{4.24b}$$

$$\frac{dz_i^{(j)}}{dt'} = A_{22}z_i^{(j)} - E_{22}q_i^{(j)} + M_z^{(j-1)}(t') \tag{4.24c}$$

$$\frac{dq_i^{(j)}}{dt'} = -D_{22}z_i^{(j)} - A_{22}'q_i^{(j)} + M_q^{(j-1)}(t') \tag{4.24d}$$

where $M_x^{(j-1)}(t'), \ldots, M_q^{(j-1)}(t')$ are known successively as exponentially decaying functions. The initial values $q_i^{(j)}(0)$, are chosen such that $q_i^{(j)}(t')$ decays expontentially and from (4.20), $z_i^{(j)}(0) = -z^{(j)}(0)$. Thus (4.24) is solved to contain exponentially decaying solutions. Then (4.20) also gives $x^{(j)}(0) = -x_i^{(j)}(0)$, which will be required for solving higher-order outer equations.

4.2.3 Final boundary-layer correction

The final BLC is determined in an analogous fashion. Here the BLC at $t = 0$ is asymptotically negligible and (4.14) and (4.9) imply that the final BLC must satisfy the set of equations

$$\frac{dx_f}{dt''} = -\left(\frac{dx}{dt} - \frac{dx_o}{dt}\right) = -\varepsilon A_{11}x_f - A_{12}z_f + \varepsilon E_{11}p_f + E_{12}q_f \tag{4.25a}$$

$$\frac{dp_f}{dt''} = -\left(\frac{dp}{dt} - \frac{dp_o}{dt}\right) = \varepsilon D_{11}x_f + D_{12}z_f + \varepsilon A_{11}'p_f + A_{21}'q_f \tag{4.25b}$$

$$\frac{dz_f}{dt''} = -\varepsilon\left(\frac{dz}{dt} - \frac{dz_o}{dt}\right) = -\varepsilon A_{21}x_f - A_{22}z_f + \varepsilon E'_{12}p_f + E_{22}q_f \qquad (4.25c)$$

$$\frac{dq_f}{dt''} = -\varepsilon\left(\frac{dq}{dt} - \frac{dq_o}{dt}\right) = \varepsilon D'_{12}x_f + D_{22}z_f + \varepsilon A'_{12}p_f + A'_{22}q_f \qquad (4.25d)$$

Now substituting asymptotic power series expansions (4.18) into (4.25) we get a series of equations. For the zeroth-order approximation,

$$\frac{dx_f^{(0)}}{dt''} = -A_{12}z_f^{(0)} + E_{12}q_f^{(0)} \qquad (4.26a)$$

$$\frac{dp_f^{(0)}}{dt''} = D_{12}z_f^{(0)} + A'_{21}q_f^{(0)} \qquad (4.26b)$$

$$\frac{dz_f^{(0)}}{dt''} = -A_{22}z_f^{(0)} + E_{22}q_f^{(0)} \qquad (4.26c)$$

$$\frac{dq_f^{(0)}}{dt''} = D_{22}z_f^{(0)} + A'_{22}q_f^{(0)} \qquad (4.26d)$$

This set of differential equations in $z_f^{(0)}(t'')$ and $q_f^{(0)}(t'')$ is solved using from (4.20) the condition $q_f^{(0)}(0) = q(T) - q^{(0)}(T)$ and using the fact that (4.26) should have exponentially decaying solutions. Once we solve for the zeroth-order corrections, (4.20) gives $p^{(1)}(T) = -p_f^{(0)}(0)$, which will be required for solving the first-order outer equations (4.21).

For higher-order $(j \geq 1)$ approximations of the final BLC, we have

$$\frac{dx_f^{(j)}}{dt''} = -A_{12}z_f^{(j)} + E_{12}q_f^{(j)} + M_x^{(j-1)}(t'') \qquad (4.27a)$$

$$\frac{dp_f^{(j)}}{dt''} = D_{12}z_f^{(j)} + A'_{21}q_f^{(j)} + M_p^{(j-1)}(t'') \qquad (4.27b)$$

$$\frac{dz_f^{(j)}}{dt''} = -A_{22}z_f^{(j)} + E_{22}q_f^{(j)} + M_z^{(j-1)}(t'') \qquad (4.27c)$$

$$\frac{dq_f^{(j)}}{dt''} = D_{22}z_f^{(j)} + A'_{22}q_f^{(j)} + M_q^{(j-1)}(t'') \qquad (4.27d)$$

where $M_x^{(j-1)}(t''), \ldots, M_q^{(j-1)}(t'')$ are known successively as exponentially decaying functions. The initial values are given from (4.20) as $q_f^{(j)}(0) = -q^{(j)}(T)$ and $z_f^{(j)}(0)$ is chosen to retain the stable solutions only. Thus all the terms in (4.27) are determined to be decaying components. From (4.20), $p^{(j)}(T)$ $= -p_f^{(j-1)}(0)$ is used for solving higher-order equations.

This completes the construction of asymptotic expansions for the state and co-state variables. To summarise, the reduced problem (4.10) and (4.11) first determines $\{x^{(0)}, p^{(0)}, z^{(0)}, q^{(0)}\}$. Using the reduced solution and (4.20), the zeroth-order BLCs for the initial and final points, i.e., $\{x_i^{(0)}, p_i^{(0)}, z_i^{(0)}, q_i^{(0)}\}$ and $\{x_f^{(0)}, p_f^{(0)}, z_f^{(0)}, q_f^{(0)}\}$, are determined from (4.23) and (4.26). Now the boundary conditions $x^{(1)}(0)$ and $p^{(1)}(T)$ are known from (4.20). Next the solution of (4.21) gives us $\{x^{(1)}, p^{(1)}, z^{(1)}, x^{(1)}\}$. Thus, proceeding recursively, the outer expansion terms $\{x^{(j)}, p^{(j)}, z^{(j)}, q^{(j)}\}$ for all $j \leq k$) allow us to calculate the kth-order boundary-layer terms $\{x_i^{(k)}, p_i^{(k)}, z_i^{(k)}, q_i^{(k)}\}$ and $\{x_f^{(k)}, p_f^{(k)}, z_f^{(k)}, q_f^{(k)}\}$. These boundary-layer terms, in turn, allow us to calculate $(k+1)$th outer expansion terms $\{x^{(k+1)}, p^{(k+1)}, z^{(k+1)}, q^{(k+1)}\}$.

4.2.4 *Asymptotic expansions for the optimal control and cost*

Having obtained unique asymptotic series expansions of the form (4.14) for the co-states p and q, the relation (4.4) for the optimal control implies that the corresponding optimal control has an analogous asymptotic expansion of the form

$$u(t,\varepsilon) = u_o(t,\varepsilon) + u_i(t',\varepsilon) + u_f(t'',\varepsilon) \tag{4.28}$$

where u_o, u_i, and u_f, all have asymptotic series expansions in ε such that the boundary-layer corrections u_i, and u_f, tend to zero as t' and t'' tend to infinity, respectively. From (4.4), (4.14) and (4.28), we can write

$$u_o(t,\varepsilon) = -R^{-1}[B_1'p_o(t,\varepsilon) + B_2'q_o(t,\varepsilon)] \tag{4.29a}$$

$$u_i(t',\varepsilon) = -R^{-1}[\varepsilon B_1'p_i(t',\varepsilon) + B_2'q_i(t',\varepsilon)] \tag{4.29b}$$

$$u_f(t'',\varepsilon) = -R^{-1}[\varepsilon B_1'p_f(t'',\varepsilon) + B_2'q_f(t'',\varepsilon)] \tag{4.29c}$$

Then using (4.2) and (4.14) for the state vectors x and z, we obtain an asymptotic expansion for the optimal cost function

$$J^*(\varepsilon) = \frac{1}{2}L_0(\varepsilon) + \frac{1}{2}\int_0^T L_1(t,\varepsilon)dt + \frac{\varepsilon}{2}\int_0^\infty L_2(t',\varepsilon)dt' + \frac{\varepsilon}{2}\int_0^\infty L_3(t'',\varepsilon)dt'' \tag{4.30}$$

where L_0 and L_j have asymptotic series expansions as ε tends to zero with integrable coefficients. In other words

$$J^*(\varepsilon) = \sum_{j=0}^\infty J_j^* \varepsilon^j \tag{4.31}$$

Here the leading term J_0^* is the optimal cost of the reduced problem, i.e.

$$J_0^* = \frac{1}{2}x^{(0)'}(T)S_{11}x^{(0)}(T) + \frac{1}{2}\int_0^T \{y^{(0)'}(t)Dy^{(0)}(t) + u^{(0)'}(t)Ru^{(0)}(t)\} \, dt \quad (4.32)$$

Higher-order terms are affected by the boundary-layer correction terms.

4.2.5 Total series solution

We shall obtain the theorem for the total series solution under the following assumptions (O'Malley 1972a,b; O'Malley and Kung, 1975):

(i)   the matrix A_{22} is invertible
(ii) the reduced problem (4.10) and (4.11a) has a unique solution
(iii) the matrix

$$\begin{bmatrix} A_{22} & -E_{22} \\ -D_{22} & -A_{22}' \end{bmatrix}$$

has nonzero real parts throughout $0 \leqslant t \leqslant T$.

Assumption (i) plays several roles. It guarantees the solvability of algebraic equations of the type (4.10c) and (4.10d) for $z^{(0)}$ and $q^{(0)}$. It rules out turning-point behaviour (Wasow, 1965). Also, as we have seen, it defines the behaviour in the boundary layers.
Theorem 4.1: Under the assumptions (i)–(iii), the optimal control problem (4.1) and (4.2) has a unique asymptotic solution for ε sufficiently small such that, for every integer $N \geqslant 0$, the optimal trajectories $x(t,\varepsilon)$ and $z(t,\varepsilon)$ satisfy

$$x(t,\varepsilon) = x^{(0)}(t) + \sum_{j=1}^N \{x^{(j)}(t) + x_i^{(j-1)}(t') + x_f^{(j-1)}(t'')\}\varepsilon^j + 0(\varepsilon^{N+1}) \quad (4.33a)$$

$$z(t,\varepsilon) = \sum_{j=0}^N \{z^{(j)}(t) + z_i^{(j)}(t') + z_f^{(j)}(t'')\}\varepsilon^j + 0(\varepsilon^{N+1}) \quad (4.33b)$$

uniformly on the interval $0 \leqslant t \leqslant T$. Further, using the control relation (4.4) and the cost functional (4.2), the optimal control and the optimal cost have the corresponding asymptotic expansions

$$u(t,\varepsilon) = \sum_{j=0}^N \{u^{(j)}(t) + u_i^{(j)}(t') + u_f^{(j)}(t'')\}\varepsilon^j + 0(\varepsilon^{N+1}) \quad (4.33c)$$

and

$$J^*(\varepsilon) = \sum_{j=1}^N J_j^* + 0(\varepsilon^{N+1}) \quad (4.33d)$$

Here the terms depending on $t' = t/\varepsilon$ (or $t'' = (T-t)/\varepsilon$) decay to zero exponentially as t' (or t'') tends to infinity; i.e. away from $t = 0$ (or T).

Example 4.1
Let us consider a second-order system (Sannuti, 1968; Naidu, 1977)

$$\frac{dx}{dt} = z \tag{4.34a}$$

$$\varepsilon\frac{dz}{dt} = -x - z + u \tag{4.34b}$$

The performance index to be minimised is

$$J = \frac{1}{2}\int_0^T (x^2 + u^2)dt \tag{4.35}$$

The initial conditions $x(0)$ and $z(0)$ are prescribed and the final states $x(T)$ and $z(T)$ are free. The singularly perturbed TPBVP corresponding to (4.9) is

$$\frac{dx}{dt} = z, \quad x(t=0) = x(0) \tag{4.36a}$$

$$\frac{dp}{dt} = -x + q, \quad p(t=T) = 0 \tag{4.36b}$$

$$\varepsilon\frac{dz}{dt} = -x - z - q, \quad z(t=0) = z(0) \tag{4.36c}$$

$$\varepsilon\frac{dq}{dt} = -p + q, \quad q(t=T) = 0 \tag{4.36d}$$

The optimal control given by (4.4) is

$$u = -q \tag{4.37}$$

The zeroth-order outer (reduced-order) problem is

$$\frac{dx^{(0)}}{dt} = z^{(0)}, \quad x^{(0)}(0) = x(0) \tag{4.38a}$$

$$\frac{dp^{(0)}}{dt} = -x^{(0)} + q^{(0)}, \quad p^{(0)}(T) = 0 \tag{4.38b}$$

$$0 = -x^{(0)} - z^{(0)} + q^{(0)} \tag{4.38c}$$

$$0 = -p^{(0)} + q^{(0)} \tag{4.38d}$$

and

$$u^{(0)} = -q^{(0)} \tag{4.39}$$

Solving

$$
\left.
\begin{aligned}
x^{(0)}(t) &= C_1 \exp(at) + C_2 \exp(-at) \\
z^{(0)}(t) &= C_3 \exp(at) + C_4 \exp(-at) \\
u^{(0)}(t) &= C_5 \exp(at) + C_6 \exp(-at)
\end{aligned}
\right\} \tag{4.40}
$$

where $a = 2^{0.5}$ and C_1 to C_6 are constants depending on $x(0)$ and T.

The initial *BLC* equations for the zeroth-order approximation are given by (4.23) as

$$\frac{dx_i^{(0)}}{dt'} = z_i^{(0)} \tag{4.41a}$$

$$\frac{dp_i^{(0)}}{dt'} = q_i^{(0)} \tag{4.41b}$$

$$\frac{dz_i^{(0)}}{dt'} = -z_i^{(0)} - q_i^{(0)} \tag{4.41c}$$

$$\frac{dq_i^{(0)}}{dt'} = q_i^{(0)} \tag{4.41d}$$

The condition is that $z_i^{(0)}(0) = z(0) - z^{(0)}(0)$ and that the solutions decay exponentially. Thus

$$x_i^{(0)}(t') = -[z(0) - z^{(0)}(0)] \exp(-t'); \quad p_i^{(0)}(t') = 0$$

$$z_i^{(0)}(t') = [z(0) - z^{(0)}(0)] \exp(-t'); \quad q_i^{(0)}(t') = 0$$

Using these solutions, the initial condition $x^{(1)}(0)$ is evaluated from (4.20) as

$$x^{(1)}(0) = -x_i^{(0)} = z(0) - z^{(0)}(0)$$

Similarly, the terminal BLC equations for the zeroth-order approximation are given by (4.26) as

$$\frac{dx_f^{(0)}}{dt''} = -z_f^{(0)} \tag{4.42a}$$

$$\frac{dp_f^{(0)}}{dt''} = -q_f^{(0)} \tag{4.42b}$$

$$\frac{dz_f^{(0)}}{dt''} = z_f^{(0)} + q_f^{(0)} \tag{4.42c}$$

$$\frac{dq_f^{(0)}}{dt''} = -q_f^{(0)} \tag{4.42d}$$

Using the conditions that $q_f^{(0)}(0) = q(T) - q^{(0)}(T) = 0$ as seen from (4.38) and that the solutions decay exponentially, we have

$$x_f^{(0)}(t'') = p_f^{(0)}(t'') = z_f^{(0)}(t'') = q_f^{(0)}(t') = 0$$

This makes $p^{(1)}(T) = -p_f^{(0)}(0) = 0$ as seen from (4.20).

Next, the equations for the first-order approximation are given by (4.21) as

$$\frac{dx^{(1)}}{dt} = z^{(1)} \tag{4.43a}$$

$$\frac{dp^{(1)}}{dt} = -x^{(1)} + q^{(1)} \tag{4.43b}$$

$$\frac{dz^{(0)}}{dt} = -x^{(1)} - z^{(1)} - q^{(1)} \tag{4.43c}$$

$$\frac{dq^{(0)}}{dt} = -p^{(1)} + q^{(1)} \tag{4.43d}$$

and $\quad u^{(1)} = -q^{(1)} \tag{4.43e}$

Using the boundary conditions $x^{(1)}(0) = z(0) - z^{(0)}(0)$ and $p^{(1)}(T) = 0$ the solutions for (4.43) are obtained as

$$\left.\begin{array}{l} x^{(1)}(t) = (C_7 + C_8 t)\exp(at) + (C_9 + C_{10}t)\exp(-at) \\[2mm] z^{(1)}(t) = (C_{11} + C_{12}t)\exp(at) + (C_{13} + C_{14}t)\exp(-at) \\[2mm] u^{(1)}(t) = (C_{15} + C_{16}t)\exp(at) + (C_{17} + C_{18}t)\exp(-at) \end{array}\right\} \tag{4.44}$$

where C_7 to C_{18} are constants depending on $x(0)$, $z(0)$, and $u^{(0)}(0)$.

The initial and final BLC equations for the first-order approximation as given by (4.24) and (4.27) are formed, respectively, and the solutions are obtained as shown below:

$$\left.\begin{array}{l} x_i^{(1)}(t') = \{z^{(1)}(0) + z^{(0)}(0) - z(0) - (z(0) - z^{(0)}(0))t'\}\exp(-t') \\[2mm] p_i^{(1)}(t') = \{z(0) - z^{(0)}(0)\}\exp(-t') \\[2mm] z_i^{(1)}(t') = \{-z^{(1)}(0) + (z(0) - z^{(0)}(0))t'\}\exp(-t') \\[2mm] q_i^{(1)}(t') = 0 \end{array}\right\} \tag{4.45a}$$

and

$$
\left.
\begin{aligned}
x_f^{(1)}(t'') &= z_f^{(1)}(t'') = -\frac{1}{2}q^{(1)}(T)\,\exp(-t'') \\[2mm]
p_f^{(1)}(t'') &= q_f^{(1)}(t'') = -q^{(1)}(T)\,\exp(-t'')
\end{aligned}
\right\}
\tag{4.45b}
$$

Now applying the theorem given by (4.43), we can form the asymptotic series solutions for x, z and u, within $0(\varepsilon^2)$ for $0 \leqslant t \leqslant T$.

The higher-order approximations are obtained in an analogous manner.

The performance index with different initial conditions and $T = 1\cdot0$ and $\varepsilon = 0\cdot25$ is shown in Table 4.1.

Table 4.1   *Performance index*

$x(0)$	$z(0)$	Zeroth order	First order	Exact solution
2	−10	0·1571	0·1348	0·1324
2	−4	0·5695	0·5478	0·5467
2	4	1·9982	1·9825	1·9815
2	10	4·5298	4·4568	4·4550

Example 4.2

In order to demonstrate the advantage of order reduction in the singular perturbation method, consider a voltage-regulator problem given by a fifth-order state equation (Sannuti, 1968; Naidu, 1977):

$$
\frac{dx_1}{dt} = -0\cdot2x_1 + 0\cdot5x_2
\tag{4.46a}
$$

$$
\frac{dx_2}{dt} = -0\cdot5x_2 + 1\cdot6z_1
\tag{4.46b}
$$

$$
\varepsilon\frac{dz_1}{dt} = -5z_1/7 + 30z_2/7
\tag{4.46c}
$$

$$
\varepsilon\frac{dz_2}{dt} = -1\cdot25z_2 + 3\cdot75z_3
\tag{4.46d}
$$

$$
\varepsilon\frac{dz_3}{dt} = -0\cdot5z_3 + 1\cdot5u
\tag{4.46e}
$$

The performance index to be minimised is

$$
J = 0\cdot5\int_0^T (x_1^2 + u^2)\,dt
\tag{4.47}
$$

where ε is the small parameter defined as the ratio of the largest of the time constants to be neglected and the smallest of the time constants to be considered. The initial states are prescribed and the final states are free. Then the co-states and optimal control equations are given by

$$\frac{dp_1}{dt} = -x_1 + 0 \cdot 2p_1 \tag{4.48a}$$

$$\frac{dp_2}{dt} = -0 \cdot 5p_1 + 0 \cdot 5p_2 \tag{4.48b}$$

$$\varepsilon \frac{dq_1}{dt} = -1 \cdot 6p_2 + 5q_2/7 \tag{4.48c}$$

$$\varepsilon \frac{dq_2}{dt} = -30q_1/7 + 1 \cdot 25q_2 \tag{4.48d}$$

$$\varepsilon \frac{dq_3}{dt} = -3 \cdot 75q_2 + 0 \cdot 5q_3 \tag{4.48e}$$

$$0 = 1 \cdot 5q_3 + u \tag{4.48f}$$

with the final conditions for the costates being zero.

The problem described by (4.46) and (4.48) constitutes a tenth-order singularly perturbed TPBVP. Using the method, the results for the control and states are obtained for zeroth-, first- and second-order approximations as shown in Fig. 4.1. The variation of performance index with ε is shown

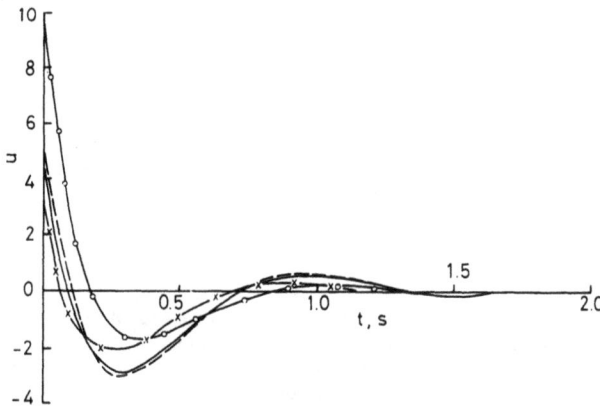

Fig. 4.1a *Exact and approximate solutions of u(t) of Example 4.2*
$\varepsilon = 0 \cdot 2$; $x_1(0) = -10 \cdot 0$; $x_2(0) = z_1(0) = z_2(0) = 0$; $z_3(0) = 10 \cdot 0$
———— exact solution
o——o——o zeroth-order solution
×——×——× first-order solution
– – – – – – second-order solution

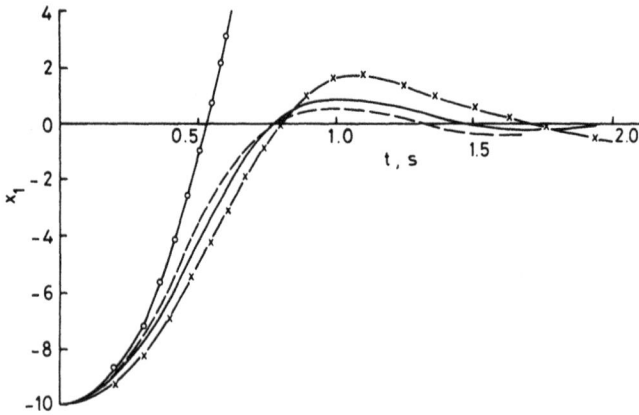

(b)

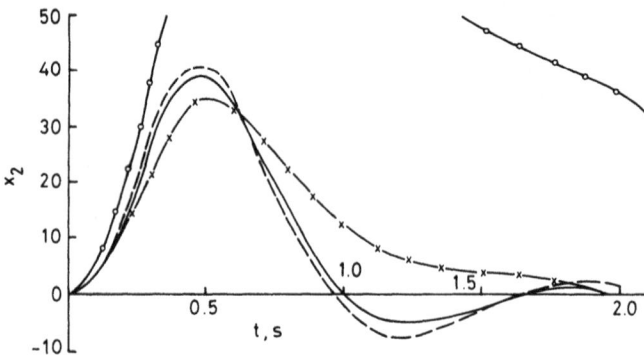

(c)

Fig. 4.1b and c Exact and approximate solutions of $x_1(t)$ and $x_2(t)$ of Example 4.2
$\varepsilon = 0 \cdot 2$; $x_1(0) = -10 \cdot 0$; $x_2(0) = z_1(0) = z_2(0) = 0$; $z_3(0) = 10 \cdot 0$
———————— exact solution
o——o——o zeroth-order solution
×——×——× first-order solution
— — — — — second-order solution

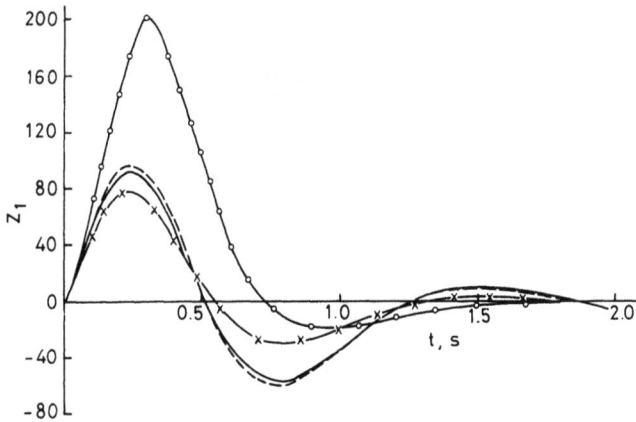

(d)

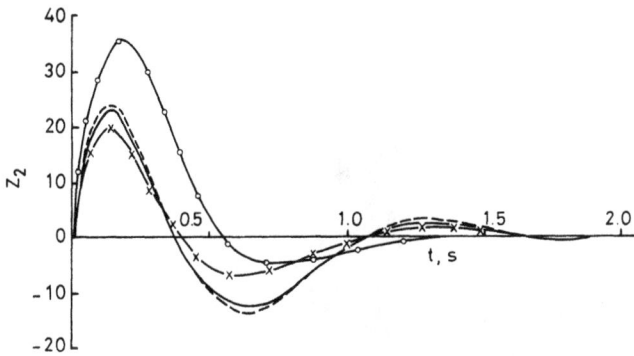

(e)

Fig. 4.1d and e   *Exact and approximate solutions of $z_1(t)$ and $z_2(t)$ of Example 4.2*
$\varepsilon = 0\cdot2$; $x_1(0) = -10\cdot0$; $x_2(0) = z_1(0) = z_2(0) = 0$; $z_3(0) = 10\cdot0$
————————— exact solution
○———○———○ zeroth-order solution
×——×——× first-order solution
– – – – – – second-order solution

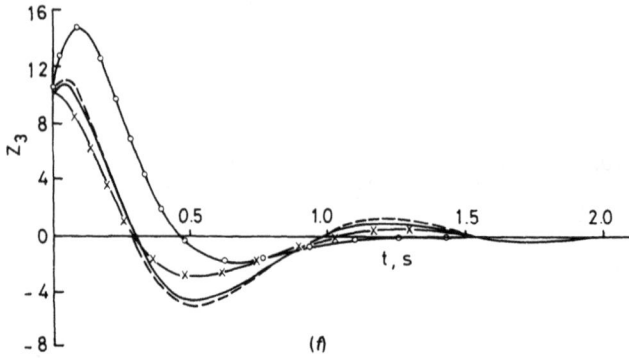

Fig. 4.1f *Exact and approximate solutions of $z_3(t)$ of Example 4.2*
$\varepsilon = 0\cdot2$; $x_1(0) = -10\cdot0$; $x_2(0) = z_1(0) = z_2(0) = 0$; $z_3(0) = 10\cdot0$
——————— exact solution
o———o———o zeroth-order solution
×——×——× first-order solution
– – – – – – second-order solution

in Fig. 4.2. The performance index for different sets of initial conditions is shown in Table 4.2.

It is clearly seen from these results that the zeroth-order approximation

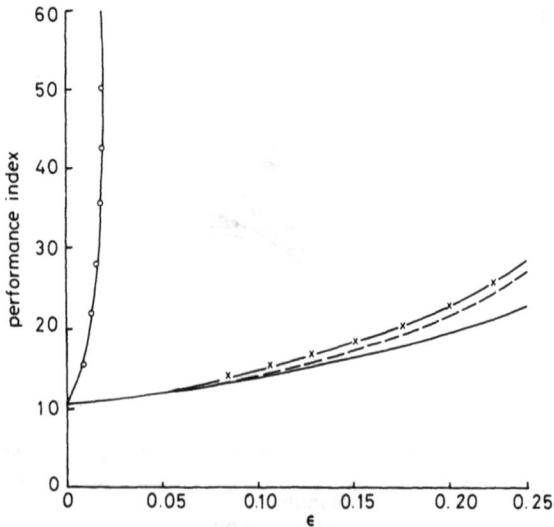

Fig. 4.2 *Performance index of Example 4.2*
$T = 2\cdot0$; $x_1(0) = -10\cdot0$; $x_2(0) = z_1(0) = z_2(0) = 0$; $z_3(0) = 10\cdot0$
——————— exact solution
o———o———o zeroth-order solution
×——×——× first-order solution
– – – – – – second-order solution

is quite unsatisfactory whereas the second-order approximation gives results closer to the optimal values over a good range of the small parameter ε.

Table 4.2   *Performance index with* T $= 2 \cdot 0$ *and* $\varepsilon = 0 \cdot 2$

$x_1(0)$	$x_2(0)$	$z_1(0)$	$z_2(0)$	$z_3(0)$	Zeroth order	First order	Second order	Exact solution
−10	10	10	10	10	876·562	22·505	21·567	19·912
−10	0	0	0	0	546·814	31·481	28·235	26·231
−10	10	0	0	10	597·504	28·165	23·568	20·450
−10	10	10	0	10	659·584	25·529	22·790	19·821
−10	0	10	0	10	604·807	28·649	26·853	24·991
−10	0	0	10	10	741·159	26·800	24·786	22·357

4.3   Fixed-end-point problem

In the previous Section, we have developed a singular perturbation method for obtaining the asymptotic series expansions for the TPBVP arising in the open-loop optimal control of a singularly perturbed, linear, free-end-point problem. Here the TPBVP has been solved using the initial conditions for the states and the final conditions for the co-states.

We now consider the optimal control problem (4.1) and (4.2) with fixed endpoints. The boundary conditions are

$$x(t{=}0) = x(0); \quad z(t{=}0) = z(0) \tag{4.49a}$$

$$x(t{=}T) = x(T); \quad z(t{=}T) = z(T) \tag{4.49b}$$

Proceeding in an identical fashion to the case of the free-end-point problem, we arrive at the optimal control (4.4) and the singularly perturbed, linear TPBVP (4.9). The only difference is that the terminal condition (4.6) is replaced by the fixed-end-point conditions (4.49).

Let us again seek an asymptotic solution of the form (4.14). The reduced problem becomes (4.10) or (4.13) with boundary conditions

$$x^{(0)}(0) = x(0); \quad x^{(0)}(T) = x(T) \tag{4.50a}$$

$$z^{(0)}(0) \neq z(0); \quad z^{(0)}(T) \neq z(T) \tag{4.50b}$$

The various equations of the outer solution, initial and final BLC remain the same as in the case of the free-end-point problem. The only difference is in the boundary conditions to be used for solving the various equations. That is, instead of (4.19) and (4.20), we have

$$x(0) = x^{(j)}(0) + \varepsilon x_i^{(j)}(0), \quad z(0) = z^{(j)}(0) + z_i^{(j)}(0) \tag{4.51a}$$

$$x(T) = x^{(j)}(T) + \varepsilon x_f^{(j)}(0), \quad z(T) = z^{(j)}(T) + z_f^{(j)}(0) \tag{4.51b}$$

Thus

$$x^{(0)}(0) = x(0); \quad z_i^{(0)}(0) = z(0) - z^{(0)}(0)$$

$$x^{(0)}(T) = x(T); \quad z_f^{(0)}(0) = z(T) - z^{(0)}(T)$$

$$x^{(j)}(0) = -x_i^{(j-1)}(0); \quad z_i^{(j)}(0) = -z^{(j)}(0); \quad j \geqslant 1 \tag{4.52a}$$

$$x^{(j)}(T) = -x_f^{(j-1)}(0); \quad z_f^{(j)}(0) = -z^{(j)}(T); \quad j \geqslant 1 \tag{4.52b}$$

Note that the boundary conditions $x^{(j)}(0)$ and $x^{(j)}(T)$ required for solving the outer equations are given in terms of the next-lower order boundary-layer corrections.

Thus the fixed-end-point problem can be solved very much like the free-end-point problem (O'Malley and Kung, 1975).

An interesting method is given by Wilde and Kokotovic (1973) for an approximate open-loop control of a fixed-end-point problem. The method uses a dichotomy transformation to separate the original optimal boundary value problem into two free-end-point problems where the latter problems, unlike the original, can be solved as singularly perturbed initial-value problems.

4.4   Nonlinear systems

In the previous Sections we have considered linear singularly perturbed systems. In this Section we are interested in obtaining asymptotic series solutions for the TPBVP arising in certain nonlinear singularly perturbed optimal control systems. Our treatment is dependent on the works of O'Malley (1974a) and Sannuti (1974b).

4.4.1   Formulation of the problem

Consider a class of nonlinear system

$$\frac{dx}{dt} = a_1(x,t,\varepsilon) + b_1(x,t,\varepsilon)z + c_1(x,t,\varepsilon)u \tag{4.53a}$$

$$\varepsilon \frac{dz}{dt} = a_2(x,t,\varepsilon) + b_2(x,t,\varepsilon)z + c_2(x,t,\varepsilon)u \tag{4.53b}$$

where x and z are m- and n-dimensional state vectors, respectively, u is an r-dimensional control vector.

The objective is to find an optimal control $u(t,\varepsilon)$ which takes the initial states

$$x(t=0) = x(0); \quad z(t=0) = z(0) \tag{4.54}$$

to the free end, with fixed final time T, while minimising the cost functional

$$J = S[x(T),z(T),\varepsilon] + \int_0^T V[x(t),z(t),u(t),t,\varepsilon]\,dt \qquad (4.55)$$

where the scalar functions S and V have the special forms

$$S(x,z,\varepsilon) = S_1(x,\varepsilon) + \frac{1}{2}\varepsilon z'S_2(x,\varepsilon)z$$

and

$$V(x,z,u,t,\varepsilon) = V_1(x,t,\varepsilon) + V_2'(x,t,\varepsilon)z + \frac{1}{2}z'V_3(x,t,\varepsilon)z + \frac{1}{2}u'R(x,t,\varepsilon)u$$

Thus, the state equations (4.53) are linear in z and u, V is quadratic in z and u, and S is quadratic in z but independent of z in the limit when $\varepsilon = 0$. Further, S and V are positive semidefinite functions of x and z while R is a positive definite matrix. Finally, R, S_3 and V_3 are symmetric matrices, and a_1, a_2, b_1, b_2, c_1, c_2, S and V all have asymptotic power series expansions as ε tends to zero with infinitely differential coefficients in an appropriate domain.

For each $\varepsilon > 0$, this problem has a unique optimal control which minimises J (Athans and Falb, 1966). We are interested, however, in obtaining the asymptotic solution of the problem as the parameter ε tends to zero. To obtain necessary conditions for an optimal control, we introduce the Hamiltonian

$$H(x,z,p,q,t,\varepsilon) = V(x,z,u,t,\varepsilon) + p'f_1(x,z,u,t,\varepsilon) + q'f_2(x,z,u,t,\varepsilon) \qquad (4.56)$$

where

$$f_i(x,z,u,t,\varepsilon) = a_i(x,t,\varepsilon) + b_i(x,t,\varepsilon)z + c_i(x,t,\varepsilon)u; \quad i = 1,2$$

and p and εq are the co-states corresponding to x and z, respectively. Note that it is conventional to use εq as the co-state variable; however, our use of q in (4.56) leads us to TPBVP in singularly perturbed form.

Along an optimal trajectory we have

$$\frac{\partial H}{\partial u} = Ru + c_1'p + c_2'q = 0$$

or $\qquad u = -R^{-1}(c_1'p + c_2'q) \qquad (4.57)$

The co-states satisfy

$$\frac{dp}{dt} = -\frac{\partial H}{\partial x} = -\frac{\partial V}{\partial x} - p'f_{1x} - q'f_{2x} \qquad (4.58a)$$

$$\varepsilon\frac{dq}{dt} = -\frac{\partial H}{\partial z} = -\frac{\partial V}{\partial z} - b_1'p - b_2'q \tag{4.58b}$$

and the boundary conditions

$$p(T) = \frac{\partial S}{\partial x}\Big|_{t=T} \tag{4.59a}$$

$$q(T) = \frac{1}{\varepsilon}\frac{\partial S}{\partial z}\Big|_{t=T} = S_2[x(T),\varepsilon]z(T) \tag{4.59b}$$

Since $\partial^2 H/\partial u^2 = 2R$ is positive definite, the control (4.57) minimises the cost functional J of (4.55).

Substituting (4.57) for u in the $(m+n)$ state equations (4.53) and $(m+n)$ co-state equations (4.58), we get a system of nonlinear equations

$$\frac{dx}{dt} = g_1(x,z,p,q,t,\varepsilon) \tag{4.60a}$$

$$\frac{dp}{dt} = g_2(x,z,p,q,t,\varepsilon) \tag{4.60b}$$

$$\varepsilon\frac{dz}{dt} = g_3(x,z,p,q,t,\varepsilon) \tag{4.60c}$$

$$\varepsilon\frac{dq}{dt} = g_4(x,z,p,q,t,\varepsilon) \tag{4.60d}$$

where

$$g_1 = a_1 + b_1 z - E_1 p - E_3 q$$

$$g_2 = -\frac{\partial H}{\partial x}[x,z,-R^{-1}(c_1'p + c_2'q),p,q,t,\varepsilon]$$

$$g_3 = a_2 + b_2 z - E_3' p - E_2 q$$

and

$$g_4 = -V_2 - V_3 z - b_1' p - b_2' q$$

$$E_i = c_i R^{-1} c_i', \quad i = 1,2, \quad E_3 = c_1 R^{-1} c_2'$$

Note that the special form (4.53) and (4.55) is selected in order to make g_3 and g_4 linear in z, p and q.

The problem (4.60) along with the $(m+n)$ initial conditions (4.54) on the states x and z and the terminal conditions (4.59) on the co-states p and q form a singularly perturbed, nonlinear $2(m+n)$th-order TPBVP, in the sense that, when $\varepsilon=0$, the order of (4.60) drops from $2(m+n)$ to $2m$. The reduced problem is then

$$\frac{dx^{(0)}}{dt} = g_1(x^{(0)},z^{(0)},p^{(0)},q^{(0)},t,0) \qquad (4.61a)$$

$$\frac{dp^{(0)}}{dt} = g_2(x^{(0)},z^{(0)},p^{(0)},q^{(0)},t,0) \qquad (4.61b)$$

$$0 = g_3(x^{(0)},z^{(0)},p^{(0)},q^{(0)},t,0) \qquad (4.61c)$$

$$0 = g_4(x^{(0)},z^{(0)},p^{(0)},q^{(0)},t,0) \qquad (4.61d)$$

with boundary conditions

$$x^{(0)}(0) = x(0); \quad p^{(0)}(T) = S_{1x}[x(T),0] \qquad (4.62)$$

Evidently the reduced problem (4.61) has dropped the other boundary conditions.

The reduced problem (4.61) and (4.62) can also be obtained by an alternative method. First we neglect ε in the state equations (4.53) and then obtain the necessary conditions for an optimal control. This leads to a TPBVP which is identical to the reduced problem (4.61) and (4.62).

4.4.2 Asymptotic expansions

Our main interest here is to obtain asymptotic expansions for the high-order TPBVP (4.60) as the scalar parameter ε tends to zero. Though the reduced problem (4.61) and (4.62) offers computational simplicity, the best it can do in general is to approximate the high-order solution only in an open interval $0<t<T$. For even when ε is very small, large discrepancies near $t=0$ and $t=T$ occur since the initial condition on z and the terminal condition on q are disregarded in the reduced problem. This loss of initial and final conditions leads to the occurrence of boundary-layer phenomena. These boundary layers are predominant only at the end points and correspond to the fast dynamics of the high-order problem. The reduced problem, where fast dynamics are neglected, corresponds to slow dynamics of the high-order system. Hence the asymptotic solution is assumed in the form of an outer solution corresponding to slow modes and the initial and final BLCs corresponding to fast modes. Thus the asymptotic solutions for the TPBVP defined by (4.60), (4.54) and (4.59) are assumed in the same form as given by (4.14)–(4.18). The boundary conditions (4.20) needed to solve the various expansion equations become

$$x^{(0)}(0) = x(0); \quad z_i^{(0)}(0) = z(0) - z^{(0)}(0)$$

$$p^{(0)}(T) = S_{1x}[x(T),0]; \quad q_f^{(0)}(0) = q(T) - q^{(0)}(T)$$

$$x^{(j)}(0) = -x_i^{(j-1)}(0); \quad z_i^{(j)}(0) = -z^{(j)}(0) \tag{4.63}$$

$$p^{(j)}(T) = S_{1xx}[x^{(0)}(T),0]x^{(j)}T + \bar{e}_1; \quad j \geq 1$$

$$q_f^{(j)}(0) = S_2[z^{(j)}(T) + z_f^{(j)}(0)] + \bar{e}_2; \quad j \geq 1$$

where $\bar{e}_1$ and $\bar{e}_2$ are successively known in terms of the preceding coefficients.

To obtain the equations for the outer expansion, we substitute (4.16) in (4.60) and collect the coefficients. For zeroth order ($\varepsilon = 0$), we have the reduced problem (4.61). For higher-order expansions, we have the following linear equations:

$$\frac{dx^{(j)}}{dt} = g_{1x}(t)x^{(j)} + b_1^{(0)}(x^{(0)},t)z^{(j)} + E_1^{(0)}(x^{(0)},t)p^{(j)} - E_3^{(0)}(x^{(0)},t)q^{(j)} + e_1(t)$$

$$\tag{4.64a}$$

$$\frac{dp^{(j)}}{dt} = g_{2x}(t)x^{(j)} + g_{2z}(t)z^{(j)} + g_{2p}(t)p^{(j)} + g_{2q}(t)q^{(j)} + e_2(t) \tag{4.64b}$$

$$0 = g_{3x}(t)x^{(j)} + b_2^{(0)}[x^{(0)},t]z^{(j)} - E_3^{(0)\prime}[x^{(0)},t]p^{(j)} - E_2^{(0)}[x^{(0)},t]q^{(j)} + e_3(t)$$

$$\tag{4.64c}$$

$$0 = g_{4x}(t)x^{(j)} - V_3^{(0)}[x^{(0)},t]z^{(j)} - b_1^{(0)\prime}[x^{(0)},t]p^{(j)} - b_2^{(0)\prime}[x^{(0)},t]q^{(j)} + e_4(t)$$

$$\tag{4.64d}$$

where $e_k(t)$, $k = 1$–4 are known in terms of preceding coefficients. Note the boundary conditions (4.63) for solving (4.64).

For obtaining the initial BLCs, we use relations similar to (4.22) using the fact that at $t = 0$, the final BLCs are negligible. Then we have

$$\frac{dx_i}{dt'} = g_{1i}(x_i, z_i, p_i, q_i, t', \varepsilon) \tag{4.65a}$$

$$\frac{dp_i}{dt'} = g_{2i}(x_i, z_i, p_i, q_i, t', \varepsilon) \tag{4.65b}$$

$$\frac{dz_i}{dt'} = g_{3i}(x_i, z_i, p_i, q_i, t', \varepsilon) \tag{4.65c}$$

$$\frac{dq_i}{dt'} = g_{4i}(x_i, z_i, p_i, q_i, t', \varepsilon) \tag{4.65d}$$

where

$$g_{1i} = g_1[x_o(\varepsilon t',\varepsilon) + \varepsilon x_i(t',\varepsilon), z_o(\varepsilon t',\varepsilon) + z_i(t',\varepsilon),$$
$$p_o(\varepsilon t',\varepsilon) + \varepsilon p_i(t',\varepsilon), q_o(\varepsilon t',\varepsilon) + q_i(t',\varepsilon), \varepsilon t',\varepsilon]$$
$$-g_1[x_o(\varepsilon t',\varepsilon), z_o(\varepsilon t',\varepsilon), p_o(\varepsilon t',\varepsilon), q_o(\varepsilon t',\varepsilon), \varepsilon t',\varepsilon] \qquad (4.66)$$

and g_{2i}, g_{3i}, g_{4i} are similarly defined.

Substituting asymptotic power series (4.17) in (4.65) and equating like coefficients of ε^j, we obtain a system of equations for the initial BLC. In particular, the lowest-order (ε^0) terms satisfy the nonlinear equations

$$\frac{dx_i^{(0)}}{dt'} = b_1^{(0)}[x^{(0)}(0),0]z_i^{(0)}(t') - E_3^{(0)}[x^{(0)}(0),0]q_i^{(0)}(t') \qquad (4.67a)$$

$$\frac{dp_i^{(0)}}{dt'} = g_2^{(0)}[x^{(0)}(0),z^{(0)}(0) + z_i^{(0)}(t'),p^{(0)}(0),q^{(0)}(0)$$
$$+ q_i^{(0)}(t'),0] - g_2^{(0)}[x^{(0)}(0),x^{(0)}(0),z^{(0)}(0),p^{(0)}(0),q^{(0)}(0),0] \qquad (4.67b)$$

$$\frac{dz_i^{(0)}}{dt'} = b_2^{(0)}[x^{(0)}(0),0]z_i^{(0)}(t') - E_2^{(0)}[x^{(0)}(0),0]q_i^{(0)}(t') \qquad (4.67c)$$

$$\frac{dq_i^{(0)}}{dt'} = V_3^{(0)}[x^{(0)}(0),0]z_i^{(0)}(t') - b_2^{(0)\prime}[x^{(0)}(0),0]q_i^{(0)}(t') \qquad (4.67d)$$

We solve (4.67) for exponentially decaying solutions using (4.63).

Likewise, from the coefficients of ε^j, $j = 1$, we obtain the linear system

$$\frac{dx_i^{(j)}}{dt'} = b_1^{(0)}[x^{(0)}(0),0]z_i^{(j)}(t') - E_3^{(0)}[x^{(0)}(0),0]q_i^{(j)}(t') + e_{1i}(t') \qquad (4.68a)$$

$$\frac{dp_i^{(j)}}{dt'} = g_{2z}^{(0)}[x^{(0)}(0),z^{(0)}(0) + z_i^{(0)}(t'),p^{(0)}(0),q^{(0)}(0)$$
$$+ q_i^{(0)}(t'),0]z_i^{(j)}(t') + g_{2q}^{(0)}[x^{(0)}(0),z^{(0)}(0)$$
$$+ z_i^{(0)}(t'),p^{(0)}(0),q^{(0)}(0) + q_i^{(0)}(t'),0]q_i^{(j)}(t') + e_{2i}(t') \qquad (4.68b)$$

$$\frac{dz_i^{(j)}}{dt'} = b_2^{(0)}[x^{(0)}(0),0]z_i^{(j)}(t') - E_2^{(0)}[x^{(0)}(0),0]q_i^{(j)}(t') + e_{3i}(t') \qquad (4.68c)$$

$$\frac{dq_i^{(j)}}{dt'} = -V_3^{(0)}[x^{(0)}(0),0]z_i^{(j)}(t') - b_2^{(0)\prime}[x^{(0)}(0),0]q_i^{(j)}(t') + e_{4i}(t') \qquad (4.68d)$$

where $e_{1i}(t'), \ldots, e_{4i}(t')$ are known successively and are exponentially decaying as t' tends to infinity. We use (4.63) to solve (4.68) for exponentially decaying solutions.

The final BLCs are determined in an analogous manner. At the final point $t = T$, the initial BLCs are negligible. Then we have

$$\frac{dx_f}{dt''} = g_{1f}(x_f, z_f, p_f, q_f, t'', \varepsilon) \tag{4.69a}$$

$$\frac{dp_f}{dt''} = g_{2f}(x_f, z_f, p_f, q_f, t'', \varepsilon) \tag{4.69b}$$

$$\frac{dz_f}{dt''} = g_{3f}(x_f, z_f, p_f, q_f, t'', \varepsilon) \tag{4.69c}$$

$$\frac{dq_f}{dt''} = g_{4f}(x_f, z_f, p_f, q_f, t'', \varepsilon) \tag{4.69d}$$

where

$$\begin{aligned}
g_{1f} = &\, g_1[x_o(t,\varepsilon), p_o(t,\varepsilon), z_o(t,\varepsilon), q_o(t,\varepsilon), t, \varepsilon] \\
&- g_1[x_o(t,\varepsilon) + \varepsilon x_f(t'',\varepsilon), p_o(t,\varepsilon) + \varepsilon p_f(t'',\varepsilon), \\
&z_o(t,\varepsilon) + z_f(t'',\varepsilon), q_o(t,\varepsilon) + q_f(t'',\varepsilon), t, \varepsilon]
\end{aligned} \tag{4.70}$$

in which t should be replaced by $T - \varepsilon t''$ and g_{2f}, g_{3f}, g_{4f}, are similarly defined. Substituting asymptotic power series (4.18) for final BLCs in (4.69) and equating like coefficients of ε^j, we obtain a system of equations. For ε^0, we have the nonlinear equations.

$$\frac{dx_f^{(0)}}{dt''} = -b_1^{(0)}[x^{(0)}(T), T]z_f^{(0)}(t'') + E_3^{(0)}[x^{(0)}(T), T]q_f^{(0)}(t'') \tag{4.71a}$$

$$\begin{aligned}
\frac{dp_f^{(0)}}{dt''} = &\, g_2^{(0)}[x^{(0)}(T), p^{(0)}(T), z^{(0)}(T), q^{(0)}(T), T, 0] \\
&- g_2^{(0)}[x^{(0)}(T), p^{(0)}(T), z^{(0)}(T) + z_f^{(0)}(t''), q^{(0)}(T) \\
&+ q_f^{(0)}(t''), T, 0]
\end{aligned} \tag{4.71b}$$

$$\frac{dz_f^{(0)}}{dt''} = b_2^{(0)}[x^{(0)}(T), T]z_f^{(0)}(t'') + E_2^{(0)}[x^{(0)}(T), T]q_f^{(0)}(t'') \tag{4.71c}$$

$$\frac{dq_f^{(0)}}{dt''} = V_3^{(0)}[x^{(0)}(T), T]z_f^{(0)}(t'') - b_2^{(0)'}[x^{(0)}(T), T]q_f^{(0)}(t'') \tag{4.71d}$$

The above problem is solved for exponentially decaying solutions using (4.63). Continuing, we equate higher-order coefficients in (4.69) to obtain a linear BVP for the terminal BLC terms and solve them.

Knowing the asymptotic expansions for the states and co-states, and using the control law (4.57), we can obtain an asymptotic series expansion for the control as given by (4.28).

We give the following assumptions for the final result (O'Malley, 1974a; Sannuti, 1974b):

(i) The reduced problem (4.61) and (4.62) has a unique continuously differentiable solution on $0 \leqslant t \leqslant T$.

(ii) The matrices $V_3(t)$, $S_{1xx}(T)$ and $S_2(T)$ are positive semi-definite.

(iii) For each fixed t, $0 \leqslant t \leqslant T$, we assume that the pair $[b_2(t), c_2(t)]$ is completely controllable and the pair $[b_2(t), d_2(t)]$ is completely observable where $V_3(t) = d_2'(t)d_2(t)$; that is, we have

$$\text{Rank } [c_2(t), b_2(t)c_2(t), \ldots, b_2^{n-1}(t)c_2(t)] = n$$

$$\text{Rank } [d_2'(t), b_2'(t)d_2'(t), \ldots, b_2^{n-1}(t)d_2'(t)] = n$$

Controllability and observability assumptions of the type (iii) are used in linear singular perturbation theory (Kokotivic and Yackel, 1972; Wilde and Kokotovic, 1973). For our purpose, it is necessarily restrictive but simplifies the presentation.

Under these assumptions, the final result is given by (4.33).

We note that the above analysis for nonlinear systems can also be done using the maximum principle, where we maximize $-J$ in (4.55). This procedure is used in the example below (Sannuti, 1974b).

Example 4.3*

Although the method is meant for high-order systems, let us take a second-order system to illustrate the method of obtaining the various expansions. Consider

$$\frac{dx}{dt} = x + x^2 + z; \quad x(t=0) = x(0) \tag{4.72a}$$

$$\varepsilon\frac{dz}{dt} = x + b(t)z + u; \quad z(t=0) = z(0) \tag{4.72b}$$

where $b(t) = 4(t - 0 \cdot 5)^2$ and the cost functional to be minimised is

$$J = S_1x^2(1) + \varepsilon S_2z^2(1) + \frac{1}{2}\int_0^1 (x^2 + z^2 + u^2)dt \tag{4.73}$$

The maximum principle gives that the state and co-state variables satisfy

* Reprinted with permission from SANNUTI, P. (1974): 'Asymptotic series solution of singularly perturbed optimal control problem', *Automatica*, **10**, pp. 183–194.

$$\frac{dp}{dt} = x - 2xp - p - q, \quad p(1) = -2S_1 x(1)$$

$$\varepsilon \frac{dq}{dt} = -p + z - bq, \quad q(1) = -2S_2 z(1)$$

$$(4.74)$$

and the optimal control is given by

$$u = q \tag{4.75}$$

The outer expansion satisfies

$$\frac{dx_o}{dt} = x_o + x_o^2 + z_o$$

$$\frac{dp_o}{dt} = x_o - 2x_o p_o - p_o - q_o$$

$$\varepsilon \frac{dz_o}{dt} = x_o + bz_o + q_o$$

$$\varepsilon \frac{dq_o}{dt} = -p_o + z_o - bq_o$$

$$(4.76)$$

The reduced problem is

$$\frac{dx^{(0)}}{dt} = x^{(0)} + x^{(0)2} + z^{(0)}$$

$$\frac{dp^{(0)}}{dt} = x^{(0)} - 2x^{(0)}p^{(0)} - p^{(0)} - q^{(0)}$$

$$0 = x^{(0)} + bz^{(0)} + q^{(0)}$$

$$0 = -p^{(0)} + z^{(0)} - bq^{(0)}$$

$$(4.77)$$

The initial BLCs satisfy

$$\frac{dx_i}{dt'} = \varepsilon x_i + z_i + \varepsilon [2x_o(\varepsilon t')x_i + \varepsilon x_i^2]$$

$$\frac{dp_i}{dt'} = \varepsilon x_i - \varepsilon p_i - q_i - 2\varepsilon [p_o(\varepsilon t')x_i + x_o(\varepsilon t')p_i + \varepsilon x_i p_i]$$

$$\frac{dz_i}{dt'} = \varepsilon x_i + b(\varepsilon t')z_i + q_i$$

$$\frac{dq_i}{dt'} = -\varepsilon p_i + z_i - b(\varepsilon t')q_i$$

$$(4.78)$$

The leading terms of the above equations satisfy

$$\frac{dx_i^{(0)}}{dt'} = z_i^{(0)}; \quad \frac{dp_i^{(0)}}{dt'} = -q_i^{(0)},$$

$$\frac{dz_i^{(0)}}{dt'} = z_i^{(0)} + q_i^{(0)}, \quad \frac{dq_i^{(0)}}{dt'} = z_i^{(0)} - q_i^{(0)}$$

(4.79)

along with the condition that $z_i^{(0)}(0) = z(0) - z^{(0)}(0)$ and that the solution decays exponentially. Thus

$$x_i^{(0)}(t') = -(1/d_0)z_i^{(0)}(t'); \quad p_i^{(0)}(t') = -(d_1/d_0)z_i^{(0)}(t') \tag{4.80}$$

$$z_i^{(0)}(t') = [z(0)-z^{(0)}(0)] \exp(-d_0 t'); \quad q_i^{(0)}(t') = -d_1 z_i^{(0)}(t')$$

where $d_0 = 2^{0.5}; \quad d_1 = 1 + d_0$

The initial condition $x^{(1)}(0)$ is

$$x^{(1)}(0) = -x_i^{(0)}(0) = [z(0) - z^{(0)}(0)]/d_0 \tag{4.81}$$

The equations defining the final BLCs are

$$\frac{dx_f}{dt''} = -\varepsilon x_f - z_f - \varepsilon[2x_o(1-\varepsilon t'')x_f + \varepsilon x_f^2]$$

$$\frac{dp_f}{dt''} = -\varepsilon x_f + \varepsilon p_f + q_f + 2\varepsilon[p_o(1-\varepsilon t'')x_f + x_o(1-\varepsilon t'')p_f + \varepsilon x_f p_f]$$

$$\frac{dz_f}{dt''} = -\varepsilon x_f - b(1-\varepsilon t'')z_f - q_f$$

$$\frac{dq_f}{dt''} = \varepsilon p_f - z_f + b(1-\varepsilon t'')q_f$$

(4.82)

The zeroth-order terms of the above satisfy

$$\frac{dx_f^{(0)}}{dt''} = -z_f^{(0)}; \quad \frac{dp_f^{(0)}}{dt''} = q_f^{(0)}$$

$$\frac{dz_f^{(0)}}{dt''} = -z_f^{(0)} - q_f^{(0)}; \quad \frac{dq_f^{(0)}}{dt''} = -z_f^{(0)} + q_f^{(0)}$$

(4.83)

along with the condition

$$q_f^{(0)}(0) = -(1 + 2S_2 d_2)^{-1}[q^{(0)}(1) + 2S_2 z^{(0)}(1)] \tag{4.84}$$

Solving

$$x^{(0)}(t'') = (d_1/d_0)q_f^{(0)}(t''); \quad p_f^{(0)}(t'') = -(1/d_0)q_f^{(0)}(t'') \atop z_f^{(0)}(t'') = d_1 q_f^{(0)}(t''); \quad q_f^{(0)}(t'') = q_f^{(0)}(0) \exp(-d_0 t'') \Bigg\}$$

(4.85)

The final condition $p^{(1)}(1)$ is

$$p^{(1)}(1) = -2S_1 x^{(1)}(1) - p_f^{(0)}(0) - 2S_1 x_f^{(0)}(0)$$

(4.86)

Note that the zeroth-order expansion enables us to calculate $x^{(1)}(0)$ and $p^{(1)}(1)$ so that the first-order outer expansion can be determined uniquely. Proceeding in an analogous manner, we can construct the asymptotic expansions up to any desired order.

More general TPBVP for singularly perturbed nonlinear systems are considered by Hadlock (1973), Chow (1977), Chow and Kokotovic (1981), Freedman and Kaplan (1976), Freedman and Granoff (1976). The nonlinear fixed-end-point problem is considered by Chow (1977, 1979).

4.5   Conclusions and discussion

In this Chapter we have mainly concentrated on obtaining asymptotic power-series expansions for the singularly perturbed TPBVP arising in open-loop optimal control systems. First, in Section 4.1 we considered in detail the linear singularly perturbed system for which the necessary optimal conditions have been applied to formulate the TPBVP in terms of state and co-state variables. The suppression of the small parameter resulted in a reduced-order TPBVP which could not satisfy all the given boundary conditions. Also the boundary layers present in the original problem have been destroyed in the reduced problem. Thus, in order to recover the boundary conditions lost during the degeneration process, the asymptotic solutions for the states, co-states and control have been constructed in the form of an outer solution and the initial and final BLCs as shown in Sections 4.2.1 and 4.2.5. The asymptotic solution for the cost functional has also been given.

We indicated in Section 4.3 the main difference in the method to be applicable to a fixed-end-point optimal control problem.

Next in Section 4.4 we presented an important class of nonlinear problems leading to the singularly perburbed TPBVP. The analysis similar to the linear problem has been given.

Singular perturbation methods were first introduced into optimal control theory by Sannuti (1968) and Kokotovic and Sannuti (1968). This first result was without any boundary-layer corrections for the singularly perturbed TPBVP. Later Hadlock (1973) indicated the method with BLCs, but the complete expansions have not been obtained. O'Malley (1972a,b,c) has successfully constructed the complete expansions for linear-state regulator problem.

The same treatment has been modified in his later works (O'Malley, 1975; O'Malley and Kung, 1975).

A class of nonlinear free-end-point optimal control problems, first considered by Sannuti (1968), Sannuti and Kokotovic (1969) without corrections has been later on thoroughly investigated by Sannuti (1974b, 1975) and O'Malley (1974a). More general nonlinear TPBVP arising in optimal control have been explored by Hadlock (1973), Freedman and Kaplan (1976) and Freedman and Granoff (1976).

The open-loop optimal-control method has been applied to nuclear reactor problem (Kao and Bankoff, 1974; Kando and Iwazumi, 1983c) and electric power system (Reddy and Sannuti, 1976).

The TPBVP resulting from the fixed-end-point linear optimal control problem has been addressed by Wilde and Kokotovic (1973), Kando and Iwazumi (1983c), whereas the nonlinear case has received the attention of Chow (1979).

The optimal control problem with three-time-scale behaviour has been investigated by Jamshidi (1976), Murthy (1978), Ozguner (1979), Kando and Iwazumi (1981) and with multiple scales by Kando and Iwazumi (1981), Dragan and Halanay (1982).

There have been many other works dealing with the open-loop and the general control problems (Dmitriev, 1978; Dontchev, 1977, 1983; Hadlock, 1977; Kurina, 1977a,b, 1983a,b; Gaitsogori and Pervozvanskii, 1979; Vasileva and Faminskaya, 1981; Eitelberg, 1982; Ardema, 1983a,b; Habets, 1983; Bensoussan *et al.*, 1984; Dmitriev and Klishevic, 1984; Gichev, 1984; Kokotovic, 1984, 1985; Sobolev, 1984; Dontchev and Veliov, 1985; Milusheva and Bainov, 1985; Saksena and Cruz, 1985a,b; Roberts, 1986; Visser and Shinar, 1986).

The open-loop optimal control problems leading to the singularly perturbed TPBVP are often called trajectory optimisation problems which have had spectacular success in the application to aerospace problems (Kokotovic *et al.*, 1976; Saksena *et al.*, 1984). The various trajectory optimisation problems considered relate to jet-engine control, missile guidance, energy management, pursuit evasion (Kelley, 1970a,b; Hadlock, 1973; Calise, 1976, 1980; Ardema, 1977; Mehra *et al.*, 1979; Sridhar and Gupta, 1980; Shinar, 1981; Cliff *et al.*, 1982; Shinar and Negrin, 1983; Rajan and Ardema, 1984, 1985; Breakwell *et al.*, 1985; Moerder and Calise, 1985; Visser and Shinar, 1986).

Dynamic programming is also used for singularly perturbed optimal control problems (Krikorian and Leondes, 1982a,b; Bensoussan *et al.*, 1984), while Rozov and Gichev (1983) analyse the perturbation of the 'minimal pulse' problem.

Chapter 5

Open-loop optimal control of discrete systems*

In this Chapter, we present a perturbation method for the TPBVP arising in the open-loop optimal control of singularly perturbed discrete systems. A state–space model with a three-time-scale property exhibiting boundary-layer behaviour at the initial and final points is formulated in Section 5.1. The solution of the model is obtained as the sum of an outer series solution and two correction series solutions for the initial and final boundary layer. In Section 5.2, the optimal control problem with a quadratic cost function is then considered. Using the discrete maximum principle, the state and co-state equations are obtained and cast in the singularly perturbed form which exhibits the three-time-scale property. In Section 5.3, a method is described to solve the resulting two-point boundary-value problem. Examples are provided to illustrate the method. Finally, conclusions and discussion are given.

5.1 Three-time-scale discrete systems

5.1.1 State–space model

The two-time-scale property of discrete models is examined in detail in Chapter 2. It is shown that in the C-model, where the small parameter ε is associated with $z(k)$, the fast modes have eigenvalues with moduli much less than unity which approach zero in the degeneration process. Consequently, the effect of the fast modes on the total solution is a fast *decay* at $k = 0$, leading to the formation of the boundary layer at the initial point $(k = 0)$. In the D-model, where the small parameter ε is associated with $z(k + 1)$, the fast modes have the eigenvalues with moduli much larger than unity which approach infinity in the degeneration process. Therefore their contribution to the total solution is a fast *rise* near $k = N$ (terminal or final

* This Chapter is based on the research monograph NAIDU, D. S., and RAO, A. K. (1985): 'Singular Perturbation Analysis of Discrete Control Systems' Lecture Notes in Mathematics, Vol. 1154 (Springer–Verlag, Berlin). The permission given by Springer–Verlag is hereby acknowledged.

point), leading to the formation of the boundary layer at the final point $(k = N)$. Keeping these points in mind, a state–space model which inherits both these features is formulated as below:

$$\begin{bmatrix} x(k+1) \\ y(k+1) \\ \varepsilon z(k+1) \end{bmatrix} = \begin{bmatrix} A_{11} & \varepsilon A_{12} & A_{13} \\ A_{21} & \varepsilon A_{22} & A_{23} \\ A_{31} & \varepsilon A_{32} & A_{33} \end{bmatrix} \begin{bmatrix} x(k) \\ y(k) \\ z(k) \end{bmatrix} + \begin{bmatrix} E \\ F \\ G \end{bmatrix} u(k) \tag{5.1}$$

where, x, y and z are n_1-, n_2-, and n_3-dimensional state vectors, respectively, u is an r-dimensional control vector, A_{ij}, $i, j = 1, 2, 3$, E, F and G are matrices of compatible dimensionality and the boundary conditions are

$$x(k = 0) = x(0) \quad \text{or} \quad x(k = N) = x(N)$$

$$y(k = 0) = y(0); \qquad z(k = N) = z(N)$$

The degenerate form of model (5.1) is

$$x^{(0)}(k+1) = A_{11}x^{(0)}(k) + A_{13}z^{(0)}(k) + Eu(k)$$

$$y^{(0)}(k+1) = A_{21}x^{(0)}(k) + A_{23}z^{(0)}(k) + Fu(k)$$

$$0 = A_{31}x^{(0)}(k) + A_{33}z^{(0)}(k) + Gu(k)$$

The above equations are rearranged as

$$x^{(0)}(k+1) = (A_{11} - A_{13}A_{33}^{-1}A_{31})x^{(0)}(k) + (E - A_{13}A_{33}^{-1}G)u(k) \tag{5.2a}$$

$$y^{(0)}(k+1) = (A_{21} - A_{23}A_{33}^{-1}A_{31})x^{(0)}(k) + (F - A_{23}A_{33}^{-1}G)u(k) \tag{5.2b}$$

$$z^{(0)}(k) = -A_{33}^{-1}A_{31}x^{(0)}(k) - A_{33}^{-1}Gu(k) \tag{5.2c}$$

In the above three equations, only (5.2a) is a difference equation of order n_1 while the other two (5.2b,c) are algebraic relationships. In other words, once $x^{(0)}(k)$ is solved from (5.2a), the solutions $y^{(0)}(k)$ and $z^{(0)}(k)$ are automatically fixed from (5.2b,c). Since the order of (5.1) drops from $(n_1 + n_2 + n_3)$ to n_1, it is in the singularly perturbed form. It follows from (5.2) that A_{33} should be non-singular to solve (5.2a).

For obtaining the basic idea of the analysis, let us consider a second-order equation with the features of (5.1) as

$$\begin{bmatrix} y(k+1) \\ \varepsilon z(k+1) \end{bmatrix} = \begin{bmatrix} a_{11} & a_{12} \\ a_{21} & a_{22} \end{bmatrix} \begin{bmatrix} \varepsilon y(k) \\ z(k) \end{bmatrix} \tag{5.3}$$

The two eigenvalues of (5.3) are

$$p_1 = \frac{a_{22}}{\varepsilon}; \quad p_2 = \varepsilon \left(\frac{a_{11}a_{22} - a_{12}a_{21}}{a_{22}} \right)$$

Given $y(0)$ and $z(N)$, the zeroth-order solution of (5.3), obtained by using

Sylvester's expansion (Hildebrand, 1968), is

$$y(k) = \underbrace{\varepsilon^{N-k+1}[a_{12}(a_{22})^{k-N-1}z(N)]}_{y_f^{(0)}(k)} + \underbrace{\varepsilon^k \left[y(0) \left\{ \frac{a_{11}a_{22} - a_{12}a_{21}}{a_{22}} \right\}^k \right]}_{y_i^{(0)}(k)} \qquad (5.4a)$$

$$z(k) = \underbrace{\varepsilon^{N-k}[(a_{22})^{k-N}z(N)]}_{z_f^{(0)}(k)} + \underbrace{\varepsilon^{k+1} \left[\frac{-a_{21}}{a_{22}} y(0) \left\{ \frac{a_{11}a_{22} - a_{12}a_{21}}{a_{22}} \right\}^k \right]}_{z_i^{(0)}(k)} \qquad (5.4b)$$

Note that the degenerate form of (5.3) is of zero order and hence has a trivial solution. A close examination of (5.4a) shows that as ε tends to zero, the uniform convergence of $y(k)$ to the trivial degenerate solution fails at $k = 0$, leading to the formation of the boundary layer at the initial point $(k = 0)$. Similarly, from (5.4b) it follows that as ε tends to zero, the uniform convergence of $z(k)$ to the trivial degenerate solution fails at $k = N$, leading to the formation of a boundary layer at the final point $(k = N)$. Thus in the process of degeneration, both the boundary conditions $y(0)$ and $z(N)$ are lost.

The structure of (5.4) suggests that the lost two boundary conditions can be recovered by incorporating two separate correction series called the initial boundary-layer correction (BLC) series and the final boundary-layer correction (BLC) series using the following transformations:

For the initial BLC,

$$y_i(k) = y(k)/\varepsilon^k; \quad z_i(k) = z(k)/\varepsilon^{k+1} \qquad (5.5a)$$

For the final BLC,

$$y_f(k) = y(k)/\varepsilon^{N-k+1}; \quad z_f(k) = z(k)/\varepsilon^{N-k} \qquad (5.5b)$$

Using (5.5a), the equation for initial BLC is

$$\begin{bmatrix} y_i(k+1) \\ \varepsilon^2 z_i(k+1) \end{bmatrix} = \begin{bmatrix} a_{11} & a_{12} \\ a_{21} & a_{22} \end{bmatrix} \begin{bmatrix} y_i(k) \\ z_i(k) \end{bmatrix} \qquad (5.6)$$

Using (5.5b), the equation for the final BLC is

$$\begin{bmatrix} y_f(k+1) \\ z_f(k+1) \end{bmatrix} = \begin{bmatrix} \varepsilon^2 a_{11} & a_{12} \\ \varepsilon^2 a_{21} & a_{22} \end{bmatrix} \begin{bmatrix} y_f(k) \\ z_f(k) \end{bmatrix} \qquad (5.7)$$

Taking $y_i^{(0)}(0) = y(0)$, if the degenerate equation (5.6) is solved, the solutions $y_i^{(0)}(k)$ and $z_i^{(0)}(k)$ obtained are the same as indicated in (5.4). Similarly, taking $z_f^{(0)}(N) = z(N)$, if the degenerate form of (5.7) is solved, the resulting solutions $y_f^{(0)}(k)$ and $z_f^{(0)}(k)$ are the same as shown in (5.4). This clearly demonstrates the recovery of the lost boundary conditions.

5.1.2 Analysis

Consider the $(n_1 + n_2 + n_3)$-order model (5.1). The degenerate form of this model (5.2) is shown to be of order n_1. Utilising the results of the second-order model (5.3), it is apparent that since ε is associated with $y(k)$, a boundary layer occurs at $k = 0$ accounting for the loss of n_2 initial conditions $y(0)$. Similarly, since ε is associated with $z(k + 1)$, another boundary layer occurs at $k = N$, accounting for the loss of n_3 terminal conditions $Z(N)$. The degenerate model of (5.2), which is of order n_1, satisfies the boundary conditions of $x(k)$ only. In order to recover these lost boundary conditions, we incorporate two separate correction series, known as initial BLC and final BLC using the transformations (5.5).

The total series solution is given as

$$
\left.
\begin{aligned}
x(k) &= x_o(k) + \varepsilon^{k+1} x_i(k) + \varepsilon^{N-k+1} x_f(k) \\
y(k) &= y_o(k) + \varepsilon^k y_i(k) + \varepsilon^{N-k+1} y_f(k) \\
z(k) &= z_o(k) + \varepsilon^{k+1} z_i(k) + \varepsilon^{N-k} z_f(k)
\end{aligned}
\right\}
\tag{5.8}
$$

where

$x_o(k)$, $y_o(k)$ and $z_o(k)$ are the outer series terms

$x_i(k)$, $y_i(k)$ and $z_i(k)$ are the initial BLC terms

$x_f(k)$, $y_f(k)$ and $z_f(k)$ are the final BLC terms

Substituting (5.8) in (5.1) and separating the outer series, the initial and final BLC series terms, three equations are obtained. The outer series equation is identical with (5.1).

The initial BLC series equation is obtained as

$$
\begin{bmatrix} \varepsilon x_i(k+1) \\ y_i(k+1) \\ \varepsilon^2 z_i(k+1) \end{bmatrix} = \begin{bmatrix} A_{11} & A_{12} & A_{13} \\ A_{21} & A_{22} & A_{23} \\ A_{31} & A_{32} & A_{33} \end{bmatrix} \begin{bmatrix} x_i(k) \\ y_i(k) \\ z_i(k) \end{bmatrix}
\tag{5.9}
$$

Note that the order of the degenerate model of (5.9) is n_2.

The final BLC series equation is obtained as

$$
\begin{bmatrix} x_f(k+1) \\ y_f(k+1) \\ z_f(k+1) \end{bmatrix} = \begin{bmatrix} \varepsilon A_{11} & \varepsilon^2 A_{12} & A_{13} \\ \varepsilon A_{21} & \varepsilon^2 A_{22} & A_{23} \\ \varepsilon A_{31} & \varepsilon^2 A_{32} & A_{33} \end{bmatrix} \begin{bmatrix} x_f(k) \\ y_f(k) \\ z_f(k) \end{bmatrix}
\tag{5.10}
$$

Note that the degenerate model of (5.10) is of order n_3.

5.1.3 Three-time-scale property

Consider the second-order model (5.3) and its eigenvalues p_1 and p_2. For sufficiently small values of ε, $|p_1| \gg |p_2|$. In the degeneration process as ε tends to zero, p_1 approaches infinity and p_2 approaches zero. Thus, for this model, p_2 is associated with a stable fast mode which corresponds to a fast decay of solution near $k = 0$ and p_1 is associated with an unstable fast mode which corresponds to a fast rise near the terminal point $k = N$. These ideas are easily extended to the $(n_1 + n_2 + n_3)$-order model (5.1). The n_2 eigenvalues of the stable fast modes have moduli much less than unity which approach zero in the degeneration process. Thus they lead to the formation of the boundary layer at the initial point $(k = 0)$. The n_3 eigenvalues of the unstable fast modes have moduli much larger than unity, which approach infinity in the degeneration process. Thus they lead to the formation of the boundary layer at the final point $(k = N)$. The n_1 eigenvalues of the slow modes have moduli in the neighbourhood of unity and approach the eigenvalues of $(A_{11} - A_{13}A_{33}^{-1}A_{31})$ in the degeneration process. In the complex plane, the n_1 eigenvalues of the slow modes lie in the neighbourhood of the circumference of the unit circle. The n_2 eigenvalues of the stable fast modes and n_3 eigenvalues of the unstable fast modes are located near the origin and far outside the circumference of the unit circle respectively. Thus the model (5.1) with three distinct groups of eigenvalues widely separated from each other and contributing to slow, fast decay and fast rise modes is said to possess the three-time-scale property.

Example 5.1
Consider a fourth-order two-point boundary-value problem (Rao and Naidu, 1981).

$$\begin{bmatrix} x_1(k+1) \\ x_2(k+1) \\ y(k+1) \\ \varepsilon z(k+1) \end{bmatrix} = \begin{bmatrix} 0 & 1 & 0 & 0 \\ 0 & 0 & 0 & 1 \\ 1 & 0 & 0 & 0 \\ -1\cdot15 & 0\cdot5 & 1\cdot5\varepsilon & 0\cdot7 \end{bmatrix} \begin{bmatrix} x_1(k) \\ x_2(k) \\ y(k) \\ z(k) \end{bmatrix}$$

with the boundary conditions

$$\begin{bmatrix} x_1(0) \\ y(0) \end{bmatrix} = \begin{bmatrix} 15\cdot0 \\ 20\cdot0 \end{bmatrix}; \quad \begin{bmatrix} x_2(8) \\ z(8) \end{bmatrix} = \begin{bmatrix} 2\cdot0 \\ 3\cdot0 \end{bmatrix}$$

and the small parameter $\varepsilon = 0\cdot1$.
    The eigenvalues of the above model are

$$p_1 = -1\cdot5367; \quad p_2 = 0\cdot9291$$
$$p_3 = 7\cdot4670; \quad p_4 = 0\cdot1407$$

The eigenvalues p_1 and p_2 correspond to slow modes and p_3 and p_4 correspond to fast rise and fast decay modes, respectively.

The eigenvalues of the degenerate model are

$$p_1 = -1{\cdot}6877; \quad p_2 = 0{\cdot}9734$$

which are nearly equal to the eigenvalues of the slow modes.

Using the method developed, the series solutions up to second-order approximation are obtained as shown in Table 5.1. It is noted that the series solutions approach the exact solution with the increased order of approximation.

Table 5.1 *Comparison of approximate and exact solutions of Example 5.1*

(a) For $x_1(k)$

k	Degenerate solution	Zeroth-order solution	First-order solution	Second-order solution	Exact solution
0	15·0000	15·0000	15·0000	15·0000	15·0000
1	14·3689	14·3689	13·2136	13·1785	13·2061
2	14·3793	14·3793	12·8239	12·5381	12·5138
3	13·3352	13·3352	11·3478	11·1958	11·1269
4	14·0980	14·0980	11·5743	11·1119	11·0872
5	11·8378	11·8378	9·3949	9·2401	9·1458
6	14·7055	14·7055	10·7245	10·2683	10·2680
7	8·9438	8·9438	7·2918	6·8189	6·7927
8	17·7707	17·7707	10·1594	10·2662	10·3428

(b) For $x_2(k)$

k	Degenerate solution	Zeroth-order solution	First-order solution	Second-order solution	Exact solution
0	14·3689	14·3689	13·2136	13·1785	13·2061
1	14·3793	14·3793	12·8239	12·5381	12·5138
2	13·3352	13·3352	11·3478	11·1958	11·1269
3	14·0980	14·0980	11·5743	11·1119	11·0872
4	11·8378	11·8378	9·3949	9·2401	9·1458
5	14·7055	14·7055	10·7245	10·2653	10·2680
6	8·9438	8·9438	7·2918	6·8189	6·7927
7	17·7707	17·7707	10·1594	10·2662	10·3428
8	2·0000	2·0000	2·0000	2·0000	2·0000

(c) For y(k)

k	Degenerate solution	Zeroth-order solution	First-order solution	Second-order solution	Exact solution
0	15·2680	20·0000	20·0000	20·0000	20·0000
1	15·0000	15·0000	15·0000	15·0000	15·0000
2	14·3689	14·3689	13·2136	13·1785	13·2061
3	14·3793	14·3793	12·8239	12·5381	12·5138
4	13·3352	13·3352	11·3478	11·1958	11·1269
5	14·0980	14·0980	11·5743	11·1119	11·0872
6	11·8378	11·8378	9·3949	9·2401	9·1458
7	14·7055	14·7055	10·7245	10·2653	10·2680
8	8·9438	8·9438	7·2918	6·8189	6·7927

(d) For z(k)

k	Degenerate solution	Zeroth-order solution	First-order solution	Second-order solution	Exact solution
0	14·3793	14·3793	12·8239	12·5381	12·5138
1	13·3352	13·3352	11·3478	11·1958	11·1268
2	14·0980	14·0980	11·5743	11·1119	11·0872
3	11·8378	11·8378	9·3949	9·2401	9·1458
4	14·7055	14·7055	10·7245	10·2652	10·2680
5	8·9438	8·9438	7·2918	6·8189	6·7927
6	17·7707	17·7707	10·1594	10·2662	10·3428
7	2·0000	2·0000	2·0000	2·0000	2·0000
8	27·7661	3·0000	3·0000	3·0000	3·0000

5.2   Open-loop optimal control problem

The open-loop optimal control problem of a singularly perturbed, linear, shift-invariant discrete-time system is considered. The resulting two-point bound-ary-value problem (TPBVP) is cast in the singularly perturbed form which exhibits the three-time-scale property. A singular perturbation method is de-scribed to obtain the approximate solution composed of an outer series, initial boundary-layer correction series and final boundary-layer correction series.

Consider the linear, shift-invariant completely controllable discrete system

$$\begin{bmatrix} y(k+1) \\ z(k+1) \end{bmatrix} = \begin{bmatrix} A_{11} & \varepsilon A_{12} \\ A_{21} & \varepsilon A_{22} \end{bmatrix} \begin{bmatrix} y(k) \\ z(k) \end{bmatrix} + \begin{bmatrix} B_1 \\ B_2 \end{bmatrix} u(k) \qquad (5.11)$$

with $y(0)$ and $z(0)$, where $y(k)$ and $z(k)$ are m and n states vectors and $u(k)$ is an r vector. A_{ij}, B_i, $i, j = 1, 2$ are real constant matrices of appropriate dimensionality.

The performance index to be minimised is

$$J = \frac{1}{2}x'(N)Sx(N) + \frac{1}{2}\sum_{k=0}^{N-1}[x'(k)Dx(k) + u'(k)Ru(k)] \tag{5.12}$$

where $x'(k) = [y'(k), \varepsilon z'(k)]$. D and S are real, positive semidefinite, symmetric matrices of order $(m + n) \times (m + n)$ given by

$$D = \begin{bmatrix} D_{11} & D_{12} \\ D'_{12} & D_{22} \end{bmatrix} \quad \text{and} \quad S = \begin{bmatrix} S_{11} & S_{12} \\ S'_{12} & S_{22} \end{bmatrix}$$

R is a real, positive definite, symmetric matrix of order $(r \times r)$. Note that the states are incorporated in (5.12) in an appropriate manner as $y(k)$ and $\varepsilon z(k)$ in order to bring the resulting TPBVP into singularly perturbed form.

The Hamiltonian of the problem is

$$H(k) = \frac{1}{2}y'(k)D_{11}y(k) + \frac{1}{2}\varepsilon z'(k)D'_{12}y(k)$$

$$+ \frac{1}{2}\varepsilon y'(k)D_{12}z(k) + \frac{1}{2}\varepsilon^2 z'(k)D_{22}z(k) + \frac{1}{2}u'(k)Ru(k)$$

$$+ p'(k+1)[A_{11}y(k) + \varepsilon A_{12}z(k) + B_1 u(k)]$$

$$+ \bar{q}'(k+1)[A_{21}y(k) + \varepsilon A_{22}z(k) + B_2 u(k)] \tag{5.13}$$

where $p(k)$ and $\bar{q}(k)$ are the co-states.

Using the results of discrete optimal control theory (Sage and White, 1977),

$$\left. \begin{aligned} \frac{\partial H(k)}{\partial p(k+1)} &= y(k+1) \\[2mm] \frac{\partial H(k)}{\partial \bar{q}(k+1)} &= z(k+1) \\[2mm] \frac{\partial H(k)}{\partial y(k)} &= p(k) \\[2mm] \frac{\partial H(k)}{\partial z(k)} &= \bar{q}(k) \\[2mm] \frac{\partial H(k)}{\partial u(k)} &= 0 \end{aligned} \right\} \tag{5.14}$$

From (5.13) and (5.14), the state and co-state equations are obtained as

$$
\begin{bmatrix} y(k+1) \\ z(k+1) \\ p(k) \\ q(k) \end{bmatrix} = \begin{bmatrix} A_{11} & \varepsilon A_{12} & -E_{11} & -\varepsilon E_{12} \\ A_{21} & \varepsilon A_{22} & -E'_{12} & -\varepsilon E_{22} \\ D_{11} & \varepsilon D_{12} & A'_{11} & \varepsilon A'_{21} \\ D'_{12} & \varepsilon D_{22} & A'_{12} & \varepsilon A'_{22} \end{bmatrix} \begin{bmatrix} y(k) \\ z(k) \\ p(k+1) \\ q(k+1) \end{bmatrix}
\tag{5.15}
$$

and the optimal control is obtained as

$$
u(k) = -R^{-1}[B'_1 p(k+1) + \varepsilon B'_2 q(k+1)]
\tag{5.16}
$$

where $\bar{q}(k) = \varepsilon q(k);$ $E_{11} = B_1 R^{-1} B'_1;$ $E_{12} = B_1 R^{-1} B'_2$

$$
E_{22} = B_2 R^{-1} B'_2
$$

The final conditions are

$$
\begin{bmatrix} p(N) \\ q(N) \end{bmatrix} = [S] \begin{bmatrix} y(N) \\ z(N) \end{bmatrix}
\tag{5.17}
$$

The equations (5.15), (5.16) and (5.17) constitute the open-loop optimal control problem. The equation (5.15) is restructured as

$$
\begin{bmatrix} y(k+1) \\ p(k) \\ z(k+1) \\ q(k) \end{bmatrix} = \begin{bmatrix} A_{11} & -E_{11} & \varepsilon A_{12} & -\varepsilon E_{12} \\ D_{11} & A'_{11} & \varepsilon D_{12} & \varepsilon A'_{21} \\ A_{21} & -E'_{12} & \varepsilon A_{22} & -\varepsilon E_{22} \\ D'_{12} & A'_{12} & \varepsilon D_{22} & \varepsilon A'_{22} \end{bmatrix} \begin{bmatrix} y(k) \\ p(k+1) \\ z(k) \\ q(k+1) \end{bmatrix}
\tag{5.18}
$$

The $2(m+n)$th-order TPBVP represented by (5.15) or (5.18) is to be solved with the boundary conditions

$$
y(0), z(0), p(N) \text{ and } q(N)
$$

5.3   Singular perturbation method

The $2(m+n)$th-order TPBVP represented by (5.18) is in the singularly perturbed form in the sense that its degenerate TPBVP

$$
\begin{bmatrix} y^{(0)}(k+1) \\ p^{(0)}(k) \end{bmatrix} = \begin{bmatrix} A_{11} & -E_{11} \\ D_{11} & A'_{11} \end{bmatrix} \begin{bmatrix} y^{(0)}(k) \\ p^{(0)}(k+1) \end{bmatrix}
\tag{5.19a}
$$

$$
\begin{bmatrix} z^{(0)}(k+1) \\ q^{(0)}(k) \end{bmatrix} = \begin{bmatrix} A_{21} & -E'_{12} \\ D'_{12} & A'_{12} \end{bmatrix} \begin{bmatrix} y^{(0)}(k) \\ p^{(0)}(k+1) \end{bmatrix}
\tag{5.19b}
$$

is of reduced order $2m$ and satisfies the boundary conditions

$$y^{(0)}(0) = y(0); \quad p^{(0)}(N) = p(N)$$

Once (5.19a) is solved with the above boundary conditions, the solutions $z^{(0)}(k)$ and $q^{(0)}(k)$ are fixed from (5.19b), and in general

$$z^{(0)}(0) \neq z(0) \quad \text{and} \quad q^{(0)}(N) \neq q(N)$$

The degenerate optimal control is obtained from (5.16) as

$$u^{(0)}(k) = -R^{-1}B_1'p^{(0)}(k+1). \tag{5.20}$$

Note that (5.18) is in the same form as the model (5.1). It is thus seen that the model (5.18) has the three-time-scale property and two boundary layers are therefore formed at the initial point ($k = 0$) and the final point ($k = N$). This situation is similar to that occurring in TPBVP arising in optimal control of continuous systems (Jamshidi, 1976). Utilising the results of the analysis in Section 5.1, the solution of (5.18) is taken as

$$\left.\begin{aligned}
y(k) &= y_o(k) + \varepsilon^{k+1}y_i(k) + \varepsilon^{N-k+1}y_f(k) \\
p(k) &= p_o(k) + \varepsilon^{k+1}p_i(k) + \varepsilon^{N-k+1}p_f(k) \\
z(k) &= z_o(k) + \varepsilon^k z_i(k) + \varepsilon^{N-k}z_f(k) \\
q(k) &= q_o(k) + \varepsilon^{k+1}q_i(k) + \varepsilon^{N-k}q_f(k)
\end{aligned}\right\} \tag{5.21}$$

where

$y_o(k)$, $p_o(k)$, $z_o(k)$, $q_o(k)$ correspond to the outer series, $y_i(k)$, $p_i(k)$, $z_i(k)$, $q_i(k)$ correspond to the initial-boundary-layer correction (BLC) series, and $y_f(k)$, $p_f(k)$, $z_f(k)$, $q_f(k)$ correspond to the final-boundary-layer correction (BLC) series.

Substituting (5.21) in (5.18) and separating the outer-series and the two correction-series terms, we get three equations. The outer series equation is found to be the same as (5.18). The equation for the initial BLC is

$$\begin{bmatrix} \varepsilon y_i(k+1) \\ z_i(k+1) \\ p_i(k) \\ q_i(k) \end{bmatrix} = \begin{bmatrix} A_{11} & A_{12} & -\varepsilon E_{11} & -\varepsilon E_{12} \\ A_{21} & A_{22} & -\varepsilon E_{12}' & -\varepsilon E_{22} \\ D_{11} & D_{12} & \varepsilon A_{11}' & \varepsilon A_{21}' \\ D_{12}' & D_{22} & \varepsilon A_{12}' & \varepsilon A_{22}' \end{bmatrix} \begin{bmatrix} y_i(k) \\ z_i(k) \\ p_i(k+1) \\ \varepsilon q_i(k+1) \end{bmatrix} \tag{5.22a}$$

The initial correction to the optimal control is

$$u_i(k) = -R^{-1}[B_1'p_i(k+1) + \varepsilon B_2'q_i(k+1)] \tag{5.22b}$$

The equation for the final BLC is

$$
\begin{bmatrix} y_f(k+1) \\ z_f(k+1) \\ \varepsilon p_f(k) \\ q_f(k) \end{bmatrix} = \begin{bmatrix} \varepsilon A_{11} & \varepsilon A_{12} & -E_{11} & -E_{12} \\ \dot{\varepsilon} A_{21} & \varepsilon A_{22} & -E'_{12} & -E_{22} \\ \varepsilon D_{11} & \varepsilon D_{12} & A'_{11} & A'_{21} \\ \varepsilon D'_{12} & \varepsilon D_{22} & A'_{12} & A'_{22} \end{bmatrix} \begin{bmatrix} y_f(k) \\ \varepsilon z_f(k) \\ p_f(k+1) \\ q_f(k+1) \end{bmatrix}
\tag{5.23}
$$

The final correction to the optimal control is given by

$$
u_f(k) = -R^{-1}[B'_1 p_f(k+1) + \varepsilon B'_2 q_f(k+1)]
\tag{5.24}
$$

Assuming series solutions for (5.18), (5.22) and (5.23), the series equations for various orders of approximations are obtained as shown.

5.3.1   Outer series

For zeroth order

$$
\begin{bmatrix} y^{(0)}(k+1) \\ p^{(0)}(k) \end{bmatrix} = \begin{bmatrix} A_{11} & -E_{11} \\ D_{11} & A'_{11} \end{bmatrix} \begin{bmatrix} y^{(0)}(k) \\ p^{(0)}(k+1) \end{bmatrix}
\tag{5.25a}
$$

$$
\begin{bmatrix} z^{(0)}(k+1) \\ q^{(0)}(k) \end{bmatrix} = \begin{bmatrix} A_{21} & -E'_{12} \\ D'_{12} & A'_{12} \end{bmatrix} \begin{bmatrix} y^{(0)}(k) \\ p^{(0)}(k+1) \end{bmatrix}
\tag{5.25b}
$$

$$
u^{(0)}(k) = -R^{-1} B'_1 p^{(0)}(k+1)
\tag{5.25c}
$$

For first order

$$
\begin{bmatrix} y^{(1)}(k+1) \\ p^{(1)}(k) \end{bmatrix} = \begin{bmatrix} A_{11} & -E_{11} \\ D_{11} & A'_{11} \end{bmatrix} \begin{bmatrix} y^{(1)}(k) \\ p^{(1)}(k+1) \end{bmatrix} + \begin{bmatrix} A_{12} & -E_{12} \\ D_{12} & A'_{21} \end{bmatrix} \begin{bmatrix} z^{(0)}(k) \\ q^{(0)}(k+1) \end{bmatrix}
\tag{5.26a}
$$

$$
\begin{bmatrix} z^{(1)}(k+1) \\ q^{(1)}(k) \end{bmatrix} = \begin{bmatrix} A_{21} & -E'_{12} \\ D'_{12} & A'_{12} \end{bmatrix} \begin{bmatrix} y^{(1)}(k) \\ p^{(1)}(k+1) \end{bmatrix} + \begin{bmatrix} A_{22} & -E_{22} \\ D_{22} & A'_{22} \end{bmatrix} \begin{bmatrix} z^{(0)}(k) \\ q^{(0)}(k+1) \end{bmatrix}
\tag{5.26b}
$$

$$
u^{(1)}(k) = -R^{-1}[B'_1 p^{(1)}(k+1) + B'_2 q^{(0)}(k+1)]
\tag{5.26c}
$$

For jth order

$$
\begin{bmatrix} y^{(j)}(k+1) \\ p^{(j)}(k) \end{bmatrix} = \begin{bmatrix} A_{11} & -E_{11} \\ D_{11} & A'_{11} \end{bmatrix} \begin{bmatrix} y^{(j)}(k) \\ p^{(j)}(k+1) \end{bmatrix} + \begin{bmatrix} A_{12} & -E_{12} \\ D_{12} & A'_{21} \end{bmatrix} \begin{bmatrix} z^{(j-1)}(k) \\ q^{(j-1)}(k+1) \end{bmatrix}
\tag{5.27a}
$$

$$\begin{bmatrix} z^{(j)}(k+1) \\ q^{(j)}(k) \end{bmatrix} = \begin{bmatrix} A_{21} & -E'_{12} \\ D'_{12} & A'_{12} \end{bmatrix} \begin{bmatrix} y^{(j)}(k) \\ p^{(j)}(k+1) \end{bmatrix} + \begin{bmatrix} A_{22} & -E_{22} \\ D_{22} & A'_{22} \end{bmatrix} \begin{bmatrix} z^{(j-1)}(k) \\ q^{(j-1)}(k+1) \end{bmatrix}$$

$$(5.27b)$$

$$u^{(j)}(k) = -R^{-1}[B'_1 p^{(j)}(k+1) + B'_2 q^{(j-1)}(k+1)] \tag{5.27c}$$

5.3.2 Initial boundary-layer correction series

For zeroth order

$$\begin{bmatrix} 0 \\ z_i^{(0)}(k+1) \end{bmatrix} = \begin{bmatrix} A_{11} & A_{12} \\ A_{21} & A_{22} \end{bmatrix} \begin{bmatrix} y_i^{(0)}(k) \\ z_i^{(0)}(k) \end{bmatrix} \tag{5.28a}$$

$$\begin{bmatrix} p_i^{(0)}(k) \\ q_i^{(0)}(k) \end{bmatrix} = \begin{bmatrix} D_{11} & D_{12} \\ D'_{12} & D_{22} \end{bmatrix} \begin{bmatrix} y_i^{(0)}(k) \\ z_i^{(0)}(k) \end{bmatrix} \tag{5.28b}$$

$$u_i^{(0)}(k) = -R^{-1}B'_1 p_i^{(0)}(k+1) \tag{5.28c}$$

For first order

$$\begin{bmatrix} y_i^{(0)}(k+1) \\ z_i^{(1)}(k+1) \end{bmatrix} = \begin{bmatrix} A_{11} & A_{12} \\ A_{21} & A_{22} \end{bmatrix} \begin{bmatrix} y_i^{(1)}(k) \\ z_i^{(1)}(k) \end{bmatrix} + \begin{bmatrix} -E_{11} & -E_{12} \\ -E'_{12} & -E_{22} \end{bmatrix} \begin{bmatrix} p_i^{(0)}(k+1) \\ 0 \end{bmatrix}$$

$$(5.29a)$$

$$\begin{bmatrix} p_i^{(1)}(k) \\ q_i^{(1)}(k) \end{bmatrix} = \begin{bmatrix} D_{11} & D_{12} \\ D'_{12} & D_{22} \end{bmatrix} \begin{bmatrix} y_i^{(1)}(k) \\ z_i^{(1)}(k) \end{bmatrix} + \begin{bmatrix} A'_{11} & A'_{21} \\ A'_{12} & A'_{22} \end{bmatrix} \begin{bmatrix} p_i^{(0)}(k+1) \\ 0 \end{bmatrix} \tag{5.29b}$$

$$u_i^{(1)}(k) = -R^{-1}[B'_1 p_i^{(1)}(k+1) + B'_2 q_i^{(0)}(k+1)] \tag{5.29c}$$

For *j*th order

$$\begin{bmatrix} y_i^{(j-1)}(k+1) \\ z_i^{(j)}(k+1) \end{bmatrix} = \begin{bmatrix} A_{11} & A_{12} \\ A_{21} & A_{22} \end{bmatrix} \begin{bmatrix} y_i^{(j)}(k) \\ z_i^{(j)}(k) \end{bmatrix} + \begin{bmatrix} -E_{11} & -E_{11} \\ -E'_{12} & -E_{22} \end{bmatrix} \begin{bmatrix} p_i^{(j-1)}(k+1) \\ q_i^{(j-2)}(k+1) \end{bmatrix}$$

$$(5.30a)$$

$$\begin{bmatrix} p_i^{(j)}(k) \\ q_i^{(j)}(k) \end{bmatrix} = \begin{bmatrix} D_{11} & D_{12} \\ D'_{12} & D_{22} \end{bmatrix} \begin{bmatrix} y_i^{(j)}(k) \\ z_i^{(j)}(k) \end{bmatrix} + \begin{bmatrix} A'_{11} & A'_{21} \\ A'_{12} & A'_{22} \end{bmatrix} \begin{bmatrix} p_i^{(j-1)}(k+1) \\ q_i^{(j-2)}(k+1) \end{bmatrix} \tag{5.30b}$$

$$u_i^{(j)}(k) = -R^{-1}[B'_1 p_i^{(j)}(k+1) + B'_2 q_i^{(j-1)}(k+1)] \tag{5.30c}$$

5.3.3 Final boundary-layer correction series

For zeroth order

$$\begin{bmatrix} y_f^{(0)}(k+1) \\ z_f^{(0)}(k+1) \end{bmatrix} = \begin{bmatrix} -E_{11} & -E_{12} \\ -E_{12}' & -E_{22} \end{bmatrix} \begin{bmatrix} p_f^{(0)}(k+1) \\ q_f^{(0)}(k+1) \end{bmatrix} \tag{5.31a}$$

$$\begin{bmatrix} 0 \\ q_f^{(0)}(k) \end{bmatrix} = \begin{bmatrix} A_{11}' & A_{21}' \\ A_{12}' & A_{22}' \end{bmatrix} \begin{bmatrix} p_f^{(0)}(k+1) \\ q_f^{(0)}(k+1) \end{bmatrix} \tag{5.31b}$$

$$u_f^{(0)}(k) = -R^{-1}B_1'p_f^{(0)}(k+1) \tag{5.31c}$$

For first order

$$\begin{bmatrix} y_f^{(1)}(k+1) \\ z_f^{(1)}(k+1) \end{bmatrix} = \begin{bmatrix} -E_{11} & -E_{12} \\ -E_{12}' & -E_{22} \end{bmatrix} \begin{bmatrix} p_f^{(1)}(k+1) \\ q_f^{(1)}(k+1) \end{bmatrix} + \begin{bmatrix} A_{11} & A_{12} \\ A_{21} & A_{22} \end{bmatrix} \begin{bmatrix} y_f^{(0)}(k) \\ 0 \end{bmatrix} \tag{5.32a}$$

$$\begin{bmatrix} p_f^{(0)}(k) \\ q_f^{(1)}(k) \end{bmatrix} = \begin{bmatrix} A_{11}' & A_{21}' \\ A_{12}' & A_{22}' \end{bmatrix} \begin{bmatrix} p_f^{(1)}(k+1) \\ q_f^{(1)}(k+1) \end{bmatrix} + \begin{bmatrix} D_{11} & D_{12} \\ D_{12}' & D_{22} \end{bmatrix} \begin{bmatrix} y_f^{(0)}(k) \\ 0 \end{bmatrix} \tag{5.32b}$$

$$u_f^{(1)}(k) = -R^{-1}[B_1'p_f^{(1)}(k+1) + B_2'q_f^{(0)}(k+1)] \tag{5.32c}$$

For the *j*th order

$$\begin{bmatrix} y_f^{(j)}(k+1) \\ z_f^{(j)}(k+1) \end{bmatrix} = \begin{bmatrix} -E_{11} & -E_{12} \\ -E_{12}' & -E_{22} \end{bmatrix} \begin{bmatrix} p_f^{(j)}(k+1) \\ q_f^{(j)}(k+1) \end{bmatrix} + \begin{bmatrix} A_{11} & A_{12} \\ A_{21} & A_{22} \end{bmatrix} \begin{bmatrix} y_f^{(j-1)}(k) \\ z_f^{(j-2)}(k) \end{bmatrix} \tag{5.33a}$$

$$\begin{bmatrix} p_f^{(j-1)}(k) \\ q_f^{(j)}(k) \end{bmatrix} = \begin{bmatrix} A_{11}' & A_{21}' \\ A_{12}' & A_{22}' \end{bmatrix} \begin{bmatrix} p_f^{(j)}(k+1) \\ q_f^{(j)}(k+1) \end{bmatrix} + \begin{bmatrix} D_{11} & D_{12} \\ D_{12}' & D_{22}' \end{bmatrix} \begin{bmatrix} y_f^{(j-1)}(k) \\ z_f^{(j-2)}(k) \end{bmatrix} \tag{5.33b}$$

$$u_f^{(j)}(k) = -R^{-1}[B_1'p_f^{(j)}(k+1) + B_2'q_f^{(j-1)}(k+1)] \tag{5.33c}$$

5.3.4 Total series solution

The total series solution is composed of the outer series, initial BLC series and the final BLC series. That is,

$$
\left.
\begin{aligned}
y(k,\varepsilon) &= \sum_{r=0}^{j} [y^{(r)}(k) + \varepsilon^{k+1} y_i^{(r)}(k) + \varepsilon^{N-k+1} y_f(k)]\varepsilon^r \\
z(k,\varepsilon) &= \sum_{r=0}^{j} [z^{(r)}(k) + \varepsilon^{k} z_i^{(r)}(k) + \varepsilon^{N-k+1} z_f(k)]\varepsilon^r \\
p(k,\varepsilon) &= \sum_{r=0}^{j} [p^{(r)}(k) + \varepsilon^{k+1} p_i^{(r)}(k) + \varepsilon^{N-k+1} p_f(k)]\varepsilon^r \\
q(k,\varepsilon) &= \sum_{r=0}^{j} [q^{(r)}(k) + \varepsilon^{k+1} q_i^{(r)}(k) + \varepsilon^{N-k} q_f^{(r)}(k)]\varepsilon^r
\end{aligned}
\right\}
\tag{5.34}
$$

The boundary conditions required to solve the series equations (5.25)–(5.33) are easily obtained from (5.34). They are

$$
\left.
\begin{aligned}
\varepsilon^0: \quad & y^{(0)}(0) = y(0); \quad p^{(0)}(N) = p(N) \\
& z_i^{(0)}(0) = z(0) - z^{(0)}(0); \quad q_f^{(0)}(N) = q(N) = q^{(0)}(N)
\end{aligned}
\right\}
\tag{5.35a}
$$

$$
\left.
\begin{aligned}
\varepsilon^1: \quad & y^{(1)}(0) = -y_i^{(0)}(0); \quad p^{(1)}(N) = -p_f^{(0)}(N) \\
& z_i^{(1)}(0) = -z^{(1)}(0); \quad q_f^{(1)}(N) = -q^{(1)}(N)
\end{aligned}
\right\}
\tag{5.35b}
$$

$$
\left.
\begin{aligned}
\varepsilon^j: \quad & y^{(j)}(0) = -y_i^{(j-1)}(0); \quad p^{(j)}(N) = p_f^{(j-1)}(N) \\
& z_i^{(j)}(0) = -z^{(j)}(0); \quad q_f^{(j)}(N) = -q^{(j)}(N)
\end{aligned}
\right\}
\tag{5.35c}
$$

Example 5.2*
Consider the second-order control system (Naidu and Rao, 1982)

$$
\begin{bmatrix} y(k+1) \\ z(k+1) \end{bmatrix} = \begin{bmatrix} 0.95 & \varepsilon \\ 1 & 0 \end{bmatrix} \begin{bmatrix} y(k) \\ z(k) \end{bmatrix} + \begin{bmatrix} 1 \\ 0 \end{bmatrix} u(k)
\tag{5.36}
$$

where the small parameter $\varepsilon = 0.12$, the initial conditions

$$
y(0) = 10.0; \quad z(0) = 15.0
$$

and the performance index

$$
J = \frac{1}{2} \sum_{k=0}^{N-1} [x'(k)Dx(k) + Ru(k)^2]
\tag{5.37}
$$

---

* Reprinted by permission from RAO, A. K. and NAIDU, D. S. (1982): 'Singular perturbation method applied to open-loop discrete optimal control problem', *Optimal Control: Applications & Methods*, **3**, pp 121–131.

where

$$D = \begin{bmatrix} 1 & 0 \\ 0 & 2 \end{bmatrix}; \quad x(k) = \begin{bmatrix} y(k) \\ \varepsilon z(k) \end{bmatrix}; \quad R = 2 \text{ and } N = 8$$

The eigenvalues of (5.27) are

$$p_1 = 0\cdot80; \quad p_2 = 0\cdot15$$

The singularly perturbed TPBVP of corresponding to (5.15) is

$$\begin{bmatrix} y(k+1) \\ z(k+1) \\ p(k) \\ q(k) \end{bmatrix} = \begin{bmatrix} 0\cdot95 & \varepsilon & -0\cdot5 & 0 \\ 1\cdot0 & 0 & 0 & 0 \\ 1\cdot0 & 0 & 0\cdot95 & \varepsilon \\ 0 & 2\cdot0\varepsilon & 1\cdot0 & 0 \end{bmatrix} \begin{bmatrix} y(k) \\ z(k) \\ p(k+1) \\ q(k+1) \end{bmatrix} \tag{5.38}$$

and the optimal control is given by

$$u(k) = -0\cdot5p(k+1) \tag{5.39}$$

The boundary conditions for solving (5.38) are

$$\begin{bmatrix} y(0) \\ z(0) \end{bmatrix} = \begin{bmatrix} 10\cdot0 \\ 15\cdot0 \end{bmatrix}; \quad \begin{bmatrix} p(N) \\ q(N) \end{bmatrix} = \begin{bmatrix} 0 \\ 0 \end{bmatrix}$$

The eigenvalues of (5.38) are

$$p_1 = 1\cdot8370; \quad p_2 = 0.5444$$
$$p_3 = -9\cdot2398; \quad p_4 = -0\cdot1082 \tag{5.40}$$

The eigenvalues of the degenerate model of (5.38) are

$$p_1 = 2.0383; \quad p_2 = 0.4906 \tag{5.41}$$

which are nearly equal to the eigenvalues of the slow modes given above.

Using the singular perturbation method described, the degenerate, zeroth-, first-, and second-order solutions are evaluated and compared with the exact solution as shown in Table 5.2 and Fig. 5.1.

Table 5.2 *Comparison of approximate and exact (optimal) solutions of Example 5.2*

k	Degenerate solution	Zeroth-order solution	First-order solution	Second-order solution	Exact solution
$y(0)$	10·0000	10·0000	10·0000	10·0000	10·0000
$z(0)$	20·3833	15·0000	15·0000	15·0000	15·0000
$p(0)$	18·7282	18·7282	21·8053	22·5680	22·5815
$q(0)$	9·1876	9·1876	15·4574	15·6991	15·7059
$u(0)$	−4·5938	−4·5938	−5·9287	−6·0496	−6·0529

Table 5.2 *continued*

k	Degenerate solution	Zeroth-order solution	First-order solution	Second-order solution	Exact solution
$y(1)$	4·9062	4·9062	5·3713	5·2504	5·2471
$z(1)$	10·0000	10·0000	10·0000	10·0000	10·0000
$p(1)$	9·1876	9·1876	11·8574	12·0991	12·1059
$q(1)$	4·5067	4·5067	8·9483	9·0096	9·0068
$u(1)$	−2·2533	−2·2533	−3·2742	−3·3048	−3·3037
$y(2)$	2·4159	2·4159	3·0369	2·8914	2·8810
$z(2)$	4·9062	4·9062	5·3713	5·2504	5·2471
$p(2)$	4·5067	4·5067	6·5483	6·6096	6·6073
$q(2)$	2·2096	2·2096	4·7459	4·8960	4·8487
$u(2)$	−1·1048	−1·7842	−1·7842	−1·8035	−1·7947
$y(3)$	1·1824	1·1824	1·6817	1·5801	1·5719
$z(3)$	2·4159	2·4159	3·0369	2·8914	2·8810
$p(3)$	2·2096	2·2096	3·5685	3·6069	3·5894
$q(3)$	1·0812	1·0812	2·4996	2·6985	2·6360
$u(3)$	−0·5406	−0·5406	−0·9599	−0·9848	−0·9723
$y(4)$	0·5827	0·5827	0·9276	0·8807	0·8668
$z(4)$	1·1824	1·1824	1·6817	1·5801	1·5720
$p(4)$	1·0812	1·0812	1·9198	1·9697	1·9445
$q(4)$	0·5248	0·5248	1·2972	1·4678	1·4180
$u(4)$	−0·2624	−0·2624	−0·5067	−0·5268	−0·5204
$y(5)$	0·2912	0·2912	0·5165	0·5064	0·4918
$z(5)$	0·5827	0·5927	0·9276	0·8807	0·8668
$p(5)$	0·5248	0·5248	1·0134	1·0536	1·0407
$q(5)$	0·2460	0·2460	0·6506	0·7737	0·7417
$u(5)$	−0·1230	−0·1230	−0·2554	−0·2755	−0·2668
$y(6)$	0·1537	0·1537	0·3053	0·3170	0·3044
$z(6)$	0·2912	0·2912	0·5165	0·5064	0·4918
$p(6)$	0·2460	0·2460	0·5108	0·5510	0·5336
$q(6)$	0·0973	0·0973	0·2864	0·3643	0·3501
$u(6)$	−0·0486	−0·0486	−0·1082	−0·1210	−0·1161
$y(7)$	0·0976	0·0976	0·2169	0·2423	0·2321
$z(7)$	0·1537	0·1537	0·3053	0·3170	0·3044
$p(7)$	0·0973	0·0973	0·2165	0·2419	0·2321
$q(7)$	0·0000	0·0000	0·0368	0·0732	0·0730
$u(7)$	0·0000	0·0000	0·0000	0·0000	0·0000
$y(8)$	0·0931	0·0931	0·2249	0·2672	0·2570
$z(8)$	0·0976	0·0976	0·2169	0·2423	0·2321
$p(8)$	0·0000	0·0000	0·0000	0·0000	0·0000
$q(8)$	−0·0973	0·0000	0·0000	0·0000	0·0000

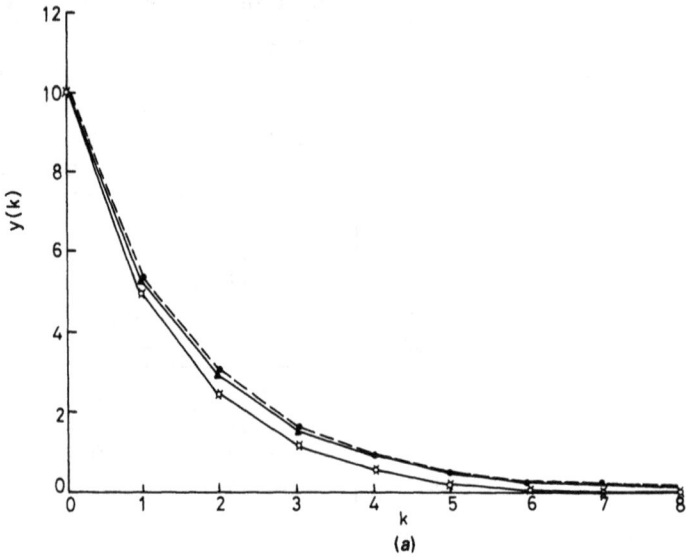

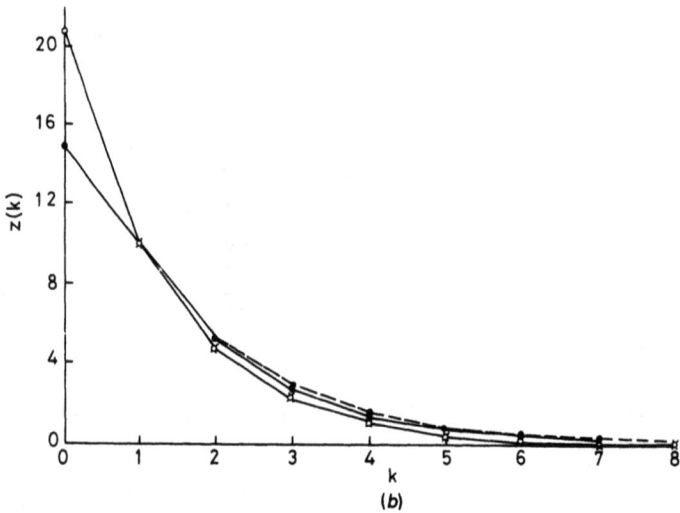

Fig. 5.1a and b   *Exact and approximate solutions of y(k) and z(k) of Example 5.2*
●————●————● exact solution
●– – –●– – –● first-order solution
×————×————× zeroth-order solution
○————○————○ degenerate solution
¤————¤————¤ solution common to degenerate and zeroth order

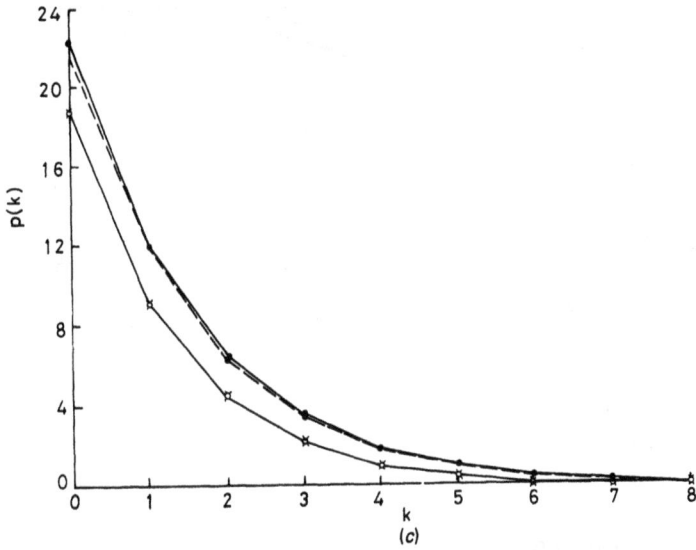

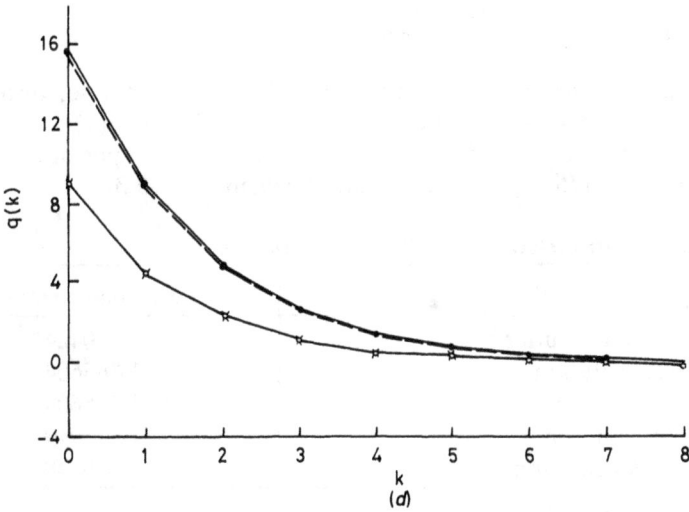

Fig. 5.1c and d *Exact and approximate solutions of p(k) and q(k) of Example 5.2*
●———●———● exact solution
●– – –●– – –● first-order solution
×———×———× zeroth-order solution
○———○———○ degenerate solution
¤———¤———¤ solution common to degenerate and zeroth order

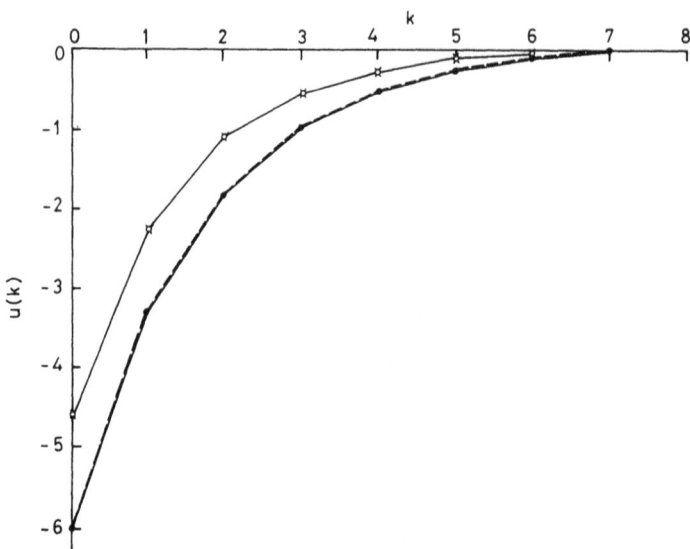

Fig. 5.1e *Exact and approximate solutions of u(k) of Example 5.2*
●——●——● exact solution
●– – –●– – –● first-order solution
¤——¤——¤ solution common to degenerate and zeroth order

Using the control $u(k)$, corresponding to exact, degenerate, zeroth, first- and second-order solutions, the resulting states $y(k)$ and $z(k)$ are calculated from the original system of (5.36) and the corresponding performance index is evaluated from (5.37). The results are shown in Table 5.3.

Table 5.3   *Comparison of performance indices*

Description of solution	Performance index
Exact (optimal solution)	127·0424
Degenerate solution	179·5604
Zeroth-order solution	179·5604
First-order solution	127·1949
Second-order solution	127·0448

The exact solution of the fourth-order singularly perturbed discrete TPBVP given by (5.38) is obtained using the method of complementary functions suggested for continuous 'stiff' problems (Roberts and Shipman, 1972).

5.4   Conclusions and discussion

A state–space model having a three-time-scale property and characterised by three distinct groups of eigenvalues has been formulated. In the process of degeneration, it is found that this model looses some of the initial and final conditions. A method of getting the approximate solution in terms of the outer series, the initial-boundary-layer correction series and final-boundary-layer correction series has been described.

The open-loop optimal control problem with a quadratic performance index has been considered. Using the discrete maximum principle, the state and co-state equations are obtained and cast in the singularly perturbed TPBVP.

The open-loop discrete optimal control problem exhibiting the three-time-scale property has the same essential features as its counterpart in continuous systems (Jamshidi, 1976).

Related problems on optimal control of singularly perturbed discrete systems are discussed by Vasileva and Dmitriev (1979), Blankenship (1981), Rajagopalan and Naidu (1981), Rao and Naidu (1982), Kando and Iwazumi (1983a,b), Litkouhi and Khalil (1984, 1985).

Other important contributions are in adaptive control (Ioannou, 1981; Ioannou and Johnson, 1983; Ioannou and Kokotovic, 1983) and stochastic control and Markov chains (Delebeque and Quadrat, 1981; Delebeque, 1983; Rao and Naidu, 1984).

Chapter 6

Closed-loop optimal control of continuous systems

Closed-loop optimal control, one of the most desirable features in system design, leads to the formulation of the well-known matrix Riccati equation. The singularly perturbed linear quadratic regulator problem is a thoroughly investigated problem (Sannuti and Kokotovic, 1969*a*; Kokotovic and Yackel, 1972; O'Malley, 1972*a,b,c*, 1975; Yackel and Kokotovic, 1973; O'Malley and Kung, 1974, 1975; Kokotovic *et al.*, 1976; Saksena *et al.*, 1984). In all these works, the analysis is based on the boundary-layer method (Vasileva, 1963; O'Malley, 1971; Hoppensteadt, 1971). Here the approximate solution is sought in terms of an outer series and a boundary-layer correction (BLC).

In this Chapter, we intend to use Vasileva's method (Vasileva, 1963; Wasow, 1965) where the approximate solution is made up of an outer series, inner series and intermediate series. It should be noted that, as shown in Chapter 2, the boundary-layer correction is equivalent to the difference between the inner and intermediate series. However, in working out a large number of control problems by Vasileva's method and the boundary-layer method (Naidu, 1977), it has been found that the former has computational advantages, especially in the case of nonlinear equations as will be apparent at the end of this Chapter. Thus, the contents of this Chapter depend mainly on results from Naidu (1977), Naidu and Rajagopalan (1980) and Rajagopalan and Naidu (1980*b*).

In Section 6.1 we pose the linear-quadratic-regulator problem and obtain the singularly perturbed matrix Riccati differential equation. Section 6.2 presents the singular perturbation method for the Riccati equation. The steady-state regulator problem, i.e. for large values of terminal time, is discussed in Section 6.3. Section 6.4 deals with the fixed-end-point problem and the associated special matrix Riccati equation. The important topic of separation of time scales and the corresponding subsystem regulator design is focused in Section 6.5. The conclusions, with a brief discussion, are given in the last Section.

6.1   Linear regulator problem

Consider a singularly perturbed, linear, time-invariant system

$$\frac{dx}{dt} = A_{11}x + A_{12}z + B_1u, \quad x(t=0) = x(0) \tag{6.1a}$$

$$\varepsilon\frac{dz}{dt} = A_{21}x + A_{22}z + B_2u, \quad z(t=0) = z(0) \tag{6.1b}$$

with performance index

$$J = \frac{1}{2}y'(T)Sy(T) + \frac{1}{2}\int_0^T (y'Qy + u'Ru)dt \tag{6.2}$$

The description of (6.1) and (6.2) is the same as that given in Chapter 4. For a free-end-point problem, the optimal control is given by (Athans and Falb, 1966)

$$u = -R^{-1}B'Py \tag{6.3a}$$

where P is an $(m+n) \times (m+n)$-dimensional positive–definite, symmetric matrix satisfying the matrix Riccati differential equation

$$\frac{dP}{dt} = -PA - A'P + PBR^{-1}B'P - Q, \quad P(t=T) = S \tag{6.3b}$$

where

$$A = \begin{bmatrix} A_{11} & A_{12} \\ \dfrac{A_{21}}{\varepsilon} & \dfrac{A_{22}}{\varepsilon} \end{bmatrix}; \quad B = \begin{bmatrix} B_1 \\ \dfrac{B_2}{\varepsilon} \end{bmatrix}; \quad Q = \begin{bmatrix} Q_{11} & Q_{12} \\ Q'_{12} & Q_{22} \end{bmatrix}$$

$$S = \begin{bmatrix} S_{11} & \varepsilon S_{12} \\ \varepsilon S'_{12} & \varepsilon S_{22} \end{bmatrix}; \quad P = \begin{bmatrix} P_{11} & \varepsilon P_{12} \\ \varepsilon P'_{12} & \varepsilon P_{22} \end{bmatrix}$$

Using the above matrices in (6.3b) and rearranging in terms of submatrices, we get

$$\frac{dP_{11}}{dt} = -P_{11}A_{11} - A'_{11}P_{11} - P_{12}A_{21} - A'_{21}P'_{12} + P_{11}E_{11}P_{11} \tag{6.4a}$$

$$+ P_{12}E'_{12}P_{11} + P_{11}E_{12}P'_{12} + P_{12}E_{22}P'_{12} - Q_{11}$$

$$\varepsilon \frac{dP_{12}}{dt} = -P_{11}A_{12} - P_{12}A_{22} - A'_{21}P_{22} + P_{11}E_{12}P_{22} \tag{6.4b}$$
$$+ P_{12}E_{22}P_{22} - Q_{12} + \varepsilon(-A'_{11}P_{12} + P_{11}E_{11}P_{12}$$
$$+ P_{12}E'_{12}P_{12})$$

$$\varepsilon \frac{dP_{22}}{dt} = -P_{22}A_{22} - A'_{22}P_{22} + P_{22}E_{22}P_{22} - Q_{22} + \varepsilon(-P'_{12}A_{12} \tag{6.4c}$$
$$- A'_{12}P_{12} + P'_{12}E_{12}P_{22} + P_{22}E'_{12}P_{12}) + \varepsilon^2 P'_{12}E_{11}P_{12}$$

with end conditions

$$P_{11}(t=T) = S_{11}; \quad P_{12}(t=T) = S_{12}; \quad P_{22}(t=T) = S_{22} \tag{6.5}$$

where

$$E_{11} = B_1 R^{-1} B'_1; \quad E_{12} = B_1 R^{-1} B'_2; \quad E_{22} = B_2 R^{-1} B'_2$$

The matrix Riccati equation (6.4) consists of $(m + n) \times (m + n + 1)/2$ first-order nonlinear differential equations which are solved backwards in time starting with the final conditions (6.5). When the small parameter ε is made zero in (6.4), the degenerate equation becomes

$$\frac{dP_{11}^{(0)}}{dt} = -P_{11}^{(0)}A_{11} - A'_{11}P_{11}^{(0)} - P_{12}^{(0)}A_{21} - A'_{21}P_{12}^{(0)} \tag{6.6a}$$
$$+ P_{11}^{(0)}E_{11}P_{11}^{(0)} + P_{12}^{(0)}E'_{12}P_{11}^{(0)} + P_{11}^{(0)}E_{12}P_{12}^{(0)'}$$
$$+ P_{12}^{(0)}E_{22}P_{12}^{(0)'} - Q_{11}$$

$$0 = -P_{11}^{(0)}A_{12} - P_{12}^{(0)}A_{22} - A'_{21}P_{22}^{(0)} + P_{11}^{(0)}E_{12}P_{22}^{(0)} + P_{12}^{(0)}E_{22}P_{22}^{(0)} - Q_{12} \tag{6.6b}$$

$$0 = -P_{22}^{(0)}A_{22} - A'_{22}P_{22}^{(0)} + P_{22}^{(0)}E_{22}P_{22}^{(0)} - Q_{22} \tag{6.6c}$$

with final conditions

$$P_{11}^{(0)}(t=T) = P_{11}(T); \quad P_{12}^{(0)}(t=T) \neq P_{12}(T); \quad P_{22}^{(0)}(t=T) \neq P_{22}(T) \tag{6.7}$$

From (6.4)–(6.7), we see that the suppression of the small parameter ε in (6.4) results in the degenerate form (6.6) with the sacrifice of some of the final conditions as shown by (6.7). Hence (6.5) is in the singularly perturbed form. The boundary layer exists at $t = T$.

An alternate way of obtaining the degenerate form (6.6) is to first make $\varepsilon = 0$ in the state equation (6.1) and then obtain the Riccati equation for the resulting degenerate state equation (Haddad and Kokotovic, 1971).

6.2 Singular perturbation method

In the previous Section we have seen that the closed-loop optimal control of a singularly perturbed linear system leads to the formulation of a singularly perturbed matrix Riccati differential equation. The solution of (6.4) has 'slow' and 'fast' modes, thereby making it computationally stiff. The degeneration has affected the order and some of the final conditions of the matrix Riccati equation.

Thus the main aim of the singular perturbation method is to retain the advantage of order reduction and to recover the final conditions sacrificed in the process of degeneration. Our development of the method is closely akin to Vasileva's results for initial-value problems (Vasileva, 1963) where the approximate solution is sought in terms of an outer, inner and intermediate series. The outer series approximates to the solution outside the boundary layer, and inner and intermediate series recover the lost auxiliary conditions and improve the solution inside the boundary layer. This Section is heavily drawn from Naidu and Rajagopalan (1980).

6.2.1   Outer-series solution

The outer-series solution (or expansion) for (6.4) is assumed in the form of

$$P(t,\varepsilon) = \sum_{i=0}^{\infty} P^{(i)}(t)\varepsilon^i, \quad P = P_{11}, P_{12}, P_{22} \tag{6.8}$$

By substituting (6.8) in (6.4) and collecting coefficients of like powers of ε, we get a set of recursive equations. For the zeroth order (ε^0) we have the same equations (6.6) and (6.7). Here we note once again that the degenerate equation (6.6) satisfies the final condition $P_{11}(T)$, and hence cannot satisfy the other two final conditions $P_{12}(T)$ and $P_{22}(T)$. $P_{22}^{(0)}(t)$ is the positive definite solution of (6.6c). Using this, (6.6b) gives a relation for $P_{11}^{(0)}(t)$ in terms of $P_{12}^{(0)}(t)$ and this in turn, used in (6.6a), finally gives a differential equation (6.6a) in $P_{11}^{(0)}(t)$ only. Thus $P_{12}^{(0)}(T)$ and $P_{22}^{(0)}(T)$ need not in general be equal to the given values $P_{12}(T)$ and $P_{22}(T)$.

For the first-order (ε^1) approximation, we have

$$
\begin{aligned}
\frac{dP_{11}^{(1)}}{dt} = &-P_{11}^{(1)}A_{11} - A_{11}'P_{11}^{(1)} - P_{12}^{(1)}A_{21} - A_{21}'P_{12}^{(1)\prime} \\
&+ P_{11}^{(1)}E_{11}P_{11}^{(0)} + P_{11}^{(0)}E_{11}P_{11}^{(1)} + P_{12}^{(1)}E_{12}'P_{11}^{(0)} \\
&+ P_{12}^{(0)}E_{12}'P_{11}^{(1)} + P_{11}^{(1)}E_{12}P_{12}^{(0)\prime} + P_{11}^{(0)}E_{12}P_{12}^{(1)\prime} \\
&+ P_{12}^{(1)}E_{22}P_{12}^{(0)\prime} + P_{12}^{(0)}E_{22}P_{12}^{(1)\prime}
\end{aligned}
\tag{6.9a}
$$

$$
\begin{aligned}
\frac{dP_{12}^{(0)}}{dt} = &-P_{11}^{(1)}A_{12} - P_{12}^{(1)}A_{22} - A_{21}'P_{22}^{(1)} \\
&+ P_{11}^{(1)}E_{12}P_{22}^{(0)} + P_{11}^{(0)}E_{12}P_{22}^{(1)} + P_{12}^{(1)}E_{22}P_{22}^{(0)} \\
&+ P_{12}^{(0)}E_{22}P_{22}^{(1)} + G_{12}^{(0)}
\end{aligned}
\tag{6.9b}
$$

$$\frac{dP_{22}^{(0)}}{dt} = -P_{22}^{(1)}A_{22} - A_{22}'P_{22}^{(1)} + P_{22}^{(1)}E_{22}P_{22}^{(0)}$$
$$+ P_{22}^{(0)}E_{22}P_{22}^{(1)} + G_{22}^{(0)}$$

(6.9c)

Similar equations can be obtained for higher-order approximations. Here $G_{12}^{(0)}$ and $G_{22}^{(0)}$ represent all the terms containing the zeroth-order coefficients. A similar notation is used in the rest of the Chapter.

Equation (6.9) can be solved only if the end condition $P_{11}^{(1)}(t = T)$ is known explicitly. The determination of this value is a vital step in the singular perturbation method and is dealt with later. Again, it is seen that, once $P_{11}^{(1)}(t)$ is evaluated $P_{12}^{(1)}(t)$ and $P_{22}^{(1)}(t)$ are fixed automatically from (6.9b) and (6.9c), respectively.

The outer-series solution (6.8) alone cannot be expected to describe the solution inside the boundary layer, but that is precisely where the final value of $P_{11}^{(1)}(t)$ is to be found. Hence there is need for additional series.

6.2.2   Inner- and intermediate-series solutions

The behaviour of the singularly perturbed matrix Riccati equation (6.4) inside the boundary layer, where the end conditions $P_{12}(T)$ and $P_{22}(T)$ are presumed to have been lost, is examined by using a stretching transformation

$$t' = \frac{(t - T)}{\varepsilon}$$

(6.10)

where t' is the stretched co-ordinate. Using (6.10) in (6.4), the stretched version becomes

$$\frac{dP_{11}(t')}{dt'} = \varepsilon[-P_{11}(t')A_{11} - A_{11}'P_{11}(t') - P_{12}(t')A_{21}$$
$$- A_{21}'P_{12}'(t') + P_{11}(t')E_{11}P_{11}(t')$$
$$+ P_{12}(t')E_{12}'P_{11}(t') + P_{11}(t')E_{12}P_{12}'(t')$$
$$+ P_{12}(t')E_{22}P_{12}'(t') - Q_{11}]$$

(6.11a)

$$\frac{dP_{12}(t')}{dt'} = -P_{11}(t')A_{12} - P_{12}(t')A_{22} - A_{21}'P_{22}(t')$$
$$+ P_{11}(t')E_{12}P_{22}(t') + P_{12}(t')E_{22}P_{22}(t') - Q_{12}$$
$$+ \varepsilon[-A_{11}'P_{12}(t') + P_{11}(t')E_{11}P_{12}(t')$$
$$+ P_{12}(t')E_{12}'P_{12}(t')]$$

(6.11b)

$$\frac{dP_{22}(t')}{dt'} = -P_{22}(t')A_{22} - A_{22}'P_{22}(t') + P_{22}(t')E_{22}P_{22}(t') - Q_{22} \quad (6.11c)$$
$$+ \varepsilon[-P_{12}'(t')A_{12} - A_{12}'P_{12}(t') + P_{12}'(t')E_{12}P_{22}(t')$$
$$+ P_{22}(t')E_{12}'P_{12}(t')] + \varepsilon^2 P_{12}'(t')E_{11}P_{12}(t')$$

The solution of the stretched equation (6.11) is sought in the form of the inner series

$$\bar{P}(t') = \sum_{i=0}^{\infty} \bar{P}^{(i)}(t')\varepsilon^i, \quad P = P_{11}, P_{12}, P_{22} \quad (6.12)$$

Substitution of (6.12) in (6.11) and collection of coefficients of like powers of ε results in a set of equations. For the zeroth-order (ε^0) approximation,

$$\frac{d\bar{P}_{11}^{(0)}}{dt'} = 0 \quad (6.13a)$$

$$\frac{d\bar{P}_{12}^{(0)}}{dt'} = -\bar{P}_{11}^{(0)}A_{12} - \bar{P}_{12}^{(0)}A_{22} - A_{21}'\bar{P}_{22}^{(0)} + \bar{P}_{11}^{(0)}E_{12}\bar{P}_{22}^{(0)} \quad (6.13b)$$
$$+ \bar{P}_{12}^{(0)}E_{22}\bar{P}_{22}^{(0)} - Q_{12}$$

$$\frac{d\bar{P}_{22}^{(0)}}{dt'} = -\bar{P}_{22}^{(0)}A_{22} - A_{22}'\bar{P}_{22}^{(0)} + \bar{P}_{22}^{(0)}E_{22}\bar{P}_{22}^{(0)} - Q_{22} \quad (6.13c)$$

For the first-order approximation

$$\frac{d\bar{P}_{11}^{(1)}}{dt'} = \bar{G}_{11}^{(0)} \quad (6.14a)$$

$$\frac{d\bar{P}_{12}^{(1)}}{dt'} = \bar{P}_{11}^{(1)}A_{12} - \bar{P}_{12}^{(1)}A_{22} - A_{21}'\bar{P}_{22}^{(1)} + \bar{P}_{11}^{(1)}E_{12}\bar{P}_{22}^{(0)} \quad (6.14b)$$
$$+ \bar{P}_{11}^{(0)}E_{12}\bar{P}_{22}^{(1)} + \bar{P}_{12}^{(1)}E_{22}\bar{P}_{22}^{(0)}$$
$$+ \bar{P}_{12}^{(0)}E_{22}\bar{P}_{22}^{(1)} + \bar{G}_{12}^{(0)}$$

$$\frac{d\bar{P}_{22}^{(1)}}{dt'} = -\bar{P}_{22}^{(1)}A_{22} - A_{22}'\bar{P}_{22}^{(1)} + \bar{P}_{22}^{(1)}E_{22}\bar{P}_{22}^{(0)} \quad (6.14c)$$
$$+ \bar{P}_{22}^{(0)}E_{22}\bar{P}_{22}^{(1)} + \bar{G}_{22}^{(0)}$$

where $\bar{G}_{11}^{(0)}$, $\bar{G}_{12}^{(0)}$ and $\bar{G}_{22}^{(0)}$ represent all the expressions with zeroth-order coefficients.

The above equations (6.13) and (6.14) have the conditions

$$\bar{P}^{(0)}(t' = 0) = P(t = T), \quad (6.15a)$$

$$\bar{P}^{(i)}(t' = 0) = 0, \qquad i > 0 \tag{6.15b}$$

The problem of determining the values $P^{(i)}(t)$, $i > 0$, is not yet resolved. In order to do so, we use the intermediate series

$$\underline{P}(t') = \sum_{i=0}^{\infty} \underline{P}^{(i)}(t')\varepsilon^i, \quad P = P_{11}, P_{12}, P_{22} \tag{6.16}$$

Insertion of (6.16) in (6.11) and collection of coefficients of like powers of ε results in a set of equations. For the zeroth order

$$\frac{d\underline{P}_{11}^{(0)}}{dt'} = 0 \tag{6.17a}$$

$$\frac{d\underline{P}_{12}^{(0)}}{dt'} = -\underline{P}_{11}^{(0)} A_{12} - \underline{P}_{12}^{(0)} A_{22} - A_{21}' \underline{P}_{22}^{(0)} + \underline{P}_{11}^{(0)} E_{12} \underline{P}_{22}^{(0)} \tag{6.17b}$$
$$+ \underline{P}_{12}^{(0)} E_{22} \underline{P}_{22}^{(0)} - Q_{12}$$

$$\frac{d\underline{P}_{22}^{(0)}}{dt'} = -\underline{P}_{22}^{(0)} A_{22} - A_{22}' \underline{P}_{22}^{(0)} + \underline{P}_{22}^{(0)} E_{22} \underline{P}_{22}^{(0)} - Q_{22} \tag{6.17c}$$

For the first-order approximation,

$$\frac{d\underline{P}_{11}^{(1)}}{dt'} = \underline{G}_{11}^{(0)} \tag{6.18a}$$

$$\frac{d\underline{P}_{12}^{(1)}}{dt'} = \underline{P}_{11}^{(0)} A_{12} - \underline{P}_{12}^{(1)} A_{22} - A_{21}' \underline{P}_{22}^{(1)} + \underline{P}_{11}^{(1)} E_{12} \underline{P}_{22}^{(0)} \tag{6.18b}$$
$$+ \underline{P}_{11}^{(0)} E_{12} \underline{P}_{22}^{(1)} + \underline{P}_{12}^{(1)} E_{22} \underline{P}_{22}^{(0)}$$
$$+ \underline{P}_{12}^{(0)} E_{22} \underline{P}_{22}^{(1)}$$
$$+ \underline{G}_{12}^{(0)}$$

$$\frac{d\underline{P}_{22}^{(1)}}{dt'} = -\underline{P}_{22}^{(1)} A_{22} - A_{22}' \underline{P}_{22}^{(1)} + \underline{P}_{22}^{(1)} E_{22} \underline{P}_{22}^{(0)} \tag{6.18c}$$
$$+ \underline{P}_{22}^{(0)} E_{22} \underline{P}_{22}^{(1)} + \underline{G}_{22}^{(0)}$$

The above equations (6.17) and (6.18) have the same subsidiary conditions as those of the outer-series (6.6) and (6.9). That is

$$\underline{P}^{(i)}(t' = 0) = P^{(i)}(t = T), \quad i \geqslant 0 \tag{6.19}$$

The end conditions $P_{11}^{(i)}(t = T)$, $i > 0$, are evaluated using the property that the contribution due to $\bar{P}^{(i)}(t') - \underline{P}^{(i)}(t')$ should be negligible as the stretched co-ordinate t' tends to $-\infty$ (Vasileva, 1963). That is

$$P_{11}^{(i)}(t = T) = \int_0^{-\infty} \left(\frac{\mathrm{d}\bar{P}_{11}^{(i)}}{\mathrm{d}t'} - \frac{\mathrm{d}\underline{P}_{11}^{(i)}}{\mathrm{d}t'} \right) \mathrm{d}t' \tag{6.20}$$

The most important feature of Vasileva's method is that the intermediate series coefficients (6.16) need not actually be solved from the corresponding equations (6.17) and (6.18), but can be formulated or generated as polynomials in the stretched co-ordinate t' as

$$\underline{P}^{(0)}(t') = P^{(0)}(T) \tag{6.21a}$$

$$\underline{P}^{(1)}(t') = P^{(1)}(T) + \dot{P}^{(0)}(T)t' \tag{6.21b}$$

$$\underline{P}^{(2)}(t') = P^{(2)}(T) + \dot{P}^{(1)}(T)t' + \ddot{P}^{(0)}(T)t'^2/2! \tag{6.21c}$$

where the 'dot' denotes differentiation with respect to t.

6.2.3 Main results

The two main results in Vasileva's method are concerned with degeneration and the asymptotic series expansions for the solutions. These are given as two theorems after stating briefly the required conditions.

(i) The boundary layer system for (6.4) is defined by

$$\begin{aligned} \frac{\mathrm{d}L_{12}(t')}{\mathrm{d}t'} &= -P_{11}(t)A_{12} + L_{12}(t')\left[E_{22}L_{22}(t') - A_{22}\right] \\ &\quad + [P_{11}(t)E_{12} - A'_{21}]L_{22}(t') - Q_{12} \end{aligned}$$
$$\tag{6.22a}$$

$$\frac{\mathrm{d}L_{22}(t')}{\mathrm{d}t'} = -L_{22}(t')A_{22} - A'_{22}L_{22}(t') + L_{22}(t')E_{22}L_{22}(t') - Q_{22} \tag{6.22b}$$

where the independent variable is t' and t is considered as a fixed parameter. Note that (6.22) can be viewed as being obtained from the stretched equation (6.11) after making $\varepsilon = 0$.

(ii) The system (6.1) is boundary-layer controllable if

$$\text{rank}\,[B_2, A_{22}B_2, \ldots, A_{22}^{n-1}B_2] = n \tag{6.23a}$$

(iii) The system (6.1) is boundary-layer observable, if

$$\text{rank}\,[C', A'_{22}C', \ldots, A_{22}^{n-1}C'] = n \tag{6.23b}$$

where C is the solution of $C'C = Q_{22}$.

Theorem 6.1
If the above conditions (i)–(iii) are satisfied, then the following convergence relations between the full (high-order) and the degenerate (low-order) solutions are valid:

$$\lim_{\varepsilon \to 0} [P_{11}(t,\varepsilon)] = P_{11}^{(0)}(t), \quad 0 \leqslant t \leqslant T \tag{6.24a}$$

$$\lim_{\varepsilon \to 0} [P_{12}(t,\varepsilon)] = P_{12}^{(0)}(t), \quad 0 \leqslant t < T \tag{6.24b}$$

$$\lim_{\varepsilon \to 0} [P_{22}(t,\varepsilon)] = P_{22}^{(0)}(t), \quad 0 \leqslant t < T \tag{6.24c}$$

This is similar to Tikhonov's theorem (Tikhonov, 1952) for initial-value problems. This aspect has been studied in detail by Kokotovic and Yackel (1972).

Theorem 6.2
If the conditions (i)–(iii) are satisfied, then the asymptotic series solutions are given by

$$P(t,\varepsilon) = \sum_{i=0}^{j} [P^{(i)}(t + \bar{P}^{(i)}(t') - \underline{P}^{(i)}(t')]\varepsilon^i + 0(\varepsilon^{j+1}) \tag{6.25}$$

where $P = P_{11}, P_{12}$ and P_{22}.

   More details are given in Yackel and Kokotovic (1973).

   Having found the asymptotic expansions for $P(t,\varepsilon)$ from (6.25), the optimal control (6.3a) is found as

$$u(t,\varepsilon) = -R^{-1}[(B_1'P_{11}(t,\varepsilon) + B_2'P_{12}'(t,\varepsilon)]x + [\varepsilon B_1'P_{12}(t,\varepsilon) + B_2'P_{22}(t,\varepsilon))z] \tag{6.26}$$

In many practical applications we are mainly interested in designing a feedback controller using singular-perturbation methodology to relieve the difficulties of dimensionality and stiffness in solving the matrix Riccati differential equation. Hence we shall be content with the evaluation of closed-loop optimal control (6.26) for the original system (6.1). The various Riccati coefficients in (6.26) are evaluated using the main result (6.25).

   Using (6.25) in the original system (6.1), we can find the asymptotic expansions for the states x and z, and hence the control u. This aspect is not of present interest and hence omitted here, but can be obtained from O'Malley (1972a,b,c), O'Malley and Kung (1974) and O'Malley (1975a).

Example 6.1*
For the sake of simplicity and clarity in demonstrating the application of the

---

* Reprinted with permission from NAIDU, D. S., and RAJAGOPALAN, P.K. (1980): 'Singular perturbation method for a closed-loop optimal control problem', *IEE Proc.*, *Pt.D.*, **127**, pp. 1–6.

method and for obtaining some explicit solutions for the various series coefficients, a second-order system is considered. In addition, by means of this example, the algorithm indicating the sequence of steps involved in the method up to second-order approximation is presented.

Consider a singularly perturbed linear system

$$\frac{dx}{dt} = z, \quad x(t=0) = x(0) \tag{6.27a}$$

$$\varepsilon \frac{dz}{dt} = -x - z + u, \quad z(t=0) = z(0) \tag{6.27b}$$

and the minimisation of a performance index

$$J = \frac{1}{2} \int_0^T (x^2 + u^2) dt \tag{6.28}$$

with free-end-point conditions.

The optimal control (6.26) is given by

$$u = -p_{12}x - p_{22}z \tag{6.29}$$

where p_{11}, p_{12} and p_{22} are the positive definite solutions of the singularly perturbed Riccati equation (6.4). That is

$$\frac{dp_{11}}{dt} = p_{12}^2 + 2p_{12} - 1, \quad p_{11}(t=T) = 0 \tag{6.30a}$$

$$\varepsilon \frac{dp_{12}}{dt} = p_{12} + p_{22} + p_{12}p_{22} - p_{11}, \quad p_{12}(t=T) = 0 \tag{6.30b}$$

$$\varepsilon \frac{dp_{22}}{dt} = p_{22}^2 + 2p_{22} - 2\varepsilon p_{12}, \quad p_{22}(t=T) = 0 \tag{6.30c}$$

The following is the sequence of steps in the application of the method.

(i) *Zeroth-order approximation*
Step 1: The zeroth-order coefficients of the outer series corresponding to (6.26) are governed by

$$\frac{dp_{11}^{(0)}}{dt} = p_{12}^{(0)^2} + 2p_{12}^{(0)} - 1, \quad p_{11}^{(0)}(T) = 0 \tag{6.31a}$$

$$0 = p_{12}^{(0)} + p_{22}^{(0)} + p_{12}^{(0)}p_{22}^{(0)} - p_{11}^{(0)}, \quad p_{12}^{(0)}(T) \neq 0 \tag{6.31b}$$

$$0 = p_{22}^{(0)^2} + 2p_{22}^{(0)}, \quad p_{22}^{(0)}(T) \neq 0 \tag{6.31c}$$

Simplifying,

$$\frac{dp_{11}^{(0)}}{dt} = p_{11}^{(0)^2} - 2p_{11}^{(0)} - 1 \tag{6.32a}$$

$$p_{12}^{(0)} = p_{11}^{(0)}; \quad p_{22}^{(0)} = 0 \tag{6.32b}$$

Step 2: The zeroth-order coefficients of the inner series are described by (6.13). That is

$$\frac{d\bar{p}_{11}^{(0)}}{dt'} = 0, \quad \bar{p}_{11}^{(0)}(t' = 0) = 0 \tag{6.33a}$$

$$\frac{d\bar{p}_{12}^{(0)}}{dt'} = \bar{p}_{12}^{(0)} + \bar{p}_{22}^{(0)} + \bar{p}_{12}^{(0)}\bar{p}_{22}^{(0)} - \bar{p}_{11}^{(0)}, \quad \bar{p}_{12}^{(0)}(t' = 0) = 0 \tag{6.33b}$$

$$\frac{d\bar{p}_{22}^{(0)}}{dt'} = \bar{p}_{22}^{(0)^2} + 2\bar{p}_{22}^{(0)}, \quad \bar{p}_{22}^{(0)}(t' = 0) = 0 \tag{6.33c}$$

Solving,

$$\bar{p}_{11}^{(0)}(t') = \bar{p}_{12}^{(0)}(t') = \bar{p}_{22}^{(0)}(t') = 0 \tag{6.34}$$

Step 3: The equation governing the intermediate-series coefficients is given by (6.17). This is identical in form to (6.33) with the conditions $\underline{p}_{11}^{(0)}$ $(t' = 0) = \underline{p}_{12}^{(0)}(t' = 0) = \underline{p}_{22}^{(0)}(t' = 0) = 0$. Solving it or formulating as per (6.21), we have

$$\underline{p}_{11}^{(0)}(t') = \underline{p}_{12}^{(0)}(t') = \underline{p}_{22}^{(0)}(t') = 0 \tag{6.35}$$

(ii) *First-order approximation*
Step 4: The final condition $\bar{p}_{11}^{(1)}(t = T)$ evaluated from (6.20) is easily seen to be

$$p_{11}^{(1)}(T) = 0 \tag{6.36}$$

Step 5: The outer-series equation corresponding to (6.9) is

$$\frac{dp_{11}^{(1)}}{dt} = 2(p_{11}^{(0)} + 1) + p_{11}^{(1)} + 2(p_{11}^{(0)^2} - 1) \tag{6.37a}$$

$$p_{12}^{(1)} = p_{11}^{(1)} + p_{11}^{(0)} - 1 \tag{6.37b}$$

$$p_{22}^{(1)} = p_{11}^{(0)} \tag{6.37c}$$

Step 6: The first-order inner-series coefficients, described by (6.13), are given by

$$\frac{d\bar{p}_{11}^{(1)}}{dt'} = \bar{p}_{12}^{(0)^2} + 2\bar{p}_{12}^{(0)} - 1, \quad \bar{p}_{11}^{(1)}(t' = 0) = 0 \tag{6.38a}$$

$$\frac{d\bar{p}_{12}^{(1)}}{dt'} = \bar{p}_{12}^{(1)} + \bar{p}_{22}^{(1)} + \bar{p}_{12}^{(1)}\bar{p}_{22}^{(0)} + \bar{p}_{12}^{(0)}\bar{p}_{22}^{(1)} - \bar{p}_{11}^{(1)}, \quad \bar{p}_{12}^{(1)}(t' = 0) = 0 \tag{6.38b}$$

$$\frac{d\bar{p}_{22}^{(1)}}{dt'} = 2\bar{p}_{22}^{(0)}\bar{p}_{22}^{(1)} + 2\bar{p}_{22}^{(1)} - 2\bar{p}_{12}^{(0)}, \quad \bar{p}_{22}^{(1)}(t' = 0) = 0 \tag{6.38c}$$

Solving

$$\bar{p}_{11}^{(1)}(t') = -t', \quad \bar{p}_{12}^{(1)}(t') = -1 - t' + \exp(t') \tag{6.39a}$$

$$\bar{p}_{22}^{(1)}(t') = 0 \tag{6.39b}$$

Step 7: The first-order intermediate-series coefficients are governed by (6.18). This is identical in form to (6.38) with the conditions $\underline{p}^{(1)}(t' = 0) = p^{(1)}(T)$. Solving it or formulating as per (6.21), we get

$$\underline{p}_{11}^{(1)}(t') = -t'; \quad \underline{p}_{12}^{(1)}(t') = -1 - t'; \quad \underline{p}_{22}^{(1)}(t') = 0 \tag{6.40}$$

(iii) *Second-order approximation*
Step 8: The end condition $p_{11}^{(2)}(t = T)$, as evaluated from (6.20), is

$$p_{11}^{(2)}(T) = -2 \tag{6.41}$$

Step 9: The second-order outer-series coefficients are obtained by solving

$$\frac{dp_{11}^{(2)}}{dt'} = 2(p_{11}^{(0)} + 1)(p_{11}^{(2)} + p_{11}^{(1)} + 2 \cdot 5 p_{11}^{(0)} - 1 \cdot 5) \tag{6.42a}$$
$$+ (p_{11}^{(1)} + p_{11}^{(0)} - 1)^2$$

$$p_{12}^{(2)} = p_{11}^{(2)} + p_{11}^{(1)} + 2 \cdot 5 p_{11}^{(0)} - 1 \cdot 5 \tag{6.42b}$$

$$p_{22}^{(2)} = p_{11}^{(1)} + 2 p_{11}^{(0)} - 1 \cdot 5 \tag{6.42c}$$

Step 10: The second-order inner-series coefficients are found to be

$$\bar{p}_{11}^{(2)}(t') = -2 - 2t' - t'^2 + 2\exp(t') \tag{6.43a}$$

$$\bar{p}_{12}^{(2)}(t') = -3 \cdot 5 - 3t' - t'^2 - 0 \cdot 5 \exp(2t') + 4\exp(t') \tag{6.43b}$$

$$\bar{p}_{22}^{(2)}(t') = -1 \cdot 5 - t' - 0 \cdot 5 \exp(2t') + 2\exp(t') \tag{6.43c}$$

Step 11: The second-order intermediate-series coefficients are either solved from the corresponding equations or formulated as given by (6.21). That is

$$\underline{p}_{11}^{(2)}(t') = -2 - 2t' - t'^2 \tag{6.44a}$$

$$\underline{p}_{12}^{(2)}(t') = -3 \cdot 5 - 3t' - t'^2 \tag{6.44b}$$

$$\underline{p}_{22}^{(2)}(t') = -1 \cdot 5 - t' \tag{6.44c}$$

Step 12: The total series solution is obtained from (6.25) up to the desired second-order approximation.

Figure 6.1 shows the effect of degeneration for the Riccati coefficients p_{11}, p_{12} and p_{22} for the two cases of without ($s_{11} = 0$) and with ($s_{11} = 1 \cdot 0$) terminal cost. For the case with terminal cost we see clearly how the degeneration affects the final conditions, in particular that of p_{12}, thus illustrating theorem 6.1 given by (6.2a).

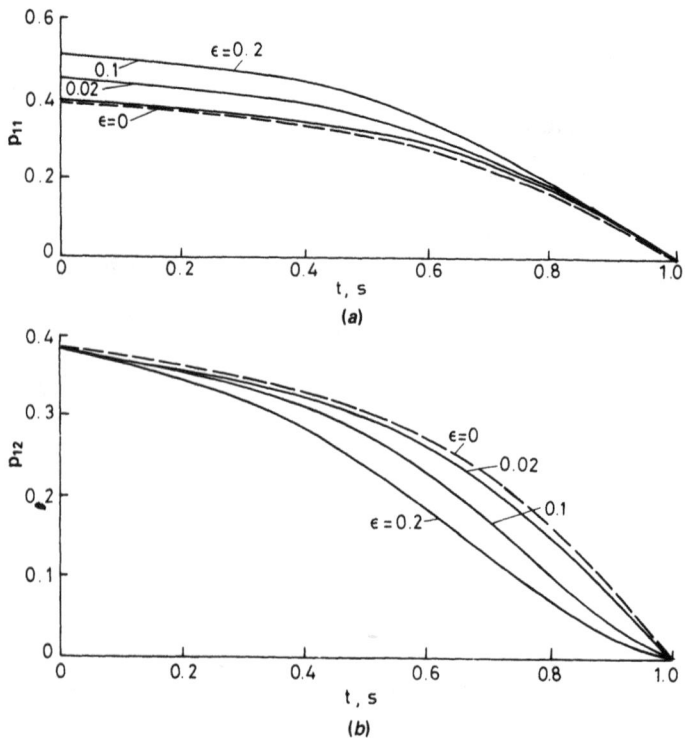

Fig. 6.1a and b   *Degeneration of solutions of $p_{11}(t)$ and $p_{12}(t)$ of Example 6.1*
$s_{11} = 0$

————— solution for $\varepsilon \neq 0$
– – – – – – solution for $\varepsilon = 0$

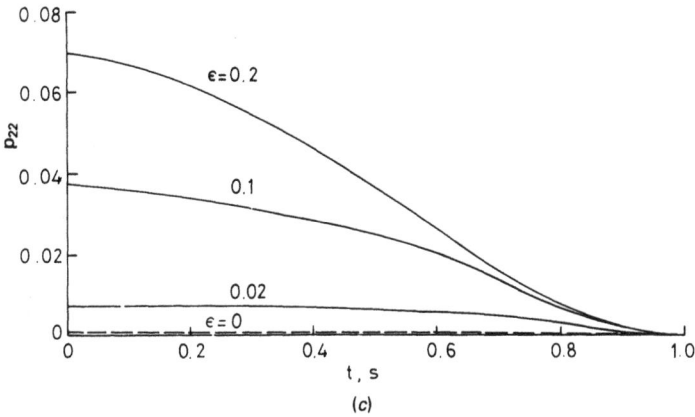

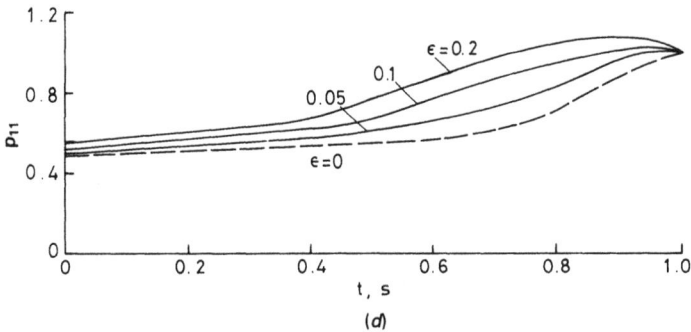

Fig. 6.1c and d *Degeneration of solutions of $p_{22}(t)$ and $p_{11}(t)$ of Example 6.1*
$c\,s_{11} = 0;\ d\,s_{11} = 1\cdot0$
————— solution for $\varepsilon \neq 0$
– – – – – – solution for $\varepsilon = 0$

Figure 6.2 shows the various approximate Riccati solutions along with their exact solutions for $\varepsilon = 0\cdot2$. Using these Riccati solutions in the control (6.29), the response of the system (6.27) can be evaluated.

6.3 Steady-state regulator problem

The steady-state regulator problem is an important case of the free-end-point regulator problem defined by (6.1) and (6.2). Under steady-state conditions the terminal time $T = \infty$ and the performance index (6.2) does not contain the terminal cost. That is

$$J = \frac{1}{2}\int_0^\infty (y'Qy + u'Ru)\mathrm{d}t \qquad (6.45)$$

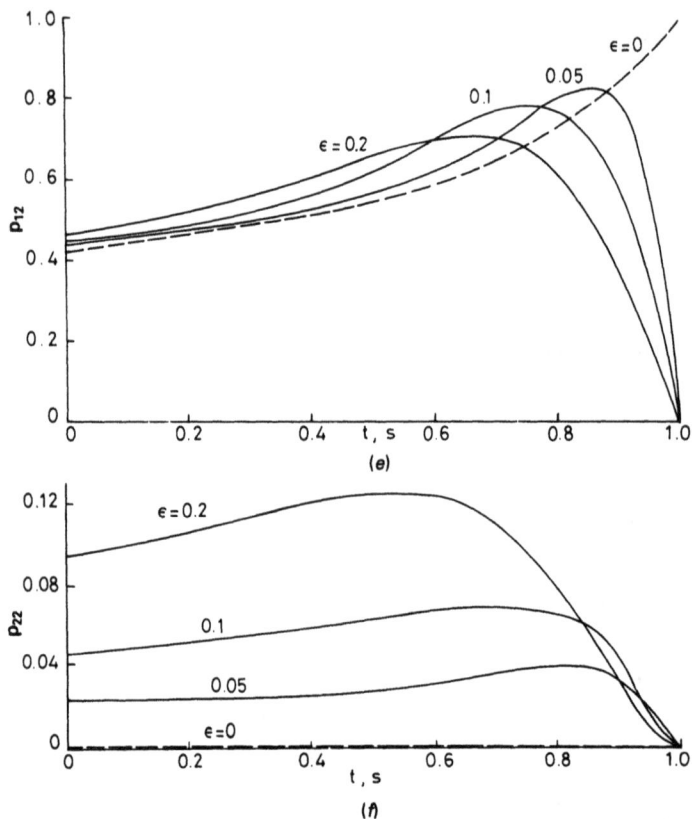

Fig. 6.1e and f *Degeneration of solutions of $p_{12}(t)$ and $p_{22}(t)$ of Example 6.1*
$s_{11} = 1 \cdot 0$
—————— solution for $\varepsilon \neq 0$
- - - - - - solution for $\varepsilon = 0$

It is well known that, with $T = \infty$, the matrix Riccati differential equation (6.4) becomes an algebraic equation (Athans and Flab, 1966):

$$-P_{11}A_{11} - A'_{11}P_{11} - P_{12}A_{21} - A'_{21}P'_{12} + P_{11}E_{11}P_{11} + P_{12}E'_{12}P_{11}$$

$$+ P_{11}E_{12}P'_{12} + P_{12}E_{22}P'_{12} - Q_{11} = 0 \tag{6.46a}$$

$$-P_{11}A_{12} - P_{12}A_{22} - A'_{21}P_{22} + P_{11}E_{12}P_{22} + P_{12}E_{22}P_{22}$$

$$- Q_{12} + \varepsilon(-A'_{11}P_{12} + P'_{12}E_{11}P_{12} + P_{12}E'_{12}P_{22}) = 0 \tag{6.46b}$$

$$-P_{22}A_{22} - A'_{22}P_{22} + P_{22}E_{22}P_{22} - Q_{22} + \varepsilon(-P'_{12}A_{12} - A'_{12}P_{12}$$

$$+ P'_{12}E_{12}P_{22} + P_{22}E'_{12}P_{12}) + \varepsilon^2 P'_{12}E_{11}P_{12} = 0 \tag{6.46c}$$

Since (6.46) is an algebraic equation instead of a differential equation, the asymptotic series solution (6.25) will be modified to contain the outer series

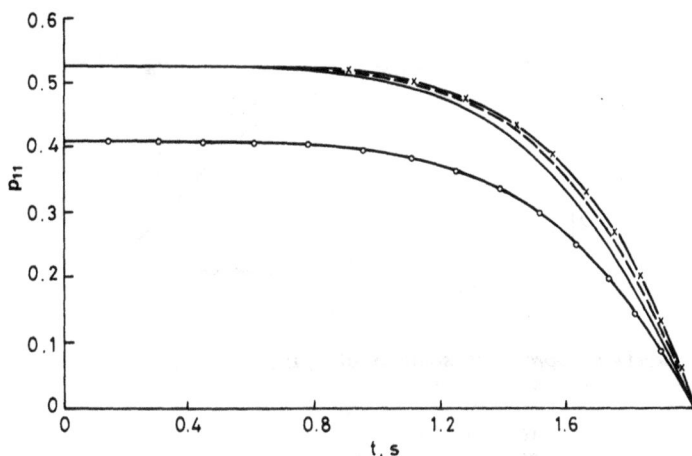

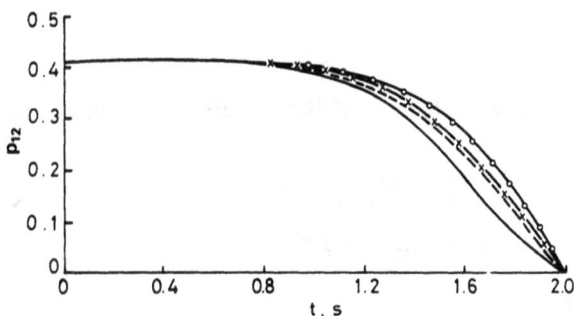

Fig. 6.2a and b *Exact and approximate solutions of $p_{11}(t)$ and $p_{12}(t)$ of Example 6.1*
———————— exact solution
o——o——o zeroth-order solution
×——×——× first-order solution
– – – – – – second-order solution

coefficients only. The inner and intermediate series do not exist, as there is no loss of end conditions and hence no need to recover them. Thus

$$P(\varepsilon) = \sum_{i=0}^{j} P^{(i)}\varepsilon^i + 0(\varepsilon^{j+1}) \qquad (6.47)$$

The rest of the procedure is identical to that of the previous case where $T \neq \infty$. However, we provide the corresponding equations. Substituting (6.47) in

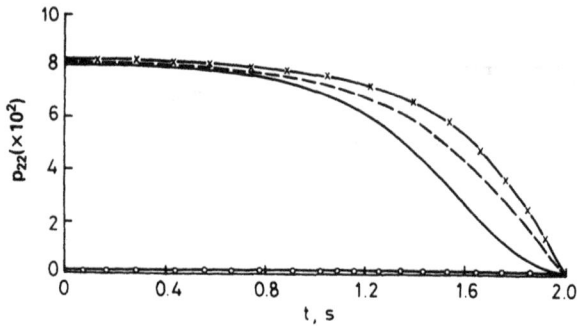

Fig. 6.2c *Exact and approximate solutions of $p_{22}(t)$ of Example 6.1*
————————— exact solution
o——o——o zeroth-order solution
×——×——× first-order solution
– – – – – – second-order solution

(6.46) and collecting coefficients of like powers of ε, we get for the zeroth-order approximation

$$-P_{11}^{(0)}A_{11} - A_{11}'P_{11}^{(0)} - P_{12}^{(0)}A_{21} - A_{21}'P_{12}^{(0)'}$$
$$+ P_{11}^{(0)}E_{11}P_{11}^{(0)} + P_{12}^{(0)}E_{12}'P_{11}^{(0)} + P_{11}^{(0)}E_{12}P_{12}^{(0)'}$$
$$+ P_{12}^{(0)}E_{22}P_{12}^{(0)'} - Q_{11} = 0 \tag{6.48a}$$

$$-P_{11}^{(0)}A_{12} - P_{12}^{(0)}A_{22} - A_{21}'P_{22}^{(0)} + P_{11}^{(0)}E_{12}P_{22}^{(0)}$$
$$+ P_{12}^{(0)}E_{22}P_{22}^{(0)} - Q_{12} = 0 \tag{6.48b}$$

$$-P_{22}^{(0)}A_{22} - A_{22}'P_{22}^{(0)} + P_{22}^{(0)}E_{22}P_{22}^{(0)} - Q_{22} = 0 \tag{6.48c}$$

Note that (6.48c) is independent of $P_{11}^{(0)}$ and $P_{12}^{(0)}$, and $P_{22}^{(0)}$ is its positive definite solution. Using this value of $P_{22}^{(0)}$ in (6.48b), the relation between $P_{11}^{(0)}$ and $P_{12}^{(0)}$ is obtained. Now (6.48a) can be solved for $P_{11}^{(0)}$. Once $P_{11}^{(0)}$ is known $P_{12}^{(0)}$ can be obtained. This procedural sequence is to be adopted for first-order and higher-order approximations. Thus for first-order approximation

$$-P_{11}^{(1)}A_{11} - A_{11}'P_{11}^{(1)} - P_{12}^{(1)}A_{21} - A_{21}'P_{12}^{(1)'}$$
$$+ P_{11}^{(1)}E_{11}P_{11}^{(0)} + P_{11}^{(0)}E_{11}P_{11}^{(1)} + P_{12}^{(1)}E_{12}'P_{11}^{(0)} + P_{12}^{(0)}E_{12}'P_{11}^{(1)} + P_{11}^{(1)}E_{12}P_{12}^{(0)'}$$
$$+ P_{11}^{(0)}E_{12}P_{12}^{(1)'} + P_{12}^{(1)}E_{22}P_{12}^{(0)'} + P_{12}^{(0)}E_{22}P_{12}^{(1)'} = 0 \tag{6.49a}$$

$$-P_{11}^{(1)}A_{12} - P_{12}^{(1)}A_{22} - A_{21}'P_{22}^{(1)} + P_{11}^{(1)}E_{12}P_{22}^{(0)}$$
$$+ P_{11}^{(0)}E_{12}P_{22}^{(1)} + P_{12}^{(1)}E_{22}P_{22}^{(0)} + P_{12}^{(0)}E_{22}P_{22}^{(1)} + G_{12}^{(0)} = 0 \tag{6.49b}$$

$$-P_{22}^{(1)}A_{22} - A_{22}'P_{22}^{(1)} + P_{22}^{(1)}E_{22}P_{22}^{(0)}$$
$$+ P_{22}^{(0)}E_{22}P_{22}^{(1)} + G_{22}^{(0)} = 0 \tag{6.49c}$$

where $G_{12}^{(0)}$ and $G_{22}^{(0)}$ represent all the terms with zeroth-order coefficients. Similar equations are obtained for higher-order approximations.

The control corresponding to (6.26) is given as

$$u(t,\varepsilon) = -R^{-1}[(B_1'P_{11} + B_2'P_{12})x + (\varepsilon B_1'P_{12} + B_2'P_{22})z] \tag{6.50}$$

Example 6.2
In order to bring out clearly the effectiveness of singular perturbation method in analysing high-order systems, a fifth-order voltage regulator problem is considered (Sannuti and Kokotovic, 1969a). The singularly perturbed system is

$$\frac{dx_1}{dt} = -0\cdot2x_1 + 0\cdot5x_2$$
$$\frac{dx_2}{dt} = -0\cdot5x_2 + 1\cdot6z_1$$
$$\varepsilon\frac{dz_1}{dt} = \frac{-10z_1 + 60z_2}{7} \tag{6.51}$$
$$\varepsilon\frac{dz_2}{dt} = -2\cdot5z_2 + 7\cdot5z_3$$
$$\varepsilon\frac{dz_3}{dt} = -z_3 + 3u$$

The cost functional to be minimised is

$$J = \frac{1}{2}\int_0^\infty (x_1^2 + u^2)dt \tag{6.52}$$

The fifth-order system (6.51) results in $5(5 + 1)/2 = 15$ nonlinear algebraic equations corresponding to (6.46). The closed-loop control is given by (6.50). In order to save space we omit the detailed procedure and give the final results regarding the coefficients in (6.47).

For the zeroth-order approximation

$$P_{11}^{(0)} = 10^{-4} \begin{bmatrix} 2\,064\cdot4 & 111\cdot0 \\ 111\cdot0 & 11\cdot5 \end{bmatrix}; \quad P_{12}^{(0)} = 10^{-4} \begin{bmatrix} 124\cdot32 & 426\cdot24 & 3\,197\cdot0 \\ 12\cdot32 & 42\cdot24 & 316\cdot8 \end{bmatrix}$$

$$P_{22}^{(0)} = 0$$

For the first-order approximation

$$P_{11}^{(1)} = 10^{-4} \begin{bmatrix} 20\,550 & 1\,964 \\ 1\,964 & 193 \end{bmatrix}; \quad P_{12}^{(1)} = 10^{-4} \begin{bmatrix} 1\,473 & 3\,614 & 44 \\ 185 & 567 & 3\,077 \end{bmatrix}$$

$$P_{22}^{(1)} = 10^{-4} \begin{bmatrix} 13\cdot8 & 47\cdot31 & 354\cdot8 \\ 47\cdot31 & 162\cdot2 & 1\,218\cdot0 \\ 354\cdot8 & 1\,218\cdot0 & 9\,177\cdot0 \end{bmatrix}$$

For the second-order approximation

$$P_{11}^{(2)} = 10^{-4} \begin{bmatrix} -12\,015 & 8\,170 \\ 8\,170 & 2\,035 \end{bmatrix}; \quad P_{12}^{(2)} = 10^{-4} \begin{bmatrix} 3\,809 & 5\,000 & 541 \\ 1\,463 & 3\,508 & -4\,310 \end{bmatrix}$$

$$P_{22}^{(2)} = 10^{-4} \begin{bmatrix} 167 & 557 & 2\,541 \\ 557 & 1\,643 & 6\,740 \\ 2\,541 & 6\,740 & 12\,600 \end{bmatrix}$$

In order to evaluate the performance of the original system (6.51), the feed-back control (6.50) is used with zeroth-, first- and second-order approximate coefficients and compared with the exact (optimal) feedback coefficients as shown in Fig. 6.3. The time behaviour of the states and control is depicted in Fig. 6.4. The variation of performance index with the small parameter ε is shown in Fig. 6.5. The performance index with different sets of initial conditions is given in Table 6.1.

It may be pointed out that the various approximate coefficients shown above have been evaluated by simple hand calculations without resorting to any numerical method on a digital computer (Sannuti, 1968; Naidu, 1977). On the other hand, the evaluation of the exact (optimal) coefficients from the 15th-order nonlinear algebraic equations like (6.46) definitely needs a numerical method on a digital computer. Also, we see from Fig. 6.5 that the range of the small parameter ε that can be used safely is quite high with the second-order approximation. Note that the zeroth-order approximation is not accep-table as it gives unstable performance.

This is a typical example which clearly shows the advantages of order reduc-tion in the singular perturbation method which provides the way between

Table 6.1   *Performance index with different initial conditions for Example 6.2*

$x_1(0)$	$x_2(0)$	$z_1(0)$	$z_2(0)$	$z_3(0)$	Zeroth order	First order	Second order	Exact solution
−10	10	10	10	10	634 581	17·29	15·00	14·57
−10	0	0	0	10	706 415	23·34	19·94	19·33
−10	10	0	0	10	598 563	18·10	16·02	15·59
−10	10	10	0	10	599 812	17·69	15·64	15·22
−10	0	10	0	10	701 401	22·69	19·42	18·82
−10	0	10	10	10	721 786	22·31	18·89	18·27
−10	0	0	0	0	536 991	26·50	23·13	22·98

(i)

u — $\dfrac{3}{0.2S+1}$ — z_3 — $\dfrac{3}{0.08S+1}$ — z_2 — $\dfrac{6}{0.14S+1}$ — z_1 — $\dfrac{3.2}{2S+1}$ — x_2 — $\dfrac{2.5}{5S+1}$ — x_1

0 | 0 | 0 | 0.0951 | 0.9591

(ii)

u — $\dfrac{3}{0.2S+1}$ — z_3 — $\dfrac{3}{0.08S+1}$ — z_2 — $\dfrac{6}{0.14S+1}$ — z_1 — $\dfrac{3.2}{2S+1}$ — x_2 — $\dfrac{2.5}{5S+1}$ — x_1

0.5515 | 0.0731 | 0.0213 | 0.1135 | 0.9618

(a)

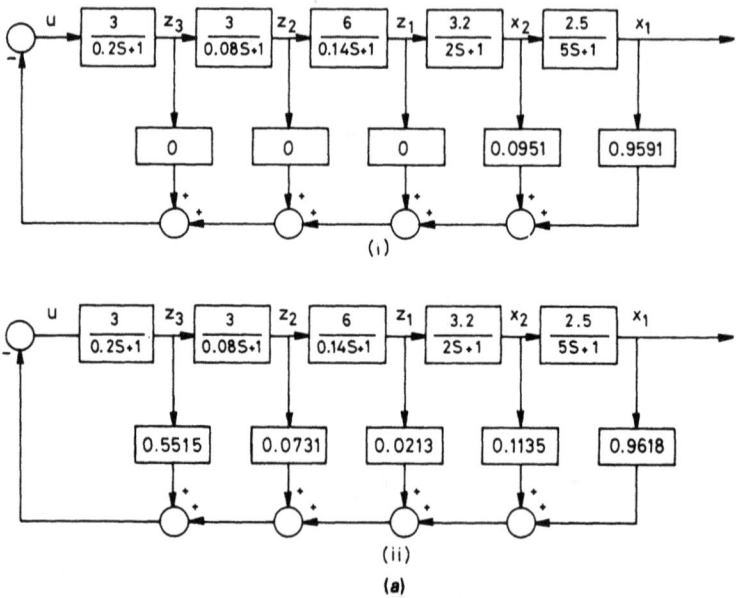

Fig. 6.3a Zeroth- and first-order feedback coefficients for the system of Example 6.2
(i) Zeroth-order solution
(ii) First-order solution

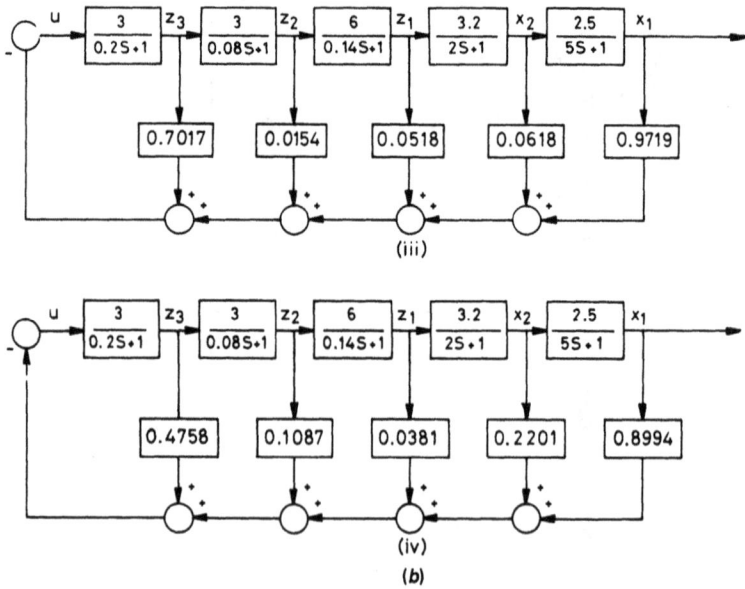

(iii)

u — $\dfrac{3}{0.2S+1}$ — z_3 — $\dfrac{3}{0.08S+1}$ — z_2 — $\dfrac{6}{0.14S+1}$ — z_1 — $\dfrac{3.2}{2S+1}$ — x_2 — $\dfrac{2.5}{5S+1}$ — x_1

0.7017 | 0.0154 | 0.0518 | 0.0618 | 0.9719

(iv)

u — $\dfrac{3}{0.2S+1}$ — z_3 — $\dfrac{3}{0.08S+1}$ — z_2 — $\dfrac{6}{0.14S+1}$ — z_1 — $\dfrac{3.2}{2S+1}$ — x_2 — $\dfrac{2.5}{5S+1}$ — x_1

0.4758 | 0.1087 | 0.0381 | 0.2201 | 0.8994

(b)

Fig. 6.3b Second-order and optimal feedback coefficients for the system of Example 6.2
(iii) Second-order solution
(iv) Optimal solution

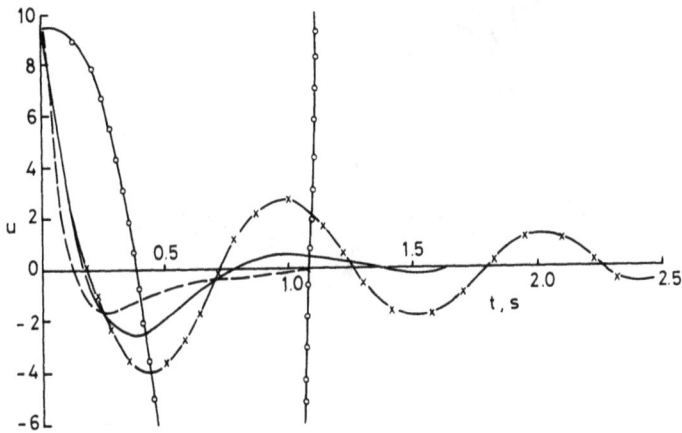

(a)

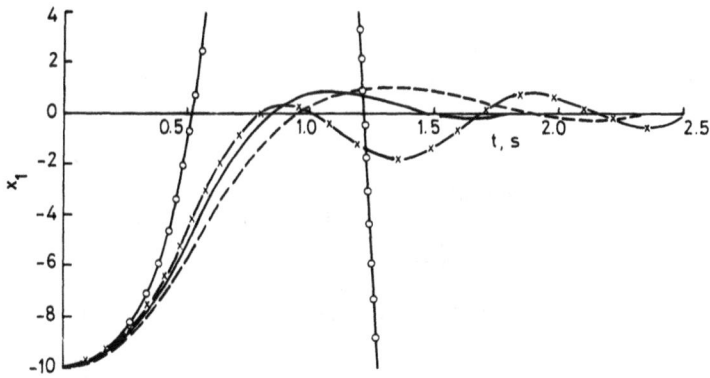

(b)

Fig. 6.4a and b *Exact and approximate solutions of u(t) and $x_1(t)$ of Example 6.2*
$x_1(0) = -10{\cdot}0$; $x_2(0) = z_1(0) = z_2(0) = z_3(0) = 0$; $\varepsilon = 0{\cdot}2$
——————— exact solution
o———o———o zeroth-order solution
×——×——× first-order solution
— — — — — — second-order solution

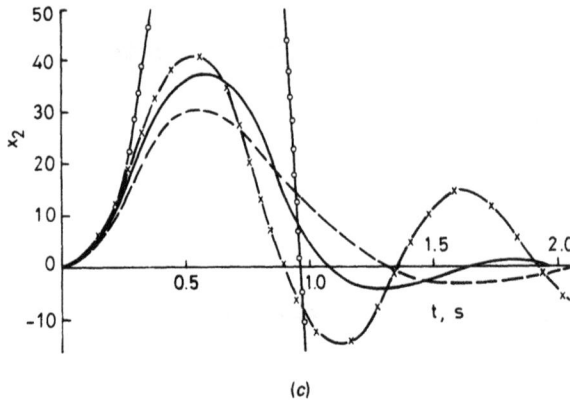

(c)

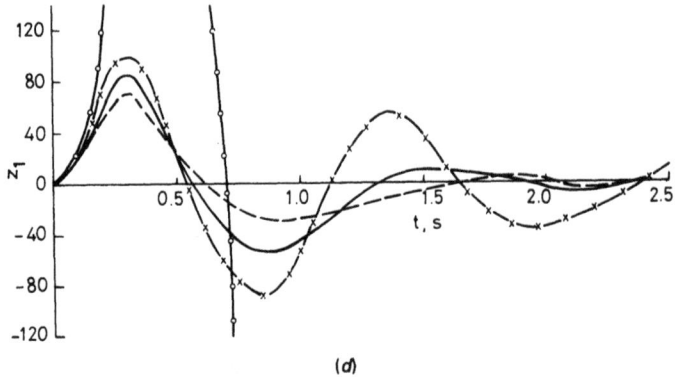

(d)

Fig. 6.4c and d   *Exact and approximate solutions of $x_2(t)$ and $z_1(t)$ of Example 6.2*
$x_1(0) = -10\cdot0;\ x_2(0) = z_1(0) = z_2(0) = z_3(0) = 0;\ \varepsilon = 0\cdot2$
——————— exact solution
o——o——o zeroth-order solution
×--×--× first-order solution
- - - - - - second-order solution

the low-order unstable performance and the high-order optimal performance requiring excessive computational effort.

6.4   Fixed-end-point problem

In the previous Sections, we were mainly concerned with the closed-loop optimal control of a singularly perturbed free-end-point problem, resulting in the normal matrix Riccati equation. On the other hand, in this Section we address the closed-loop optimal control of a singularly perturbed fixed-end-

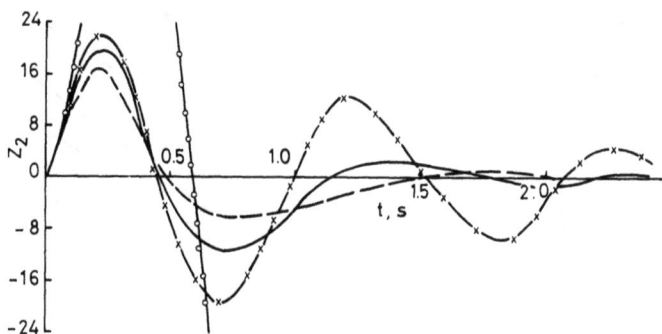

(e)

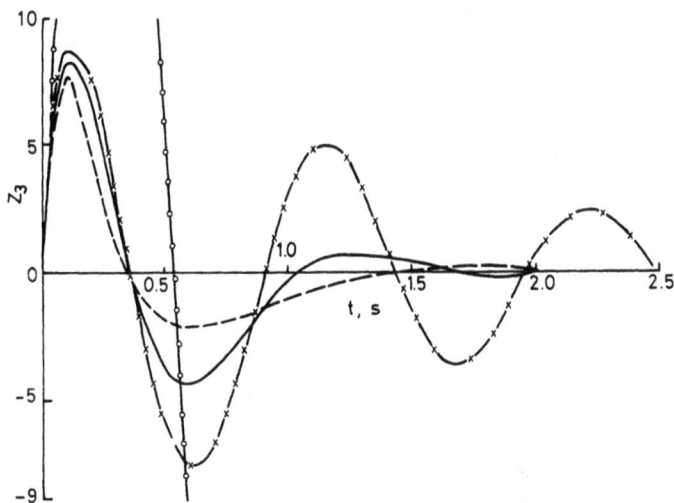

(f)

Fig. 6.4e and f   *Exact and approximate solutions of $z_2(t)$ and $z_3(t)$ of Example 6.2*
$x_1(0) = -10 \cdot 0$; $x_2(0) = z_1(0) = z_2(0) = z_3(0) = 0$; $\varepsilon = 0 \cdot 2$
——————— exact solution
○——○——○ zeroth-order solution
×——×——× first-order solution
— — — — — second-order solution

point problem. This leads to a special matrix Riccati equation (Asatani, 1974).
Based on the original works of Vasileva (1963), we present a singular perturbation method in which the asymptotic series solution is composed of an outer-,
inner-, and intermediate-series solutions. This Section is heavily dependent
on the results of Naidu (1977) and Rajagopalan and Naidu (1980b).

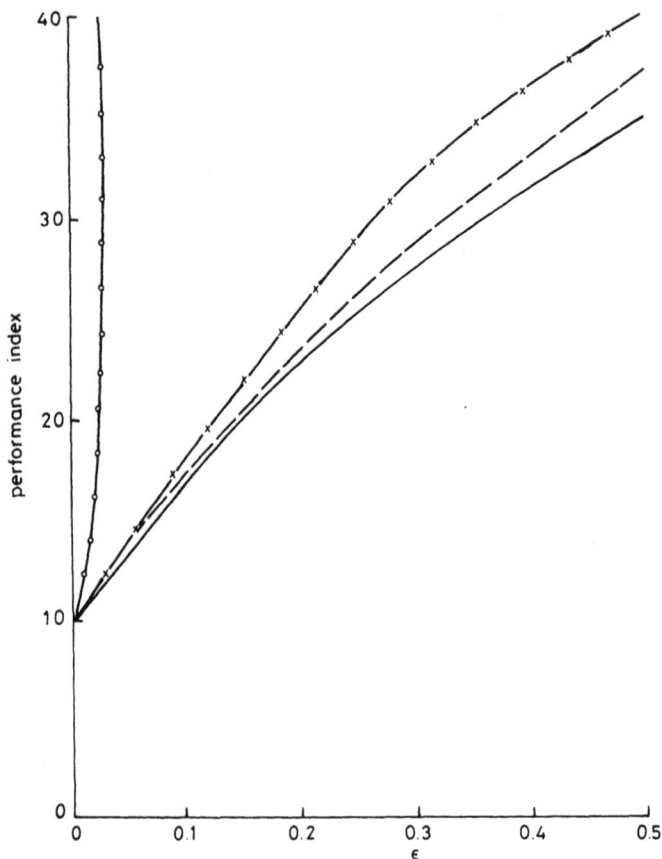

Fig. 6.5 *Performance index of Example 6.2*
$x_1(0) = -10\cdot0$; $x_2(0) = z_1(0) = z_2(0) = z_3(0) = 0\cdot0$
——————— exact solution
○———○———○ zeroth-order solution
×———×———× first-order solution
– – – – – – second-order solution

6.4.1 Problem formulation

Consider the singularly perturbed, linear, time-invariant system (6.1) and the performance index (6.2) without terminal cost ($S = 0$) and the boundary conditions

$$x(t = 0) = x(0); \quad x(t = T) = x(T) \tag{6.53a}$$

$$z(t = 0) = z(0); \quad z(t = T) = z(T) \tag{6.53b}$$

The optimal control is given by

$$u = -R^{-1}B'p \tag{6.54}$$

and the state y and the co-state p satisfy

$$\begin{bmatrix} \dot{y} \\ \dot{p} \end{bmatrix} = \begin{bmatrix} A & -BR^{-1}B' \\ -Q & -A' \end{bmatrix} \begin{bmatrix} y \\ p \end{bmatrix} \qquad (6.55)$$

where

$$A = \begin{bmatrix} A_{11} & A_{12} \\[1mm] \dfrac{A_{21}}{\varepsilon} & \dfrac{A_{22}}{\varepsilon} \end{bmatrix}; \quad B = \begin{bmatrix} B_1 \\[1mm] \dfrac{B_2}{\varepsilon} \end{bmatrix}; \quad Q = \begin{bmatrix} Q_{11} & Q_{12} \\[1mm] Q'_{12} & Q_{22} \end{bmatrix}$$

The canonical system (6.65) under the boundary conditions (6.53) is ill-conditioned; i.e. the boundary conditions on the co-state p are not directly specified (Mufti, 1970). To deal with this situation, a special Riccati transformation is introduced (Asatani, 1974) as

$$y(t) = M(t)p(t) + v(t) \qquad (6.56)$$

where $M(t)$ is an $(m + n) \times (m + n)$-dimensional matrix and $v(t)$ is an $(m + n)$-dimensional vector. Note that transformation (6.56) is different from the normal or conventional Riccati transformation

$$p(t) = \hat{M}(t)y(t) + \hat{v}(t) \qquad (6.57)$$

used for free-end-point problems (Athans and Falb, 1966). By using (6.56) in (6.55) we obtain the matrix differential equation of Riccati type for M and the corresponding differential equation for v as

$$\dot{M} = AM + MA' + MQM - BR^{-1}B' \qquad (6.58)$$

$$\dot{v} = (A + MQ)v \qquad (6.59)$$

Equations (6.58) and (6.59) are solved by using the initial conditions

$$M(t = 0) = 0 \quad \text{and} \quad v(t = 0) = y(0) \qquad (6.60)$$

or using the terminal conditions

$$M(t = T) = 0 \quad \text{and} \quad v(t = T) = y(T) \qquad (6.61)$$

The singularly perturbed structure of (6.1) suggests that the corresponding equations (6.58) and (6.59) should be cast in the singularly perturbed form. This necessitates M and v to have the following partitioned forms with appropriate dimensions:

$$M = \begin{bmatrix} M_{11} & M_{12} \\[1mm] M'_{12} & \dfrac{M_{22}}{\varepsilon} \end{bmatrix}; \quad v = \begin{bmatrix} v_1 \\[1mm] v_2 \end{bmatrix}$$

Then (6.58) and (6.59) become

$$\frac{dM_{11}}{dt} = A_{11}M_{11} + M_{11}A'_{11} + A_{12}M'_{12} + M_{12}A'_{12} + M_{11}Q_{11}M_{11} \quad (6.62a)$$
$$+ M_{12}Q'_{12}M_{11} + M_{11}Q_{12}M'_{12} + M_{12}Q_{22}M'_{12} - E_{11}$$

$$\varepsilon\frac{dM_{12}}{dt} = M_{11}A'_{21} + M_{12}A'_{22} + A_{12}M_{22} + M_{11}Q_{12}M_{22} \quad (6.62b)$$
$$+ M_{12}Q_{22}M_{22} - E_{12} + \varepsilon(A_{11}M_{12} + M_{11}Q_{11}M_{12}$$
$$+ M_{12}Q'_{12}M_{12})$$

$$\varepsilon\frac{dM_{22}}{dt} = A_{22}M_{22} + M_{22}A'_{22} + M_{22}Q_{22}M_{22} - E_{22} + \varepsilon(A_{21}M_{22} \quad (6.62c)$$
$$+ M'_{12}A'_{21} + M_{22}Q_{12}M_{12} + M'_{12}Q_{12}M_{22})$$
$$+ \varepsilon^2 M'_{12}Q_{11}M_{12}$$

$$\frac{dv_1}{dt} = (A_{11} + M_{11}Q_{11} + M_{12}Q'_{12})v_1 + (A_{12} + M_{11}Q_{12} \quad (6.63a)$$
$$+ M_{12}Q_{12})v_2$$

$$\varepsilon\frac{dv_2}{dt} = (A_{21} + M_{22}Q'_{12} + \varepsilon M'_{12}Q_{11})v_1 + (A_{22} + M_{22}Q_{22} \quad (6.63b)$$
$$+ \varepsilon M'_{12}Q_{12})v_2$$

where

$$E_{11} = B_1R^{-1}B_1, \quad E_{22} = B_2R^{-1}B_2, \quad E_{12} = B_1R^{-1}B_2$$

The boundary conditions (6.60) and (6.61) become

$$M_{11}(t=0) = M_{12}(t=0) = M_{22}(t=0) = 0 \quad (6.64a)$$
$$v_1(t=0) = x(0), \quad v_2(t=0) = z(0) \quad (6.64b)$$

or

$$M_{11}(t=T) = M_{12}(t=T) = M_{22}(t=T) = 0 \quad (6.65a)$$
$$v_1(t=T) = x(T), \quad v_2(t=T) = z(T) \quad (6.65b)$$

The system (6.62) and (6.64) is called the 'forward full system', and the system (6.63) and (6.65) is called the 'backward full system'. Here we intend to present the analysis of the forward full system, and then the analysis of the backward full system is straightforward.

6.4.2   Outer series

We know that (6.62) and (6.63) are in the singularly perturbed form since the small parameter ε multiplies some of the state variables. Also, the boundary layer exists at the initial or final point. As a first step we assume the outer series as

$$M(t,\varepsilon) = \sum_{i=0}^{\infty} M^{(i)}(t)\varepsilon^i, \quad M = M_{11}, M_{12}, M_{22} \tag{6.66}$$

$$v(t,\varepsilon) = \sum_{i=0}^{\infty} v^{(i)}(t)\varepsilon^i, \qquad v = v_1, v_2 \tag{6.67}$$

By the substitution of (6.66) and (6.67) in (6.62) and (6.63) and collection of coefficients of like powers of ε on either side, a set of equations is obtained. For the zeroth-order approximation

$$\frac{dM_{11}^{(0)}}{dt} = A_{11}M_{11}^{(0)} + M_{11}^{(0)}A_{11}' + A_{12}M_{12}^{(0)'} + M_{12}^{(0)}A_{12}' \tag{6.68a}$$
$$+ M_{11}^{(0)}Q_{11}M_{11}^{(0)} + M_{12}^{(0)}Q_{12}'M_{11}^{(0)} + M_{11}^{(0)}Q_{12}M_{12}^{(0)'}$$
$$+ M_{12}^{(0)}Q_{22}M_{12}^{(0)'} - E_{11}$$

$$0 = M_{11}^{(0)}A_{21}' + M_{12}^{(0)}A_{22}' + A_{12}M_{22}^{(0)} + M_{11}^{(0)}Q_{12}M_{22}^{(0)} \tag{6.68b}$$
$$+ M_{12}^{(0)}Q_{22}M_{22}^{(0)} - E_{12}$$

$$0 = A_{22}M_{22}^{(0)} + M_{22}^{(0)}A_{22}' + M_{22}^{(0)}Q_{22}M_{22}^{(0)} - E_{22} \tag{6.68c}$$

and

$$\frac{dv_1^{(0)}}{dt} = (A_{11} + M_{11}^{(0)}Q_{11} + M_{12}^{(0)}Q_{12}')v_1^{(0)} \tag{6.69a}$$
$$+ (A_{12} + M_{11}^{(0)}Q_{12} + M_{12}^{(0)}Q_{12})v_2^{(0)}$$

$$0 = (A_{21} + M_{22}^{(0)}Q_{12}')v_1^{(0)} + (A_{22} + M_{22}^{(0)}Q_{22})v_2^{(0)} \tag{6.69b}$$

These are the same as those obtained by letting $\varepsilon = 0$ in (6.62) and (6.63). Hence the system of (6.68) and (6.69) is called the degenerate (low-order) system. These are solved with the conditions

$$\left.\begin{array}{l} M_{11}^{(0)}(t = 0) = M_{11}(0) = 0 \\[2mm] v_1^{(0)}(t = 0) = v_1(0) = x(0) \end{array}\right\} \tag{6.70}$$

Note that $M_{12}^{(0)}(t=0)$, $M_{22}^{(0)}(t=0)$ and $v_2^{(0)}(t=0)$ need not, in general, be equal to $M_{12}(0)$, $M_{22}(0)$ and $v_2(0)$, respectively. These conditions are lost in the process of degeneration. Note that once $M_{11}^{(0)}(t)$ and $v_1^{(0)}(t)$ are determined, $M_{12}^{(0)}(t)$, $M_{22}^{(0)}(t)$ and $v_2^{(0)}(t)$ are automatically fixed by (6.68b), and (6.68c) and (6.69b).

The first-order approximation

$$
\begin{aligned}
\frac{dM_{11}^{(1)}}{dt} =\ & A_{11}M_{11}^{(1)} + M_{11}^{(1)}A_{11}' + A_{12}M_{12}^{(1)\prime} + M_{12}^{(1)}A_{12}' \\
& + M_{11}^{(1)}Q_{11}M_{11}^{(0)} + M_{11}^{(0)}Q_{11}M_{11}^{(1)} + M_{12}^{(1)}Q_{12}'M_{11}^{(0)} \\
& + M_{12}^{(0)}Q_{12}'M_{11}^{(1)} + M_{11}^{(1)}Q_{12}M_{12}^{(0)\prime} + M_{11}^{(0)}Q_{12}M_{12}^{(1)\prime} \\
& + M_{12}^{(1)}Q_{22}M_{12}^{(0)} + M_{12}^{(0)}Q_{22}M_{12}^{(1)\prime}
\end{aligned}
\tag{6.71a}
$$

$$
\begin{aligned}
\frac{dM_{12}^{(0)}}{dt} =\ & M_{11}^{(1)}A_{21}' + M_{12}^{(1)}A_{22}' + A_{12}M_{22}^{(1)} + M_{11}^{(1)}Q_{12}M_{22}^{(0)} \\
& + M_{11}^{(0)}Q_{12}M_{22}^{(1)} + M_{12}^{(1)}Q_{22}M_{22}^{(0)} \\
& + M_{12}^{(0)}Q_{22}M_{22}^{(1)} + G_{12}^{(0)}
\end{aligned}
\tag{6.71b}
$$

$$
\frac{dM_{22}^{(0)}}{dt} = A_{22}M_{22}^{(1)} + M_{22}^{(1)}A_{22}' + M_{22}^{(1)}Q_{22}M_{22}^{(0)} + M_{22}^{(0)}Q_{22}M_{22}^{(1)} + G_{22}^{(0)}
\tag{6.71c}
$$

and

$$
\begin{aligned}
\frac{dv_1^{(1)}}{dt} =\ & (A_{11} + M_{11}^{(0)}Q_{11} + M_{12}^{(0)}Q_{12}')v_1^{(1)} \\
& + (M_{11}^{(1)}Q_{11} + M_{12}^{(1)}Q_{12}')v_1^{(0)} + (A_{12} + M_{11}^{(0)}Q_{12} \\
& + M_{12}^{(0)}Q_{12})v_2^{(1)} + (M_{11}^{(1)}Q_{12} + M_{12}^{(1)}Q_{12})v_2^{(0)}
\end{aligned}
\tag{6.72a}
$$

$$
\begin{aligned}
\frac{dv_2^{(0)}}{dt} =\ & (A_{21} + M_{22}^{(0)}Q_{12}')v_1^{(1)} + M_{22}^{(1)}Q_{12}'v_1^{(0)} \\
& + (A_{22} + M_{22}^{(0)}Q_{22})v_2^{(1)} + M_{22}^{(1)}Q_{22}v_2^{(0)} + G_2^{(0)}
\end{aligned}
\tag{6.72b}
$$

where $G_{12}^{(0)}$, $G_{22}^{(0)}$ and $G_2^{(0)}$ represent terms containing zeroth-order coefficients.

The solutions of (6.71) and (6.72) require the initial conditions $M_{11}^{(1)}(0)$ and $v^{(1)}(0)$. The determination of these values is a vital step in singular perturbation analysis, and this is dealt with later.

6.4.3   Inner and intermediate series

The inner and intermediate series are formed from the stretched system obtained by using the initial transformation

$$t' = \frac{t}{\varepsilon} \tag{6.73}$$

in (6.62) and (6.63). The inner series is

$$\bar{M}(t',\varepsilon) = \sum_{i=0}^{\infty} \bar{M}^{(i)}(t')\varepsilon^i \tag{6.74}$$

$$\bar{v}(t',\varepsilon) = \sum_{i=0}^{\infty} \bar{v}^{(i)}(t')\varepsilon^i \tag{6.75}$$

For the zeroth-order approximation

$$\frac{d\bar{M}_{11}^{(0)}}{dt'} = 0 \tag{6.76a}$$

$$\frac{d\bar{M}_{12}^{(0)}}{dt'} = \bar{M}_{11}^{(0)}A_{21}' + \bar{M}_{12}^{(0)}A_{22}' + A_{12}\bar{M}_{22}^{(0)} \\ + \bar{M}_{11}^{(0)}Q_{12}\bar{M}_{22}^{(0)} + \bar{M}_{12}^{(0)}Q_{22}\bar{M}_{22}^{(0)} - E_{12} \tag{6.76b}$$

$$\frac{d\bar{M}_{22}^{(0)}}{dt'} = A_{22}\bar{M}_{22}^{(0)} + \bar{M}_{22}^{(0)}A_{22}' + \bar{M}_{22}^{(0)}Q_{22}\bar{M}_{22}^{(0)} - E_{22} \tag{6.76c}$$

$$\frac{d\bar{v}_1^{(0)}}{dt'} = 0 \tag{6.77a}$$

$$\frac{d\bar{v}_2^{(0)}}{dt'} = (A_{21} + \bar{M}_{22}^{(0)}Q_{12}')\bar{v}_1^{(0)} + (A_{22} + \bar{M}_{22}^{(0)}Q_{22})\bar{v}_2^{(0)} \tag{6.77b}$$

For the first-order approximation

$$\frac{d\bar{M}_{11}^{(1)}}{dt'} = \bar{F}_{11}^{(0)} \tag{6.78a}$$

$$\frac{d\bar{M}_{12}^{(1)}}{dt'} = \bar{M}_{11}^{(1)}A_{21}' + \bar{M}_{12}^{(1)}A_{22}' + A_{12}\bar{M}_{22}^{(1)} \\ + \bar{M}_{11}^{(1)}Q_{12}\bar{M}_{22}^{(0)} + \bar{M}_{11}^{(0)}Q_{12}\bar{M}_{22}^{(0)} \\ + \bar{M}_{12}^{(1)}Q_{22}\bar{M}_{22}^{(0)} + \bar{M}_{12}^{(0)}Q_{22}\bar{M}_{22}^{(1)} + \bar{F}_{12}^{(0)} \tag{6.78b}$$

$$\frac{d\bar{M}_{22}^{(1)}}{dt'} = A_{22}\bar{M}_{22}^{(1)} + \bar{M}_{22}^{(1)}A_{22}' + \bar{M}_{22}^{(1)}Q_{22}\bar{M}_{22}^{(0)} \\ + \bar{M}_{22}^{(0)}Q_{22}\bar{M}_{22}^{(1)} + \bar{F}_{22}^{(0)} \tag{6.78c}$$

and

$$\frac{d\bar{v}_1^{(1)}}{dt'} = \bar{F}_1^{(0)} \tag{6.79a}$$

$$\frac{d\bar{v}_2^{(1)}}{dt'} = (A_{21} + \bar{M}_{22}^{(0)}Q_{12}')\bar{v}_1^{(1)} + \bar{M}_{22}^{(1)}Q_{12}'\bar{v}_1^{(0)} \tag{6.79b}$$
$$+ (A_{22} + \bar{M}_{22}^{(0)}Q_{22})\bar{v}_2^{(1)} + \bar{M}_{22}^{(1)}Q_{22}\bar{v}_2^{(0)} + \bar{F}_2^{(0)}$$

The solutions of (6.76)–(6.79) have the initial conditions as

$$\bar{M}^{(0)}(t'=0) = M(t=0); \quad \bar{v}^{(0)}(t'=0) = v(0) \tag{6.80}$$
$$\bar{M}^{(i)}(t'=0) = 0; \qquad \bar{v}^{(i)}(t'=0) = 0; \quad i \geqslant 1 \tag{6.81}$$

Next we have the intermediate series as

$$\underline{M}(t',\varepsilon) = \sum_{i=0}^{\infty} \underline{M}^{(i)}(t')\varepsilon^i \tag{6.82}$$

$$\underline{v}(t',\varepsilon) = \sum_{i=0}^{\infty} \underline{v}^{(i)}(t')\varepsilon^i \tag{6.83}$$

The controlling equations for the various intermediate series coefficients are identical to those of (6.76)–(6.79) of the corresponding inner-series coefficients; the only difference is their initial conditions, i.e.

$$\underline{M}^{(i)}(t'=0) = M^{(i)}(t=0); \quad \underline{v}^{(i)}(t'=0) = v^{(i)}(t=0), \quad i \geqslant 0 \tag{6.84}$$

The initial values are evaluated from (Wasow, 1965)

$$M^{(i)}(t=0) = \int_0^\infty \left(\frac{d\bar{M}_{11}^{(i)}}{dt'} - \frac{d\underline{M}_{11}^{(i)}}{dt'}\right)dt', \quad i \geqslant 1 \tag{6.85}$$

$$v^{(i)}(t=0) = \int_0^\infty \left(\frac{d\bar{v}_1^{(i)}}{dt'} - \frac{d\underline{v}_1^{(i)}}{dt'}\right)dt', \quad i \geqslant 1 \tag{6.86}$$

A very special feature of Vasileva's method is that the intermediate-series coefficients $\underline{M}^{(i)}(t')$ and $\underline{v}^{(i)}(t')$ can be easily formulated or generated as polynomials in t' as

$$\left.\begin{array}{l} \underline{M}^{(0)}(t') = M^{(0)}(0) \\ \underline{M}^{(1)}(t') = M^{(1)}(0) + \dot{M}^{(0)}(0)t' \\ \underline{M}^{(2)}(t') = M^{(2)}(0) + \dot{M}^{(1)}(0)t' + \ddot{M}^{(0)}(0)t'^2/2! \end{array}\right\} \tag{6.86}$$

and

$$\underline{v}^{(0)}(t') = v^{(0)}(0)$$

$$\underline{v}^{(1)}(t') = v^{(1)}(0) + \dot{v}^{(0)}(0)t'$$

$$\underline{v}^{(2)}(t') = v^{(2)}(0) + \dot{v}^{(1)}(0)t' + \ddot{v}^{(0)}(0)t'^2/2!$$

$$\left.\right\} \qquad (6.87)$$

where the 'dot' denotes differentiation with respect to t.

6.4.4 *Main results*

The two important results in the singular perturbation methods are concerned with the degeneration and the asymptotic expansions of the solution. These results require the following basic conditions:

(i) the system (6.1) is boundary-layer controllable, i.e. rank $[B_2, A_{22}B_2, \ldots, A_{22}^{n-1}B_2] = n$

(ii) the system (6.1) is boundary-layer observable, i.e. rank $[Q_{22}, A_{22}Q_{22}, \ldots, A_{22}^{n-1}Q_{22}] = n$

(iii) A_{22} is a stable matrix

(iv) $(A_{22} + M_{22}Q_{22})$ is a stable matrix.

Theorem 6.3

If conditions (i)–(iv) are satisfied for the forward system, then the following convergence relations between the full (high-order) and degenerate (low-order) solutions are valid:

$$\underset{\varepsilon \to 0}{\text{limit}} \left[M_{11}(t,\varepsilon) \right] = M_{11}^{(0)}(t), \quad 0 \leqslant t \leqslant T \qquad (6.89a)$$

$$\underset{\varepsilon \to 0}{\text{limit}} \left[M(t,\varepsilon) \right] = M^{(0)}(t), \quad 0 < t \leqslant T, \quad M = M_{12}, M_{22} \qquad (6.89b)$$

and

$$\underset{\varepsilon \to 0}{\text{limit}} \left[v_1(t,\varepsilon) \right] = v_1^{(0)}(t), \quad 0 \leqslant t \leqslant T \qquad (6.90a)$$

$$\underset{\varepsilon \to 0}{\text{limit}} \left[v_2(t,\varepsilon) \right] = v_2^{(0)}(t), \quad 0 < t \leqslant T \qquad (6.90b)$$

This is similar to Tikhonov's convergence theorem for initial-value problems (Tikhonov, 1952).

Theorem 6.4

If conditions (i)–(iv) are valid for the forward system, then the asymptotic series solutions are given by (Wasow, 1965)

$$M(t,\varepsilon) = \sum_{i=0}^{j} [M^{(i)}(t) + \bar{M}^{(i)}(t') - \underline{M}^{(i)}(t')]\varepsilon^i + S_m(\varepsilon^{j+1}) \qquad (6.91)$$

$$v(t,\varepsilon) = \sum_{i=0}^{j} [v^{(i)}(t) + \bar{v}^{(i)}(t') - \underline{v}^{(i)}(t')]\varepsilon^i + S_v(\varepsilon^{j+1}) \tag{6.92}$$

where $M = M_{11}, M_{12}, M_{22}$ and $v = v_1, v_2$ and S_m and S_v are uniformly bounded in a given interval.

Similar results can be obtained for the 'backward' system (6.62) and (6.63) along with system (6.65). The corresponding terminal stretching transformation to be used is $t'' = (t - T)/\varepsilon$.

Example 6.3*

In order to illustrate the method and to obtain explicit solutions to various inner- and intermediate-series functions, let us consider a simple second-order system described by

$$\left. \begin{aligned} \frac{dx}{dt} &= z \\ \varepsilon \frac{dz}{dt} &= -x - z + u \end{aligned} \right\} \tag{6.93}$$

The performance index to be minimised is

$$J = \frac{1}{2} \int_0^T (x^2 + u^2)dt \tag{6.94}$$

The specified boundary conditions are $x(0)$, $z(0)$, $x(T)$ and $z(T)$.

Using (6.62) and (6.63) for the present second-order system

$$\left. \begin{aligned} \frac{dm_{11}}{dt} &= m_{11}^2 + 2m_{12} \\ \varepsilon \frac{dm_{12}}{dt} &= -m_{11} - m_{12} + m_{22} + \varepsilon m_{11}m_{12} \\ \varepsilon \frac{dm_{22}}{dt} &= -2\varepsilon m_{12} - 2m_{22} + \varepsilon^2 m_{12}^2 - 1 \end{aligned} \right\} \tag{6.95}$$

and

$$\left. \begin{aligned} \frac{dv_1}{dt} &= m_{11}v_1 + v_2 \\ \varepsilon \frac{dv_2}{dt} &= -v_1 - v_2 + \varepsilon m_{12}v_1 \end{aligned} \right\} \tag{6.96}$$

* Reprinted with permission from RAJAGOPÀLAN, P. K., and NAIDU, D. S. (1980): 'Singular perturbation analysis of a closed-loop fixed-end-point optimal control problem', *IEE Proc., Pt.D.* **127**, pp. 194–203

The 'forward' procedure is used with

$$m_{11}(t=0) = m_{12}(t=0) = m_{22}(t=0) = 0$$
$$v_1(t=0) = x(0); \quad v_2(t=0) = z(0)$$

(6.97)

Using the singular perturbation method, various solutions are obtained as summarised below.

(i) *Zeroth-order approximation*
Step 1: The outer-series equations are

$$\frac{dm_{11}^{(0)}}{dt} = m_{11}^{(0)^2} - 2m_{11}^{(0)} - 1; \quad m_{11}^{(0)}(t=0) = 0$$
$$m_{12}^{(0)} = -0{\cdot}5 - m_{11}^{(0)}$$
$$m_{22}^{(0)} = -0{\cdot}5$$

(6.98)

$$\frac{dv_1^{(0)}}{dt} = (m_{11}^{(0)} - 1)v_1^{(0)}; \quad v_1^{(0)}(t=0) = x(0)$$
$$v_2^{(0)} = -v_1^{(0)}$$

(6.99)

Step 2: The inner-series solutions are

$$\bar{m}_{11}^{(0)}(t') = 0$$
$$\bar{m}_{12}^{(0)}(t') = -0{\cdot}5 + \exp(-t') - 0{\cdot}5\exp(-2t')$$
$$\bar{m}_{22}^{(0)}(t') = -0{\cdot}5 + 0{\cdot}5\exp(-2t')$$

(6.100)

and

$$\bar{v}_1^{(0)}(t') = x(0)$$
$$\bar{v}_2^{(0)}(t') = -x(0) + [x(0) + z(0)]\exp(-t')$$

(6.101)

Step 3: The intermediate-series solutions are

$$\underline{m}_{11}^{(0)}(t') = 0; \quad \underline{m}_{12}^{(0)}(t') = \underline{m}_{22}^{(0)}(t') = -0{\cdot}5$$

(6.102)

(ii) *First-order approximation*
Step 4: The initial values evaluated are

$$m_{11}^{(1)}(0) = 1{\cdot}5; \quad v_1^{(1)}(0) = x(0) + z(0)$$

(6.103)

Step 5: The outer-series equations to be solved are

$$
\left.
\begin{aligned}
\frac{dm_{11}^{(1)}}{dt} &= 2(m_{11}^{(0)} - 1)m_{11}^{(1)} - 3m_{11}^{(0)} - 1 \\
m_{12}^{(1)} &= -m_{11}^{(1)} - 1{\cdot}5m_{11}^{(0)} - 0{\cdot}5 \\
m_{22}^{(1)} &= m_{11}^{(0)} + 0{\cdot}5
\end{aligned}
\right\}
\tag{6.104}
$$

and

$$
\left.
\begin{aligned}
\frac{dv_1^{(1)}}{dt} &= (m_{11}^{(0)} - 1)v_1^{(1)} + (m_{11}^{(1)} - 1{\cdot}5)v_1^{(0)} \\
v_2^{(1)} &= -v_1^{(1)} - 1{\cdot}5v_1^{(0)}
\end{aligned}
\right\}
\tag{6.105}
$$

Step 6: The inner-series solutions are

$$
\left.
\begin{aligned}
\bar{m}_{11}^{(1)}(t') &= 1{\cdot}5 - t' - 2\exp(-t') + 0{\cdot}5\exp(-2t') \\
\bar{m}_{12}^{(1)}(t') &= -2 + t' + 4\exp(-t') - (2 + t')\exp(-2t') \\
\bar{m}_{22}^{(1)}(t') &= 0{\cdot}5 - 2\exp(-t') + (1{\cdot}5 + t')\exp(-2t')
\end{aligned}
\right\}
\tag{6.106}
$$

and

$$
\left.
\begin{aligned}
\bar{v}_1^{(1)}(t') &= x(0) + z(0) - x(0)t' - [x(0) + z(0)]\exp(-t') \\
\bar{v}_2^{(1)}(t') &= -2{\cdot}5x(0) - z(0) + x(0)t' + [2x(0) + z(0)](1 + t') \\
&\qquad \exp(-t') + 0{\cdot}5x(0)\exp(-2t')
\end{aligned}
\right\}
\tag{6.107}
$$

Step 7: The intermediate-series solutions are

$$
\underline{m}_{11}^{(1)}(t') = 1{\cdot}5 - t'; \qquad \underline{m}_{12}^{(1)}(t') = -2 - t'; \qquad \underline{m}_{22}^{(1)}(t') = 0{\cdot}5
\tag{6.108}
$$

and

$$
\underline{v}_1^{(1)}(t') = x(0) + z(0) - x(0)t'; \qquad \underline{v}_2^{(1)}(t') = -2{\cdot}5x(0) - z(0) - x(0)t'
\tag{6.109}
$$

Proceeding in a similar manner, the second-order approximation is also evaluated.

Figure 6.6 shows the degeneration of high-order solution to low-order solution, thereby illustrating Theorem 6.3 given by (6.89) and (6.90).

The solutions for zeroth-, first- and second-order approximations along with the exact solutions are shown in Fig. 6.7. A relatively high value of ε is chosen to demonstrate clearly how the series solutions improve with the increased order of approximation, thereby illustrating Theorem 6.4 given by (6.91) and

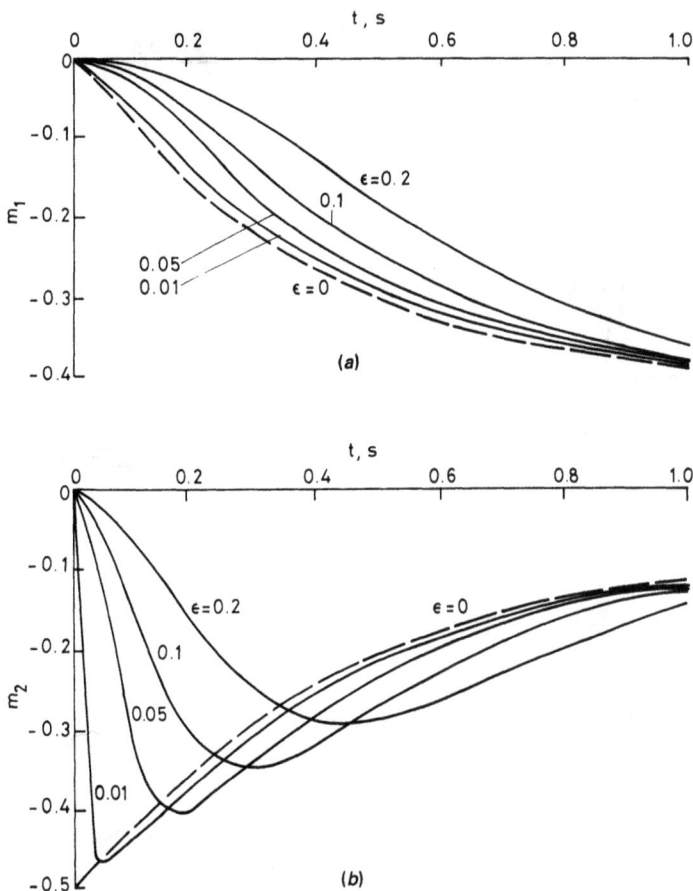

Fig. 6.6a and b   *Degeneration of solutions of $m_1(t)$ and $m_2(t)$ of Example 6.3*
——————— solution for $\varepsilon \neq 0$
- - - - - - - solution for $\varepsilon = 0$

(6.92). The error estimation based on the integral-squared-error for the solutions is shown in Fig. 6.8.

6.5   Time-scale analysis

In this Section we propose a procedure for a complete separation of slow and fast regulator designs (Chow and Kokotovic, 1976). We recall that, in Chapter 3, we have seen that a singularly perturbed system can be decoupled into slow and fast subsystems. For the sake of continuity and completeness, we reconsider the problem

$$\frac{dx}{dt} = A_{11}x + A_{12}z + B_1 u \qquad (6.110a)$$

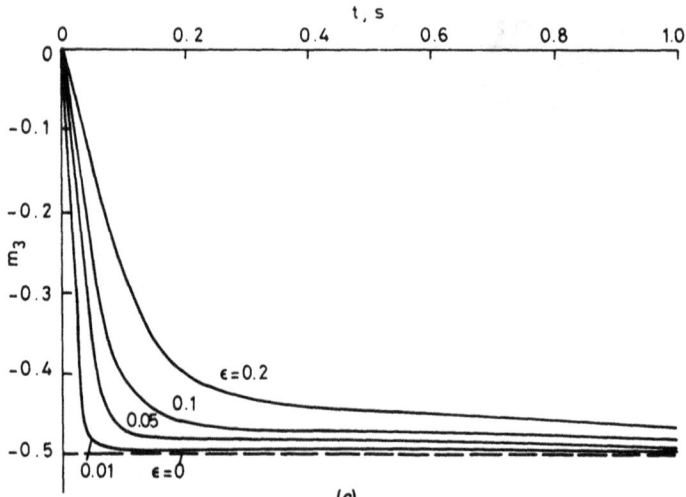

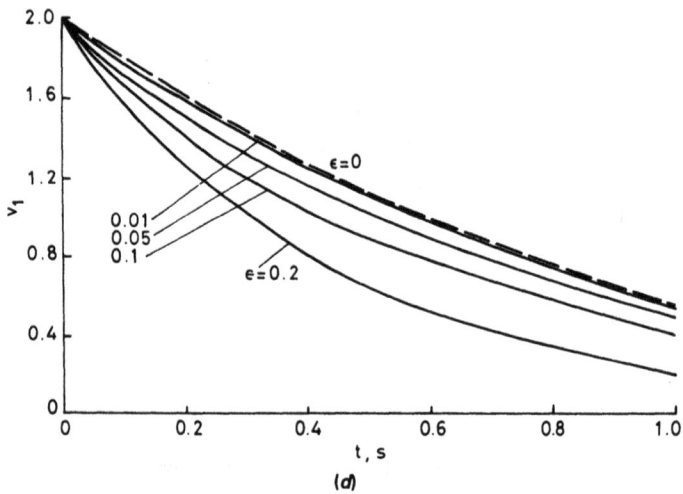

Fig. 6.6c and d Degeneration of solutions of $m_3(t)$ and $v_1(t)$ of Example 6.3
——— solution for $\varepsilon \neq 0$
- - - - - - solution for $\varepsilon = 0$

$$\varepsilon \frac{dz}{dt} = A_{21}x + A_{22}z + B_2u \tag{6.110b}$$

$$y = C_1x + C_2z \tag{6.110c}$$

6.5.1 Slow and fast subsystems

The system (6.110) possesses a two-time-scale property, i.e. it has m small

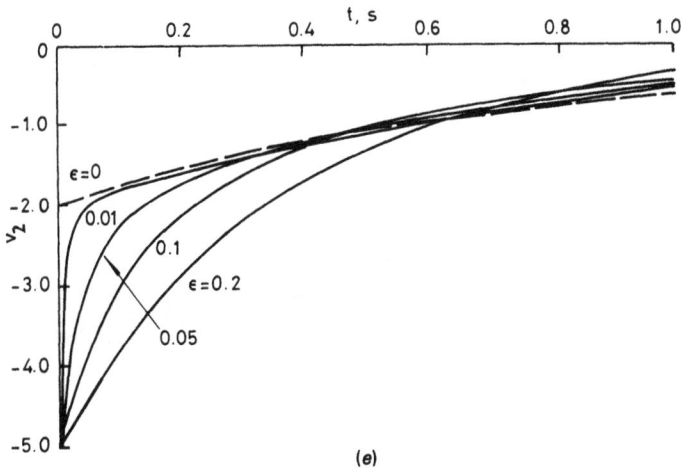

(e)

Fig. 6.6e *Degeneration of solutions of $v_2(t)$ of Example 6.3*
————— solution for $\varepsilon \neq 0$
------- solution for $\varepsilon = 0$

eigenvalues of magnitude $0(1)$ and n large eigenvalues of magnitude $0(1/\varepsilon)$. Preliminary to separation of slow and fast regulator designs, the system (6.110) is approximately decomposed into a slow subsystem with m small eigenvalues and a fast subsystem with n large eigenvalues. In an asymptotically stable system the fast modes corresponding to the large eigenvalues are important only during a short initial period. After that period they are negligible and the behaviour of the system can be described by its slow modes. Neglecting the fast modes is equivalent to assuming that they are infinitely fast, i.e. letting $\varepsilon \rightarrow 0$ in system (6.110). Hence without the fast modes, the system (6.110) reduces to

$$\dot{x}^{(0)} = A_{11}x^{(0)} + A_{12}z^{(0)} + B_1u^{(0)} \tag{6.111a}$$

$$0 = A_{21}x^{(0)} + A_{22}z^{(0)} + B_2u^{(0)} \tag{6.111b}$$

$$y^{(0)} = C_1x^{(0)} + C_2z^{(0)} \tag{6.111c}$$

Assuming that A_{22} is nonsingular, we have

$$z^{(0)} = -A_{22}^{-1}(A_{21}x^{(0)} + B_2u^{(0)}) \tag{6.112}$$

Using (6.112) in (6.111), we get the slow system as

$$\dot{x}_s = A_0x_s + B_0u_s \tag{6.113a}$$

$$y_s = C_0x_s + D_0u_s \tag{6.113b}$$

(a)

(b)

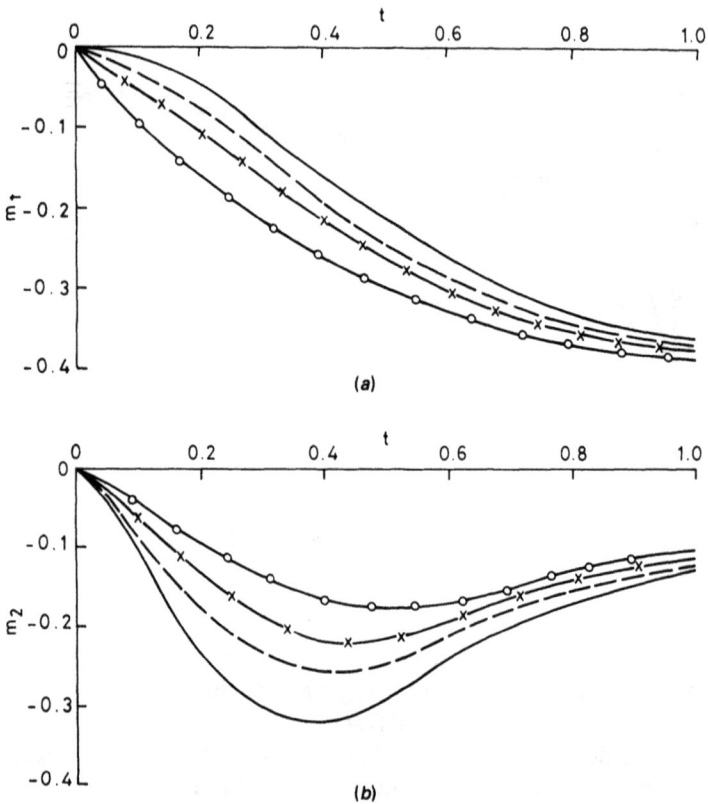

Fig. 6.7a and b *Exact and approximate solutions of $m_1(t)$ and $m_2(t)$ of Example 6.3*
$x(0) = 2 \cdot 0;\ z(0) = -5 \cdot 0;\ \varepsilon = 0 \cdot 15$
——————— exact solution
o——o——o zeroth-order solution
x——x——x first-order solution
– – – – – – second-order solution

where

$$A_0 = A_{11} - A_{12}A_{22}^{-1}A_{21}; \qquad B_0 = B_1 - A_{12}A_{22}^{-1}B_2$$

$$C_0 = C_1 - C_2A_{22}^{-1}A_{21}; \qquad D_0 = -C_2A_{22}^{-1}B_2$$

$$x^{(0)} = x_s;\quad y^{(0)} = y_s; \qquad u^{(0)} = u_s$$

To obtain the fast subsystem, we assume that the slow variables are constant during the fast transients, i.e. $\dot{z}^{(0)} = 0$ and $x^{(0)} = x_s = $ constant. From systems (6.110b) and (6.112), we get

$$\varepsilon(\dot{z} - \dot{z}^{(0)}) = A_{22}(z - z^{(0)}) + B_2(u - u^{(0)}) \qquad (6.114)$$

Letting $z_f = z - z^{(0)}$, $u_f = u - u^{(0)}$, $y_f = y - y^{(0)}$, the fast subsystem of (6.110) becomes

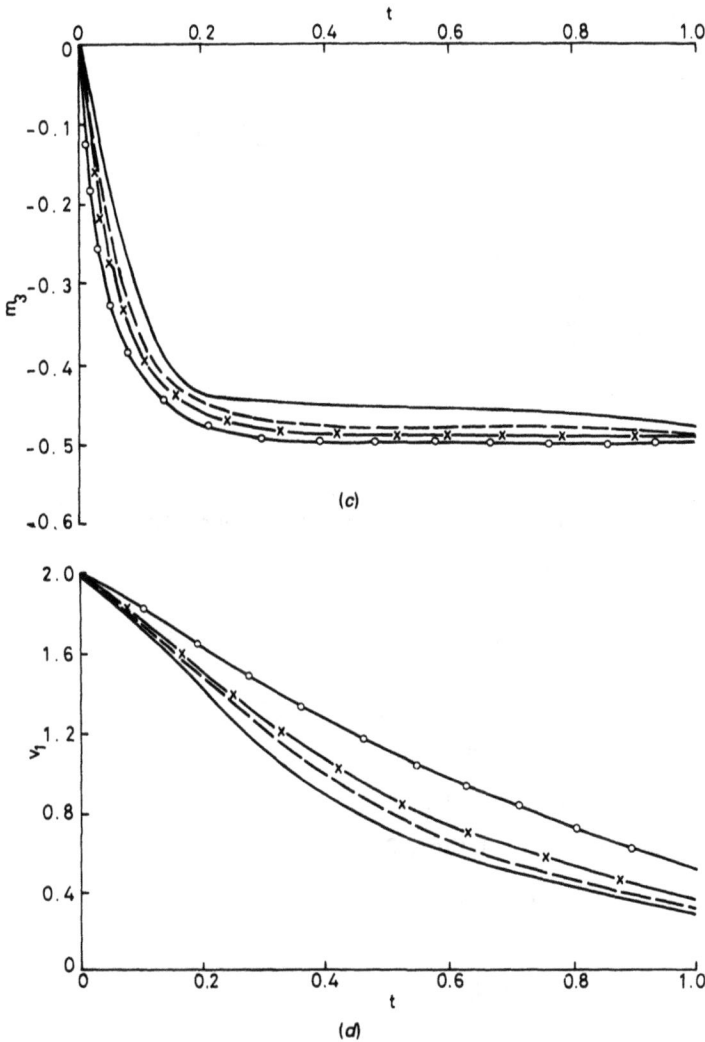

Fig. 6.7c and d Exact and approximate solutions of $m_3(t)$ and $v_1(t)$ of Example 6.3
$x(0) = 2 \cdot 0$; $z(0) = -5 \cdot 0$; $\varepsilon = 0 \cdot 15$
——————— exact solution
o———o———o zeroth-order solution
×——×——× first-order solution
- - - - - - - second-order solution

$$\varepsilon \dot{z}_f = \frac{dz_f}{dt'} = A_{22} z_f + B_2 u_f, \quad z_f(0) = z(0) - z^{(0)}(0) \qquad (6.115a)$$

$$y_f = C_2 z_f; \quad t' = \frac{t}{\varepsilon} \qquad (6.115b)$$

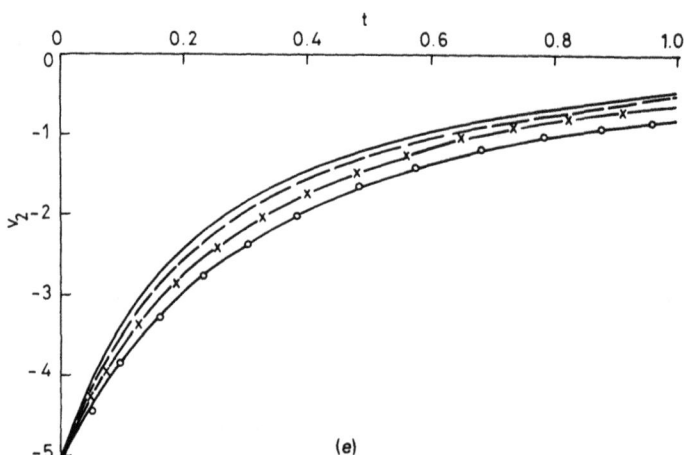

Fig. 6.7e *Exact and approximate solutions of $v_2(t)$ of Example 6.3*
  $x(0) = 2 \cdot 0$; $z(0) = -5 \cdot 0$; $\varepsilon = 0 \cdot 15$
  ——————— exact solution
  o——o——o zeroth-order solution
  x——x——x first-order solution
  – – – – – – second-order solution

Suppose now that $u_s = G_s x_s$ and $u_f = G_f x_f$ are designed such that x_s and z_f meet some specifications. In view of

$$z_s = -A_{22}^{-1}(A_{21} + B_2 G_s)x_s \qquad (6.116)$$

which follows from (6.112), the 'composite control'

$$u_c = u_s + u_f = G_s x_s + G_f z_f \qquad (6.117)$$

can be written down as

$$u_c = [(I + G_f A_{22}^{-1} B_2)G_s + G_f A_{22}^{-1} A_{21}]x_s$$
$$\qquad + G_f[-A_{22}^{-1}(A_{21} + B_2 G_s)x_s + z_f] \qquad (6.118a)$$

Now, if x_s approximates x and $z_f + z^{(0)}$ approximates z, we have

$$u_c = [(I + G_f A_{22}^{-1} B_2)G_s + G_f A_{22}^{-1} A_{21}]x + G_s z \qquad (6.118b)$$

6.5.2 Subsystem regulator design

Now, we decompose the optimum regulator problem defined by (6.110) and the performance index

$$J = \frac{1}{2}\int_0^\infty (y'y + u'Ru)\,dt \qquad (6.119)$$

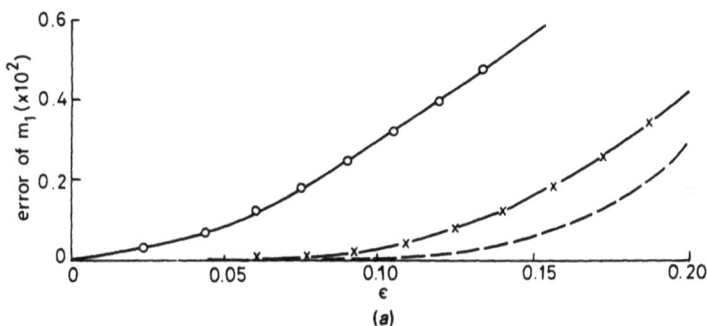

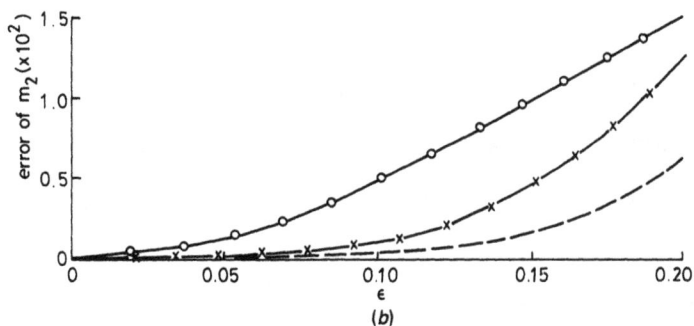

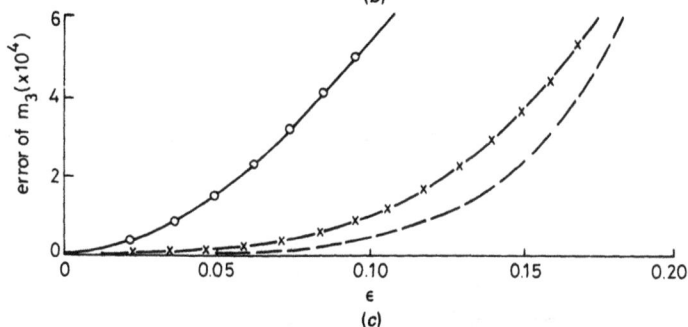

Fig. 6.8a, b and c   *Integral-squared error of $m_1(t)$, $m_2(t)$ and $m_3(t)$ of Example 6.3*
○———○———○ zeroth-order solution
×———×———× first-order solution
– – – – – – second-order solution

into two subsystem regulator problems. For that, we represent J as two quadratic performance indices, one for the slow subsystem (6.113) and the other for the fast subsystem (6.115).

For the slow subsystem (6.113), we need to find u_s to minimise

$$J = \frac{1}{2}\int_0^\infty (y_s' y_s + u_s' R u_s)\,dt \qquad (6.120)$$

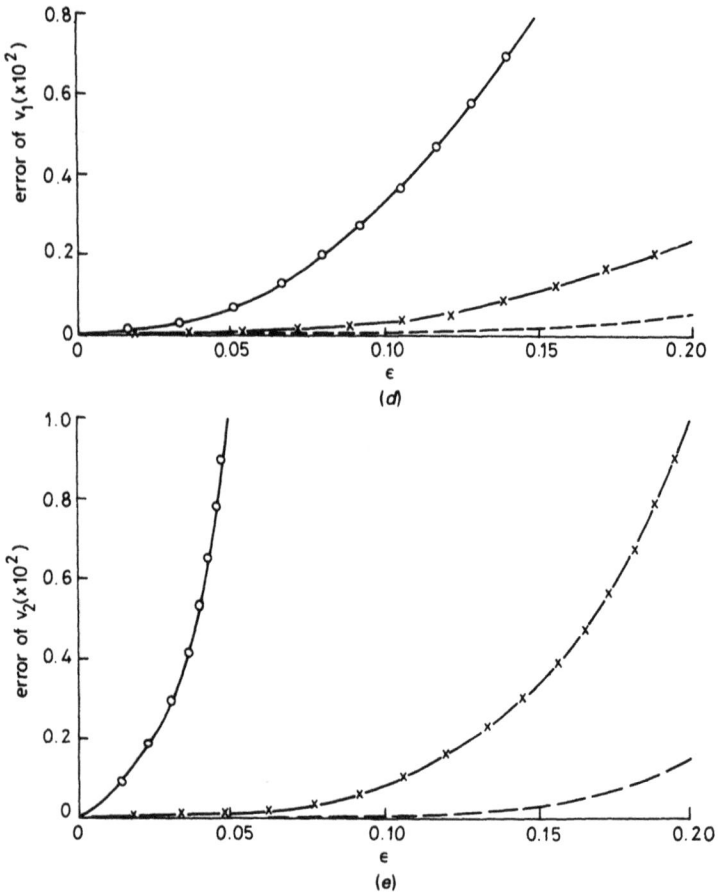

Fig. 6.8d and e   *Integral-squared error of $v_1(t)$ and $v_2(t)$ of Example 6.3*
o——o——o zeroth-order solution
×——×——× first-order solution
– – – – – – second-order solution

Using (6.113b) in (6.120),

$$J_s = \frac{1}{2}\int_0^\infty (x_s' C_0' C_0 x_s + 2u_s' D_0' D_0 x_s + u_s' R_0 u_s)dt \tag{6.121}$$

where $R_0 = R + D_0' D_0$. It is well known that the Riccati equation

$$0 = -P_s(A_0 - B_0 R_0^{-1} D_0' C_0) - (A_0 - B_0 R_0^{-1} D_0' D_0)' P_s$$

$$+ P_s B_0 R_0^{-1} B_0' P_s - C_0'(I - D_0 R_0^{-1} D_0') C_0 \tag{6.122}$$

has a positive definite solution P_s if the triple (A_0, B_0, C_0) is stabilisable-detectable. Then the optimal control for (6.113) and (6.120) is

$$u_s = -R_0^{-1}(D_0'C_0 + B_0'P_s)x_s \qquad (6.123)$$

Similarly, the optimal control u_f to minimise

$$J_f = \frac{1}{2}\int_0^\infty (y_f'y_f + u_f'Ru_f)dt \qquad (6.124)$$

with the fast subsystem (6.115), is given by

$$u_f = -R^{-1}B_2'P_f x_f \qquad (6.125)$$

where P_f is the positive definite solution of the Riccati equation

$$0 = -P_f A_{22} - A_{22}'P_f + P_f B_2 R^{-1}B_2'P_f - C_2'C_2 \qquad (6.126)$$

if the triple (A_{22}, B_2, C_2) is stabilisable-detectable.

It is of interest to find the composite control u_c in terms of x and z using (6.123) and (6.125) in (6.118b). With $G_s = -R_0^{-1}(D_0'C_0 + B_0'P_s)$ and $G_f = -R^{-1}B_2'P_f$, we have

$$\begin{aligned}u_c = &[(I - R^{-1}B_2'P_f A_{22}^{-1}B_2)R_0^{-1}(D_0'C_0 + B_0'P_s) \\ &+ R^{-1}B_2'P_f A_{22}^{-1}A_{21}]x - R^{-1}B_2'P_f z\end{aligned} \qquad (6.127)$$

Thus we have seen the complete separation of slow and fast regulator designs. An important property of the control law (6.127) is that it does not explicitly depend on the small parameter ε. But it has been shown (Chow and Kokotovic, 1976) that, even without the knowledge of ε the control (6.127) results in an $0(\varepsilon^2)$ approximation of the optimal performance. However, we note that the separation of the original system (6.110) into a slow subsystem (6.113) and fast subsystem (6.115) is only a first approximation and ε would be needed for higher-order approximations.

In addition, we see that the slow (reduced) subsystem design alone with control (6.123) has $0(\varepsilon)$ approximation of the optimal performance (Chow and Kokotovic, 1976).

Example 6.4*
Consider a magnetic-tape control system

$$\begin{bmatrix} \dot{x}_1 \\ \dot{x}_2 \\ \varepsilon \dot{z}_1 \\ \varepsilon \dot{z}_2 \end{bmatrix} = \begin{bmatrix} 0 & 0.4 & 0 & 0 \\ 0 & 0 & 0.345 & 0 \\ 0 & -0.524 & -0.465 & 0.262 \\ 0 & 0 & 0 & -1.0 \end{bmatrix} \begin{bmatrix} x_1 \\ x_2 \\ z_1 \\ z_2 \end{bmatrix} + \begin{bmatrix} 0 \\ 0 \\ 0 \\ 1.0 \end{bmatrix} u \qquad (6.128a)$$

---

* Reprinted with permission from CHOW, J. H. and KOKOTOVIC, P. V. (1976): 'A decomposition of near optimum regulators for systems with slow and fast modes', *IEEE Trans.*, **AC-21**, pp. 701–705.

$$\begin{bmatrix} y_1 \\ y_2 \end{bmatrix} = \begin{bmatrix} 1 & 0 & 0 & 0 \\ 0 & 0 & 1 & 0 \end{bmatrix} \begin{bmatrix} x_1 \\ x_2 \\ z_1 \\ z_2 \end{bmatrix}$$

(6.128b)

with performance index

$$J = \frac{1}{2}\int_0^\infty (y'y + u'u)dt$$

(6.129)

It is easily seen that the triples (A_0, B_0, C_0) and (A_{22}, B_2, C_2) are stabilisable-detectable.

From (6.126), P_f is found to be

$$P_f = \begin{bmatrix} 1 \cdot 04030 & 0 \cdot 18036 \\ 0 \cdot 18036 & 0 \cdot 04619 \end{bmatrix}$$

From (6.122), P_s is found to be

$$P_s = \begin{bmatrix} 7 \cdot 3840 & 5 \cdot 9048 \\ 5 \cdot 9048 & 7 \cdot 1516 \end{bmatrix}$$

The composite control (6.127) is given by

$$u_c = [-1 \quad -0 \cdot 86123]x + [-0 \cdot 18036 \quad -0 \cdot 046185]z$$

(6.130)

The reduced (slow) control (6.123) is given by

$$u_s = [-0 \cdot 87122 \quad -0 \cdot 57325]x$$

(6.131)

For $\varepsilon = 0 \cdot 1$, the optimal feedback solution is

$$u^* = [-1 \quad -0 \cdot 89292]x + [-0 \cdot 24396 \quad -0 \cdot 061996]z$$

(6.132)

For the initial conditions $[x'(0), z'(0)] = [1, 0, 1, 0]$, the values of the performance index are

$$J^* = 4 \cdot 2406; \quad J_c = 4 \cdot 2428; \quad J_s = 4 \cdot 2506$$

(6.133)

Hence the performance loss with u_c is less than $0 \cdot 052\%$ and with u_s less than $0 \cdot 24\%$.

6.6   Conclusions and discussion

This Chapter has dealt with the important aspect of obtaining closed-loop control for singularly perturbed linear, time-invariant systems. In Section 6.1

we posed the standard regulator problem leading to the singularly perturbed matrix Riccati differential equation to be solved with the given final conditions. Accordingly, in Section 6.2, the asymptotic series solution for the Riccati functions has been formulated in terms of the outer series based on the degenerate Riccati equation and the inner and intermediate series based on the Riccati equation obtained by using the final stretching transformation. An example has been provided to clearly illustrate the method.

The steady-state regulator problem has been considered in Section 6.3. Here we have arrived at the matrix Riccati algebraic equation. Accordingly, the asymptotic solution consisted of only the outer-series functions. In order to demonstrate the powerfulness of the singular perturbation method, a fifth-order example has been given and approximate coefficients have been evaluated by simple hand calculation.

In Section 6.4, we have addressed the closed-loop optimal control of a singularly perturbed fixed-end-point problem, leading to the special matrix Riccati equation to be solved either as a forward problem or backward problem. An asymptotic series solution has been sought in terms of the outer, inner- and intermediate-series solutions. An example clearly illustrated the method.

Time-scale analysis formed the content of Section 6.5. Here we considered the closed-loop optimal control of a singularly perturbed system decomposed into slow and fast subsystems. Then we associated the subsystems with appropriate performance indices. We obtained a composite sub-optimal control composed of slow and fast feedback controls.

In this Chapter, we have made the choice of using Vasileva's method in preference to the boundary-layer method owing to the reasons mentioned in Section 2.3 and the fact that the singularly perturbed linear regulator problem using the boundary-layer method is a thoroughly investigated problem (Kokotovic *et al.*, 1976; Saksena *et al.*, 1984) whereas that using Vasileva's method has not received sufficient attention (Naidu, 1977; Naidu and Rajagopalan, 1980; Rajagopalan and Naidu, 1980*a,b*).

The closed-loop optimal control of singularly perturbed linear systems was first considered by Sannuti (1968) and Sannuti and Kokotovic (1969). In this first work the asymptotic solution for the resulting matrix Riccati equation consisted of outer series only without any corrections. With the introduction of the boundary-layer method for linear and nonlinear initial and boundary-value problems by Butuzov *et al.* (1970), O'Malley (1971) and Hoppensteadt (1971), there was an attempt to use the same for the linear state-regulator problem. Consequently several authors have studied this problem (Kokotovic and Yackel, 1972; O'Malley, 1972*a,b,c*, 1975, 1976; Yackel and Kokotovic, 1973; Wilde and Kokotovic, 1973; O'Malley and Kung, 1974; Kokotovic *et al.*, 1976; Suzuki and Miura, 1976; Glizer and Dmitriev, 1975*a,b*, 1976, 1977*a,b*, 1978, 1979*a*; Womble *et al.*, 1976; O'Reilly, 1979*b*, Dragan and Halanay, 1982; Garden and Cruz, 1980; Pandolfi, 1981; Kano and Iwazumi, 1978,

1981, 1983*c*; Gaitsogori and Pervozvanskii, 1979; Eitelberg and Rode, 1982; Anderson *et al.*, 1983; Longchamp, 1983; Gichev, 1984; Kokotovic, 1984, 1985; Sobolev, 1984; Calise and Moerder, 1985; Dontchev and Veliov, 1985; Haraux and Murat, 1985; Saberi and Khalil, 1985; Young, 1985).

The linear quadratic regulator problem with three-time-scale behaviour has been investigated by Jamshidi (1976), Murthy (1978), Ozguner (1979), Kando and Iwazumi (1981), Mahmoud *et al.* (1982), and with multiple time scales by Dragan and Halanay (1982), and Sannuti and Wason (1985).

Another interesting investigation has been the closed-loop optimal control of linear and nonlinear systems decomposed into slow and fast subsystems (Chow and Kokotovic, 1976, 1978*a*,*b*, 1979; Gardner and Cruz, 1980; Suzuki and Miura, 1976; Saberi and Khalil, 1982; Suzuki, 1981; Kando and Iwazumi, 1983*c*; Syrcos and Sannuti, 1983, 1984).

Chapter 7

Closed-loop optimal control of discrete systems*

In this Chapter, first in Section 7.1, a method is described to analyse the singularly perturbed nonlinear difference equations for initial and boundary-value problems. The approximate solution is obtained in the form of an outer series and a correction series. It is seen that considerable care has to be taken in formulating the equations for the boundary-layer correction series in the case of nonlinear equations. Then, in Section 7.2, the closed-loop optimal-control problem is formulated, resulting in the singularly perturbed nonlinear matrix Riccati difference equation. It is seen that the degeneration (the process of suppressing a small parameter) affects some of the final conditions of the Riccati equation. In Section 7.3, a method is given to obtain approximate solutions in terms of an outer series and a terminal boundary-layer correction series. A method is also discussed in Section 7.4 for the important case of the steady-state solution of the matrix Riccati equation. The time-scale analysis of the regulator problem is also given. It is found that these methods, with the special feature of order reduction, offer considerable computational simplicity in evaluating the inverse of a matrix associated with the solution of the Riccati equation. Examples are given to illustrate these methods.

7.1 Nonlinear difference equations

Consider a second-order nonlinear difference equation

$$y(k+2) + ay(k+1) + \varepsilon y^2(k) = 0 \tag{7.1}$$

where ε is the small parameter.

Letting $\varepsilon = 0$ in (7.1), the degenerate equation

$$y^{(0)}(k+2) + ay^{(0)}(k+1) = 0$$

* Portions of this Chapter are based on the Research Monograph, NAIDU, D. S., and RAO, A. K. (1985): 'Singular perturbation analysis of discrete control systems'. Lecture Notes in Mathematics, Vol. 1154 (Springer–Verlag, Berlin, 1985). The permission given by Springer–Verlag is hereby acknowledged.

is of reduced order, and thus (7.1) is said to be in the singularly perturbed form. There can be many other types of singularly perturbed nonlinear difference equations.

7.1.1   Initial-value problem

For (7.1), the given initial conditions are $y(0)$ and $y(1)$. Since (7.1) belongs to the R-type equation dealt with in detail in Chapter 2, the degenerate equation of (7.1) satisfies $y(1)$ only and looses the other initial condition $y(0)$. The boundary layer occurs at $k = 0$. A correction is incorporated by using the transformation

$$w(k) = y(k)/\varepsilon^k$$

The total solution is given by

$$y(k) = y_o(k) + \varepsilon^k w(k) \tag{7.2}$$

where $y_o(k)$ refers to the outer solution and $w(k)$ refers to the boundary-layer-correction solution.

Using (7.2) in (7.1),

$$[y_o(k+2) + \varepsilon^{k+2} w(k+2)] + a[y_o(k+1) $$
$$+ \varepsilon^{k+1} w(k+1)] + \varepsilon[y_o(k) + \varepsilon^k w(k)]^2 = 0$$

From the above equation, separating the outer-series terms, we get

$$y_o(k+2) + ay_o(k+1) + \varepsilon y_o^2(k) = 0 \tag{7.3}$$

The remaining terms for the correction series are obtained as

$$\varepsilon w(k+2) + aw(k+1) + \varepsilon^k w^2(k) + 2y_o(k)w(k) = 0 \tag{7.4}$$

Note that (7.3) is identical to (7.1) and therefore (7.1) is also referred to as the equation of the outer series.

Outer series
Assume the outer series solution as

$$y(k) = y^{(0)}(k) + \varepsilon y^{(1)}(k) + \varepsilon^2 y^{(2)}(k) + \ldots \tag{7.5}$$

Substituting the above solution in (7.1) and equating terms with like powers of ε on either side, we get the following equations:

$$\begin{aligned}
\varepsilon^0: \quad & y^{(0)}(k+2) + ay^{(0)}(k+1) = 0 \\
\varepsilon^1: \quad & y^{(1)}(k+2) + ay^{(1)}(k+1) = -y^{(0)^2}(k) \\
\varepsilon^2: \quad & y^{(2)}(k+2) + ay^{(2)}(k+1) = -2y^{(0)}(k)y^{(1)}(k)
\end{aligned} \tag{7.6}$$

Equations for higher-order approximations are obtained, proceeding on similar lines.

Correction series
Assume the correction-series solution as

$$w(k) = w^{(0)}(k) + \varepsilon w^{(1)}(k) + \varepsilon^2 w^{(2)}(k) + \ldots \tag{7.7}$$

By substitution of the above in (7.4) and comparison of coefficients, we get the equations corresponding to zeroth-, first- and higher-order approximations. These equations require careful handling in view of the term ε^k in (7.4). However, since the correction series needs to be evaluated for only a few values of k, say $k = 0$ and 1, it does not pose any serious problems as shown below.

Substituting $k = 0$ for ε^k in (7.4),

$$\varepsilon w(k+2) + aw(k+1) + w^2(k) + 2y(k)w(k) = 0 \tag{7.8a}$$

For the above equation, the zeroth-order approximation is

$$aw^{(0)}(k+1) + w^{(0)^2}(k) + 2y^{(0)}(k)w^{(0)}(k) = 0 \tag{7.8b}$$

Using $k = 1$ for ε^k in (7.4),

$$\varepsilon w(k+2) + aw(k+1) + \varepsilon w^2(k) + 2y(k)w(k) = 0 \tag{7.8c}$$

For the above equation, the zeroth-order approximation is

$$aw^{(0)}(k+1) + 2y^{(0)}(k)w^{(0)}(k) = 0 \tag{7.8d}$$

Thus (7.8b) and (7.8d) are the two equations for the zeroth-order approximation only and should be used only for $k = 0$ and $k = 1$, respectively.

For the first-order approximation, (7.8a) and (7.8c) become, for $k = 0$,

$$aw^{(1)}(k+1) + 2w^{(0)}(k)w^{(1)}(k) + 2y^{(0)}(k)w^{(1)}(k)$$
$$= -2y^{(1)}(k)w^{(0)}(k) - w^{(0)}(k+2) \tag{7.9a}$$

for $k = 1$

$$aw^{(1)}(k+1) + 2y^{(0)}(k)w^{(1)}(k) = -w^{(0)^2}(k)$$
$$- 2y^{(1)}(k)w^{(0)}(k) - w^{(0)}(k+2) \tag{7.9b}$$

Similar equations are obtained for higher-order approximations and for higher values of k.

Total series solution
The total series solution consisting of the outer series solution and correction series solution is given by

$$y(k) = [y^{(0)}(k) + \varepsilon y^{(1)}(k) + \varepsilon^2 y^{(2)}(k) + \ldots]$$
$$+ \varepsilon^k[w^{(0)}(k) + \varepsilon w^{(1)}(k) + \ldots] \tag{7.10}$$

The initial conditions required for solving the series (7.8) and (7.9) are obtained from the above equation as

$$\left.\begin{array}{ll} \varepsilon^0: & y^{(0)}(1) = y(1); \quad w^{(0)}(0) = y^{(0)} - y^{(0)}(0) \\[4pt] \varepsilon^1: & y^{(1)}(1) = -w^{(0)}(1); \quad w^{(1)}(0) = -y^{(1)}(0) \\ \cdots & \cdots\cdots\cdots\cdots\cdots\cdots\cdots\cdots\cdots\cdots \\[4pt] \varepsilon^j: & y^{(j)}(1) = -w^{(j-1)}(1); \quad w^{(j)}(0) = -y^{(j)}(0) \end{array}\right\} \tag{7.11}$$

It is to be noted that, for a desired second-order approximation, the correction-series terms $w^{(0)}(1)$, $w^{(0)}(2)$, and $w^{(1)}(1)$ are obtained by solving the corresponding correction-series equations. On the other hand, the terms $w^{(0)}(0)$ and $w^{(1)}(0)$ are simply evaluated from the outer-series coefficients $y(0) - y^{(0)}(0)$ and $y^{(1)}(0)$, respectively.

Example 7.1
Consider a second-order nonlinear equation

$$y(k+2) - 0{\cdot}75y(k+1) + \varepsilon y(k)^2 = 0 \tag{7.12}$$

with initial conditions

$$y(0) = 1{\cdot}0; \quad y(1) = 2{\cdot}0$$

and

$$\varepsilon = 0{\cdot}05.$$

Using the method presented in Section 7.1, the degenerate, zeroth, first- and second-order series equations are obtained and solved. The results are shown in Table 7.1 which indicate that the series solutions approach the exact solution with the increased order of approximation.

Table 7.1   *Comparison of approximate and exact solutions of Example 7.1*

$y(k)$	Degenerate solution	Zeroth-order solution	First-order solution	Second-order solution	Exact solution
$y(0)$	2·6667	1·0000	1·0000	1·0000	1·0000
$y(1)$	2·0000	2·0000	2·0000	2·0000	2·0000
$y(2)$	1·5000	1·5000	1·4500	1·4500	1·4500
$y(3)$	1·1250	1·1250	0·8875	0·8875	0·8875
$y(4)$	0·8438	0·8438	0·5531	0·5606	0·5605
$y(5)$	0·6328	0·6328	0·3516	0·3839	0·3810
$y(6)$	0·4746	0·4746	0·2281	0·2769	0·2700
$y(7)$	0·3560	0·3560	0·1510	0·2054	0·1953
$y(8)$	0·2670	0·2670	0·1020	0·1545	0·1429
$y(9)$	0·2002	0·2002	0·0701	0·1168	0·1052
$y(10)$	0·1502	0·1502	0·0491	0·0885	0·0779

The degenerate and zeroth-order solutions are the same except at $k = 0$.

7.1.2   Boundary-value problem

For the BVP, given $y(0)$ and $y(N)$, the procedure is almost similar to the case of IVP. The selection of the boundary conditions is, however, different and is furnished below:

$$
\left.\begin{aligned}
&\varepsilon^0: \quad y^{(0)}(N) = y(N); \quad w^{(0)}(0) = y(0) - y^{(0)}(0) \\
&\varepsilon^1: \quad y^{(1)}(N) = 0; \qquad w^{(1)}(0) = -y^{(1)}(0) \\
&\cdots \quad \cdots\cdots\cdots\cdots\cdots\cdots\cdots\cdots\cdots\cdots \\
&\varepsilon^j: \quad y^{(j)}(N) = 0; \qquad w^{(j)}(0) = -y^{(j)}(0)
\end{aligned}\right\}
\tag{7.13}
$$

7.2   Closed-loop optimal control problem

Consider a linear, shift-invariant, completely controllable singularly perturbed discrete system described by

$$
\begin{bmatrix} y(k+1) \\ z(k+1) \end{bmatrix} = \begin{bmatrix} A_{11} & \varepsilon A_{12} \\ A_{21} & \varepsilon A_{22} \end{bmatrix} \begin{bmatrix} y(k) \\ z(k) \end{bmatrix} + \begin{bmatrix} B_1 \\ B_2 \end{bmatrix} u(k)
\tag{7.14}
$$

with initial conditions

$$
y(k=0) = y(0); \quad z(k=0) = z(0)
$$

where $y(k)$ and $z(k)$ are n_1- and n_2-dimensional state vectors, respectively; $u(k)$ is an r control vector; A_{ij}, B_i, $i, j = 1, 2$ are real constant matrices of compatible dimensionality.

The performance index to be minimised is

$$
J = \frac{1}{2} x'(N) S x(N) + \frac{1}{2} \sum_{k=0}^{N-1} [x'(k)Qx(k) + u'(k)Ru(k)]
\tag{7.15}
$$

where

$$
x'(k) = [y'(k), z'(k)]
$$

S and Q are real, positive semidefinite $(n_1 + n_2) \times (n_1 + n_2)$ order symmetric matrices; R is a real, positive definite $(r \times r)$-order symmetric matrix.

Using the well-known results of optimal control theory (Sage and White, 1977), the closed-loop optimal control is given by

$$
\begin{aligned}
u(k) &= -R^{-1}B'[P^{-1}(k+1) + BR^{-1}B']^{-1}Ax(k) \\
&= -R^{-1}B'P(k+1)[I + BR^{-1}B'P(k+1)]^{-1}Ax(k)
\end{aligned}
\tag{7.16}
$$

where the positive definite symmetric matrix $P(k+1)$ of order $(n_1 + n_2) \times (n_1 + n_2)$ satisfies the matrix Riccati difference equation

$$
P(k) = Q + A'P(k+1)[I + BR^{-1}B'P(k+1)]^{-1}A
\tag{7.17}
$$

with the final condition $P(N) = S$

$$A = \begin{bmatrix} A_{11} & \varepsilon A_{12} \\ A_{21} & \varepsilon A_{22} \end{bmatrix}; \quad B = \begin{bmatrix} B_1 \\ B_2 \end{bmatrix}; \quad q = \begin{bmatrix} Q_1 & \varepsilon Q_2 \\ \varepsilon Q_2' & \varepsilon^2 Q_3 \end{bmatrix}$$

$$P(k) = \begin{bmatrix} P_1(k) & \varepsilon P_2(k) \\ \varepsilon P_2'(k) & \varepsilon^2 P_3(k) \end{bmatrix}; \quad S = \begin{bmatrix} S_1 & \varepsilon S_2 \\ \varepsilon S_2' & \varepsilon^2 S_3 \end{bmatrix}$$

The Riccati equation (7.17) is solved backward in time with the condition $P(N) = S$, and certain gain functions called 'Kalman gains' are obtained. These precomputed gains are stored and applied to the system (7.14) as it runs forward in time. Thus we have a closed-loop optimal control problem.

Rearranging (7.17)

$$P_1(k) = Q_1 + A_{11}'[F_1(k+1)A_{11} + F_2(k+1)A_{21}]$$
$$+ A_{21}'[F_2'(k+1)A_{11} + F_3(k+1)A_{21}] \tag{7.18a}$$

$$P_2(k) = Q_2 + A_{11}'[F_1(k+1)A_{12} + F_2(k+1)A_{22}]$$
$$+ A_{21}'[F_2'(k+1)A_{12} + F_3(k+1)A_{22}] \tag{7.18b}$$

$$P_3(k) = Q_3 + A_{12}'[F_1(k+1)A_{12} + F_2(k+1)A_{22}]$$
$$+ A_{22}'[F_2'(k+1)A_{12} + F_3(k+1)A_{22}] \tag{7.18c}$$

where

$$\begin{bmatrix} F_1(k+1) & F_2(k+1) \\ F_2'(k+1) & F_3(k+1) \end{bmatrix} = \begin{bmatrix} P_1(k+1) & \varepsilon P_2(k+1) \\ \varepsilon P_2'(k+1) & \varepsilon^2 P_3(k+1) \end{bmatrix}$$

$$\times \left[\begin{bmatrix} I_1 & 0 \\ 0 & I_2 \end{bmatrix} + \begin{bmatrix} E_1 & E_2 \\ E_2' & E_3 \end{bmatrix} \begin{bmatrix} P_1(k+1) & \varepsilon P_2(k+1) \\ \varepsilon P_2'(k+1) & \varepsilon^2 P_3(k+1) \end{bmatrix} \right]^{-1} \tag{7.19}$$

$$\begin{bmatrix} E_1 & E_2 \\ E_2' & E_3 \end{bmatrix} = \begin{bmatrix} B_1 R^{-1} B_1' & B_1 R^{-1} B_2' \\ B_2 R^{-1} B_1' & B_2 R^{-1} B_2' \end{bmatrix}$$

Using $\varepsilon = 0$ in (7.19),

$$\begin{bmatrix} F_1^{(0)}(k+1) & F_2^{(0)}(k+1) \\ F_2^{(0)\prime}(k+1) & F_3^{(0)}(k+1) \end{bmatrix} = \begin{bmatrix} P_1^{(0)}(k+1)\{I_1 + E_1 P_1^{(0)}(k+1)\}^{-1} & 0 \\ 0 & 0 \end{bmatrix}$$

$$\tag{7.20}$$

Substituting (7.20) in (7.18), the degenerate matrix Riccati equation becomes

$$P_1^{(0)}(k) = Q_1 + A_{11}' F_1^{(0)}(k+1)A_{11} \tag{7.21a}$$

$$P_2^{(0)}(k) = Q_2 + A_{11}' F_1^{(0)}(k+1)A_{12} \tag{7.21b}$$

$$P_3^{(0)}(k) = Q_3 + A_{12}' F_1^{(0)}(k+1)A_{12} \tag{7.21c}$$

It is seen that (7.21a) is a difference equation of order n_1 whereas (7.21b) and (7.21c) are algebraic equations. That is, if (7.21a) is solved with

$$P_1^{(0)}(N) = P_1(N) = S_1 \tag{7.22a}$$

the solutions for $P_2^{(0)}(k)$ and $P_3^{(0)}(k)$ are automatically fixed from (7.21b) and (7.21c), respectively. This results in

$$P_2^{(0)}(N) \neq P_2(N); \quad P_3^{(0)}(N) \neq P_3(N) \tag{7.22b}$$

Thus in the degeneration process, the $(n_1 + n_2)$-order matrix Riccati equation (7.18) is reduced to the n_1-order matrix Riccati equation (7.21) and in the process loses the final conditions (7.22b). Thus (7.18) is said to be in the singularly perturbed form.

Note: It is seen that the solution of the original matrix Riccati (7.18) involves the inverse of the matrix (7.19) of order $n_1 + n_2$, whereas the degenerate matrix Riccati (7.21) requires the inverse of the matrix (7.20) of order n_1.

7.3 Singular perturbation method

7.3.1 Boundary-layer correction

The degenerate or reduced equation (7.21) offers the advantage of order reduction, but suffers from the inability to satisfy all the given final conditions. This loss of some of the final conditions (7.22b) is attributed to the existence of the boundary layer corresponding to the final point $k = N$. In order to recover these final conditions that are lost in the process of degeneration, a boundary-layer correction is incorporated by using certain transformations as explained below.

The total solution $P(k)$ is written as

$$P(k) = \bar{P}(k) + \bar{W}(k) \tag{7.23}$$

where $\bar{P}(k)$ refers to the outer solution and $\bar{W}(k)$ refers to the boundary-layer-correction solution:

$$P(k) = \begin{bmatrix} \bar{P}_1(k) & \varepsilon\bar{P}_2(k) \\ \varepsilon\bar{P}_2'(k) & \varepsilon^2\bar{P}_3(k) \end{bmatrix}; \quad \bar{W}(k) = \begin{bmatrix} \bar{W}_1(k) & \varepsilon\bar{W}_2(k) \\ \varepsilon\bar{W}_2'(k) & \varepsilon^2\bar{W}_3(k) \end{bmatrix}$$

Based on the results of Chapter 2, where the boundary layer occurs at the final point $k = N$, the transformations are given by

$$W_1(k) = \bar{W}_1(k)/\varepsilon^{N-k+1}; \quad W_2(k) = \bar{W}_2(k)/\varepsilon^{N-k}; \quad W_3(k) = \bar{W}_3(k)/\varepsilon^{N-k} \tag{7.24a}$$

Therefore

$$\bar{W}(k) = \varepsilon^{N-k+1}\begin{bmatrix} W_1(k) & W_2(k) \\ W_2'(k) & \varepsilon W_3(k) \end{bmatrix} = \varepsilon^{N-k+1}W(k) \tag{7.24b}$$

$$\bar{W}(k+1) = \varepsilon^{N-k}W(k+1) \tag{7.24c}$$

Using the above, (7.23) becomes

$$P(k+1) = \bar{P}(k+1) + \varepsilon^{N-k}W(k+1) \tag{7.25}$$

Substituting (7.25) in (7.19) and omitting the argument $(k+1)$ for simplicity, we get

$$F = [\bar{P} + \varepsilon^{N-k}W][I + E\bar{P} + \varepsilon^{N-k}EW]^{-1} \tag{7.26a}$$

Rearranging the terms, we get (Appendix 7.7.1)

$$F = \bar{F} + \varepsilon^{N-k}G \tag{7.26b}$$

where

$$\bar{F} = \bar{P}C; \quad C = [I + E\bar{P}]^{-1}$$

$$G = WC - \bar{P}CEWC - \varepsilon^{N-k}WCEWC$$

Substituting (7.26b) and (7.25) in (7.18), we get

$$\begin{bmatrix} \bar{P}_1(k) + \varepsilon^{N-k+1}W_1(k) & \bar{P}_2(k) + \varepsilon^{N-k}W_2(k) \\ \bar{P}_2'(k) + \varepsilon^{N-k}W_2'(k) & \bar{P}_3(k) + \varepsilon^{N-k}W_3(k) \end{bmatrix}$$

$$= \begin{bmatrix} Q_1 & Q_2 \\ Q_2' & Q_3 \end{bmatrix} + \begin{bmatrix} A_{11}' & A_{21}' \\ A_{12}' & A_{22}' \end{bmatrix}\begin{bmatrix} \bar{F}_1(k+1) & \bar{F}_2(k+1) \\ \bar{F}_2'(k+1) & \bar{F}_3(k+1) \end{bmatrix}$$

$$+ \varepsilon^{N-k}\begin{bmatrix} G_1(k+1) & G_2(k+1) \\ G_2'(k+1) & G_3(k+1) \end{bmatrix}\begin{bmatrix} A_{11} & A_{12} \\ A_{21} & A_{22} \end{bmatrix} \tag{7.27}$$

Separating the outer-series terms from (7.27), we get

$$\begin{bmatrix} \bar{P}_1(k) & \bar{P}_2(k) \\ \bar{P}_2'(k) & \bar{P}_3(k) \end{bmatrix} = \begin{bmatrix} Q_1 & Q_2 \\ Q_2' & Q_3 \end{bmatrix} + \begin{bmatrix} A_{11}' & A_{21}' \\ A_{12}' & A_{22}' \end{bmatrix}$$

$$\times \begin{bmatrix} \bar{F}_1(k+1) & \bar{F}_2(k+1) \\ \bar{F}_2'(k+1) & \bar{F}_3(k+1) \end{bmatrix}\begin{bmatrix} A_{11} & A_{12} \\ A_{21} & A_{22} \end{bmatrix} \tag{7.28}$$

Separating the correction-series terms from (7.27) and dividing all the terms by ε^{N-k}

$$\begin{bmatrix} \varepsilon W_1(k) & W_2(k) \\ W_2'(k) & W_3(k) \end{bmatrix} = \begin{bmatrix} A_{11}' & A_{21}' \\ A_{12}' & A_{22}' \end{bmatrix}$$

$$\times \begin{bmatrix} G_1(k+1) & G_2(k+1) \\ G_2'(k+1) & G_3(k+1) \end{bmatrix} \begin{bmatrix} A_{11} & A_{12} \\ A_{21} & A_{22} \end{bmatrix} \tag{7.29}$$

Note that (7.28) is identical to (7.18) when $\bar{P}$ and $\bar{F}$ are replaced by P and F, respectively. Hence (7.18) is referred to as the outer-series equation. The equation for the boundary-layer correction series is given by (7.29).

It is shown that the expressions for $\bar{F}(k+1)$ and $\bar{G}(k+1)$ can be expressed as power series in ε (Appendix 7.7.2).

Once the equations for the outer series (7.28) and for the BLC series (7.29) are formulated, the method of developing the series solutions is straight-forward. Firstly, the outer solution is assumed as a power series in ε and the controlling equations are obtained. Secondly, a similar procedure is adopted for BLC solution. Finally, the total series solution and the terminal conditions for solving the series equations are obtained. The detailed procedure is given below.

7.3.2 Outer and corrections series

Let the outer series be represented as

$$\bar{P}_i(k) = \bar{P}_i^{(0)}(k) + \varepsilon \bar{P}_i^{(1)}(k) + \ldots \; i = 1, 2, 3 \tag{7.30}$$

Substituting (7.30) in (7.28) and collecting coefficients of like powers of ε, the following series equations are obtained.

For the zeroth-order approximation

$$\begin{bmatrix} \bar{P}_1^{(0)}(k) & \bar{P}_2^{(0)}(k) \\ \bar{P}_2^{(0)'}(k) & \bar{P}_3^{(0)}(k) \end{bmatrix} = \begin{bmatrix} Q_1 & Q_2 \\ Q_2' & Q_3 \end{bmatrix} + \begin{bmatrix} A_{11}' & A_{21}' \\ A_{12}' & A_{22}' \end{bmatrix}$$

$$\times \begin{bmatrix} \bar{F}_1^{(0)}(k+1) & \bar{F}_2^{(0)}(k+1) \\ \bar{F}_2^{(0)'}(k+1) & \bar{F}_3^{(0)}(k+1) \end{bmatrix} \begin{bmatrix} A_{11} & A_{12} \\ A_{21} & A_{22} \end{bmatrix} \tag{7.31a}$$

For the first-order approximation

$$\begin{bmatrix} \bar{P}_1^{(1)}(k) & \bar{P}_2^{(1)}(k) \\ \bar{P}_2^{(1)'}(k) & \bar{P}_3^{(1)}(k) \end{bmatrix} = \begin{bmatrix} A_{11}' & A_{21}' \\ A_{12}' & A_{22}' \end{bmatrix}$$

$$\times \begin{bmatrix} \bar{F}_1^{(1)}(k+1) & \bar{F}_2^{(1)}(k+1) \\ \bar{F}_2^{(1)'}(k+1) & \bar{F}_3^{(1)}(k+1) \end{bmatrix} \begin{bmatrix} A_{11} & A_{12} \\ A_{21} & A_{22} \end{bmatrix} \tag{7.31b}$$

Similar equations are obtained for higher-order approximations.

In (7.31), $\bar{F}^{(0)}(k+1)$ and $\bar{F}^{(1)}(k+1)$ represent the zeroth and first-order coefficient matrices.

Let the correction series be represented as

$$W_i(k) = W_i^{(0)}(k) + \varepsilon W_i^{(1)}(k) + \ldots, \, i = 1, 2, 3 \tag{7.32}$$

Inserting (7.32) in (7.29) and collecting coefficients of like powers of ε we get for the zeroth-order approximation

$$\begin{bmatrix} 0 & W_2^{(0)}(k) \\ W_2^{(0)\prime}(k) & W_3^{(0)}(k) \end{bmatrix} = \begin{bmatrix} A_{11}' & A_{21}' \\ A_{12}' & A_{22}' \end{bmatrix}$$

$$\times \begin{bmatrix} G_1^{(0)}(k+1) & G_2^{(0)}(k+1) \\ G_2^{(0)\prime}(k+1) & G_3^{(0)}(k+1) \end{bmatrix} \begin{bmatrix} A_{11} & A_{12} \\ A_{21} & A_{22} \end{bmatrix} \tag{7.33a}$$

For the first-order approximation

$$\begin{bmatrix} W_1^{(0)}(k) & W_2^{(1)}(k) \\ W_2^{(1)\prime}(k) & W_3^{(1)}(k) \end{bmatrix} = \begin{bmatrix} A_{11}' & A_{21}' \\ A_{12}' & A_{22}' \end{bmatrix}$$

$$\times \begin{bmatrix} G_1^{(1)}(k+1) & G_2^{(1)}(k+1) \\ G_2^{(1)\prime}(k+1) & G_3^{(1)}(k+1) \end{bmatrix} \begin{bmatrix} A_{11} & A_{12} \\ A_{21} & A_{22} \end{bmatrix} \tag{7.33b}$$

Higher-order equations are obtained in a similar way.

In (7.33), $G^{(0)}(k+1)$ and $G^{(1)}(k+1)$ represent the zeroth and first-order-coefficient matrices.

7.3.3   Total series solution

The total series solution is obtained by proper summing of the outer series (7.30) and the correction series (7.32). That is

$$P_1(k) = [\bar{P}_1^{(0)}(k) + \varepsilon \bar{P}_1^{(1)}(k) + \ldots] + \varepsilon^{N-k+1}[W_1^{(0)}(k)$$
$$+ \varepsilon W_1^{(1)}(k) + \ldots] \tag{7.34a}$$

$$P_i(k) = [\bar{P}_i^{(0)}(k) + \varepsilon \bar{P}_i^{(1)}(k) + \ldots]$$
$$+ \varepsilon^{N-k}[W_i^{(0)}(k) + \varepsilon W_i^{(1)}(k) + \ldots], \quad i = 2, 3 \tag{7.34b}$$

The final conditions required to solve the outer-series (7.31) and the final BLC series (7.33) are obtained based on the fact that the total series solution (7.34) should satisfy the given final conditions. That is

$$\begin{aligned} \varepsilon^0: \quad & \bar{P}_1^{(0)}(N) = P_1(N); \quad W^{(0)}(N) = P_i(N) - \bar{P}_i^{(0)}(N), \quad i = 2, 3 \\ \varepsilon^1: \quad & \bar{P}_1^{(1)}(N) = -W_1^{(0)}(N); \quad W_i^{(1)}(N) = -\bar{P}_i^{(1)}(N), \quad i = 2, 3 \end{aligned} \tag{7.35}$$

Important Note: Owing to the association of coefficients ε^{N-k+1} and ε^{N-k} with the correction series in the total series solution (7.34), the correction-series (7.33) needs to be solved for only a few values of k near the final point, utilising the inverse of the reduced order matrix [for $G^{(j)}(k+1)$], which is already evaluated for the solution of the outer-series (7.31). Thus the solution of the correction-series (7.33) does not require any extra matrix inversions and is obtained by simple recursion. This aspect of the correction series offers a considerable reduction in the overall computation for obtaining the approximate solutions.

Example 7.2
Consider a second-order discrete control system

$$\begin{bmatrix} y(k+1) \\ z(k+1) \end{bmatrix} = \begin{bmatrix} 0{\cdot}95 & \varepsilon \\ 1{\cdot}0 & 0 \end{bmatrix} \begin{bmatrix} y(k) \\ z(k) \end{bmatrix} + \begin{bmatrix} 1 \\ 0 \end{bmatrix} u(k) \qquad (7.36)$$

where the small parameter $\varepsilon = 0{\cdot}12$.
  The initial conditions are

$$y(0) = 10{\cdot}0; \quad z(0) = 15{\cdot}0$$

and the performance index is

$$J = \frac{1}{2} \sum_{k=0}^{N-1} [x'(k)Qx(k) + Ru(k)^2] \qquad (7.37)$$

where $\quad Q = \begin{bmatrix} 1 & 0 \\ 0 & 2\varepsilon^2 \end{bmatrix}; \quad x(k) = \begin{bmatrix} y(k) \\ z(k) \end{bmatrix}; \quad R = 2 \text{ and } N = 8$

$$BR^{-1}B' = \begin{bmatrix} 0{\cdot}5 & 0 \\ 0 & 0 \end{bmatrix}, \quad \text{where,} \quad B = \begin{bmatrix} 1 \\ 0 \end{bmatrix}$$

The singularly perturbed matrix Riccati difference equation becomes

$$\begin{bmatrix} p_1(k) & \varepsilon p_2(k) \\ \varepsilon p_2'(k) & \varepsilon^2 p_3(k) \end{bmatrix} = \begin{bmatrix} 1 & 0 \\ 0 & 2\varepsilon^2 \end{bmatrix} + \begin{bmatrix} 0{\cdot}95 & 1{\cdot}0 \\ \varepsilon & 0 \end{bmatrix}$$

$$\left[\left\{ \begin{bmatrix} p_1(k+1) & \varepsilon p_2(k+1) \\ \varepsilon p_2'(k+1) & \varepsilon^2 p_3(k+1) \end{bmatrix} \right\}^{-1} + \begin{bmatrix} 0{\cdot}5 & 0 \\ 0 & 0 \end{bmatrix} \right]^{-1} \begin{bmatrix} 0{\cdot}95 & \varepsilon \\ 1 & 0 \end{bmatrix} \qquad (7.38)$$

The closed-loop optimal control is given by

$$u(k) = -0\cdot5[1 \quad 0]\left[\left\{\begin{matrix} p_1(k+1) & \varepsilon p_2(k+1) \\ \varepsilon p_2'(k+1) & \varepsilon^2 p_3(k+1) \end{matrix}\right\}^{-1} + \left\{\begin{matrix} 0\cdot5 & 0 \\ 0 & 0 \end{matrix}\right\}\right]^{-1}$$

$$\times \begin{bmatrix} 0\cdot95 & \varepsilon \\ 1 & \varepsilon \end{bmatrix}\begin{bmatrix} y(k) \\ z(k) \end{bmatrix} \tag{7.39}$$

Proceeding on the lines suggested above, the series equations for zeroth- and first-order approximations are obtained.

Outer Series
For zeroth order

$$p_1^{(0)}(k) = \frac{2 + 2\cdot805 p_1^{(0)}(k+1)}{2 + p_1^{(0)}(k+1)} \tag{7.40a}$$

$$p_2^{(0)}(k) = \frac{1\cdot9 p_1^{(0)}(k+1)}{2 + p_1^{(0)}(k+1)} \tag{7.40b}$$

$$p_3^{(0)}(k) = \frac{4 + 4 p_1^{(0)}(k+1)}{2 + p_1^{(0)}(k+1)} \tag{7.40c}$$

For first order

$$p_1^{(1)}(k) = \frac{p_1^{(1)}(k+1)[2\cdot805 - p_1^{(0)}(k)] + 3\cdot8 p_2^{(0)}(k+1)}{2 + p_1^{(0)}(k+1)} \tag{7.41a}$$

$$p_2^{(1)}(k) = \frac{p_1^{(1)}(k+1)[1\cdot9 - p_2^{(0)}(k)] + 2 p_2^{(0)}(k+1)}{2 + p_1^{(0)}(k+1)} \tag{7.41b}$$

$$p_3^{(1)}(k) = \frac{p_1^{(1)}(k+1)[4 - p_3^{(0)}(k)]}{2 + p_1^{(0)}(k+1)} \tag{7.41c}$$

Correction series
For zeroth order

$$0 = w_1^{(0)}(k+1)[2\cdot805 - p_1^{(0)}(k)] + 3\cdot8 w_2^{(0)}(k+1) \tag{7.42a}$$

$$w_2^{(0)}(k) = \frac{w_1^{(0)}(k+1)[1\cdot 9 - p_2^{(0)}(k) - \varepsilon^{N-k}w_2^{(0)}(k)] + 2w_2^{(0)}(k+1)}{2 + p_1^{(0)}(k+1)}$$

(7.42b)

$$w_3^{(0)}(k) = \frac{w_1^{(0)}(k+1)[4 - \varepsilon^{N-k}w_3^{(0)}(k)] - w_1^{(0)}(k)p_3^{(0)}(k)}{2 + p_1^{(0)}(k+1)}$$

(7.42c)

The total series solution is given by

$$p_1(k) = [p_1^{(0)}(k) + \varepsilon p_1^{(1)}(k) + \varepsilon^2 p_1^{(2)}(k) + \ldots]$$
$$+ \varepsilon^{N-k+1}[w_1^{(0)}(k) + \varepsilon w_1^{(1)}(k) + \ldots]$$

(7.43a)

$$p_i(k) = [p_i^{(0)}(k) + \varepsilon p_i^{(1)}(k) + \varepsilon^2 p_i^{(2)}(k) + \ldots]$$
$$+ \varepsilon^{N-k}[w_i^{(0)}(k) + \varepsilon w_i^{(1)}(k) + \ldots], \quad i = 2, 3$$

(7.43b)

Note: In (7.42), the terms associated with ε^{N-k} are to be handled carefully, as explained in Section 7.1. It is evident that the correction-series equations are to be solved for k less than N only. Therefore those terms associated with ε^{N-k} should not be considered for the zeroth-order approximation.

The Riccati coefficients $p(k)$ corresponding to degenerate-, zeroth- and first-order solutions are evaluated. The results are compared with the exact solution obtained from (7.38) in Table 7.2. Using the series solutions obtained, the closed-loop optimal control and the state trajectories are evaluated from (7.39) and (7.36), respectively. The results are compared with the exact solution as shown in Table 7.3. The performance indices are shown in Table 7.4.

Example 7.3*
Consider the fourth-order discrete control problem

$$\begin{bmatrix} y_1(k+1) \\ y_2(k+1) \\ z_1(k+1) \\ z_2(k+1) \end{bmatrix} = \begin{bmatrix} 1\cdot 0 & 1\cdot 0 & 1\cdot 0\varepsilon & 1\cdot 0\varepsilon \\ 0\cdot 0 & 1\cdot 0 & 0\cdot 0\varepsilon & 0\cdot 1\varepsilon \\ 1\cdot 0 & -1\cdot 0 & 0\cdot 1\varepsilon & 0\cdot 5\varepsilon \\ 0\cdot 2 & 0\cdot 0 & 0\cdot 0\varepsilon & 1\cdot 0\varepsilon \end{bmatrix} \begin{bmatrix} y_1(k) \\ y_2(k) \\ z_1(k) \\ z_2(k) \end{bmatrix} + \begin{bmatrix} 1 \\ 1 \\ 0 \\ 0 \end{bmatrix} u(k) \quad (7.44)$$

where the small parameter $\varepsilon = 0\cdot 1$.
The initial conditions are

$$\begin{bmatrix} y_1(0) \\ y_2(0) \end{bmatrix} = \begin{bmatrix} 10\cdot 0 \\ 8\cdot 0 \end{bmatrix}; \quad \begin{bmatrix} z_1(0) \\ z_2(0) \end{bmatrix} = \begin{bmatrix} 5\cdot 0 \\ 2\cdot 0 \end{bmatrix}$$

* Reprinted by permission from NAIDU, D. S., and RAO, A. K., 'Singular perturbation analysis of closed-loop discrete optimal control problem', *Optimal Control: Applications & Methods*, **55**, pp. 19–38.

Table 7.2 *Comparison of the approximate and exact solutions of the matrix Riccati equation of Example 7.2*

$p(k)$	Degenerate solution	Zeroth-order solution	First-order solution	Exact solution
$p_1(0)$	1·8728	1·8728	2·0151	2·0734
$p_2(0)$	0·9192	0·9192	1·0117	1·0275
$p_3(0)$	2·9653	2·953	3·0069	3·0208
$p_1(1)$	1·8726	1·8726	2·0146	2·0724
$p_2(1)$	0·9183	0·9183	1·0108	1·0267
$p_3(1)$	2·9653	2·9653	3·0069	3·0278
$p_1(2)$	1·8719	1·8719	2·0125	2·0695
$p_2(2)$	0·9175	0·9175	1·0092	1·0242
$p_3(2)$	2·9653	2·9653	3·0000	3·0139
$p_1(3)$	1·8687	1·8687	2·0005	2·0601
$p_2(3)$	0·9142	0·9142	1·0008	1·0158
$p_3(3)$	2·9653	2·9653	2·9931	3·0069
$p_1(4)$	1·8557	1·8557	1·9770	2·0285
$p_2(4)$	0·9008	0·9008	0·9750	0·9867
$p_3(4)$	2·9514	2·9514	2·9722	2·9792
$p_1(5)$	1·8027	1·8027	1·8829	1·9270
$p_2(5)$	0·8450	0·8450	0·8875	0·8950
$p_3(5)$	2·8889	2·8889	2·8889	2·8958
$p_1(6)$	1·6017	1·6017	1·6017	1·6305
$p_2(6)$	0·6333	0·6333	0·6333	0·6333
$p_3(6)$	2·6667	2·6667	2·6667	2·6667
$p_1(7)$	1·0000	1·0000	1·0000	1·0000
$p_2(7)$	0·0000	0·0000	0·0000	0·0000
$p_3(7)$	2·0000	2·0000	2·0000	2·0000
$p_1(8)$	0·0000	0·0000	0·0000	0·0000
$p_2(8)$	−1·0525	0·0000	0·0000	0·0000
$p_3(8)$	0·8889	0·0000	0·0000	0·0000

and the performance index is

$$J = \frac{1}{2} \sum_{k=0}^{N-1} [x'(k)Qx(k) + u'(k)Ru(k)]$$

(7.45)

where

$$Q = \begin{bmatrix} 1 & 0 & 0 & 0 \\ 0 & 1 & 0 & 0 \\ 0 & 0 & 0 & 0 \\ 0 & 0 & 0 & 0 \end{bmatrix}; \quad x(k) = \begin{bmatrix} y_1(k) \\ y_2(k) \\ z_1(k) \\ z_2(k) \end{bmatrix}$$

Table 7.3 *Comparison of approximate and exact solutions of states and optimal control of Example 7.2*

$y(k)/z(k)$ $u(k)$	Degenerate solution	Zeroth-order solution	First-order solution	Exact solution
$y(0)$	10·0000	10·0000	10·0000	10·0000
$z(0)$	15·0000	15·0000	15·0000	15·0000
$u(0)$	− 5·7488	− 5·7488	− 5·9727	− 6·0529
$y(1)$	5·5512	5·5512	5·3273	5·2471
$z(1)$	10·0000	10·0000	10·0000	10·0000
$u(1)$	− 3·2876	− 3·2876	− 3·3009	− 3·3037
$y(2)$	3·1860	3·1860	2·9600	2·8811
$z(2)$	5·5512	5·5512	5·3273	5·2461
$u(2)$	− 1·8741	− 1·8741	− 1·8147	− 1·7947
$y(3)$	1·8187	1·8187	1·6365	1·5720
$z(3)$	3·1860	3·1860	2·9600	2·8811
$u(3)$	− 1·0665	− 1·0665	− 0·9976	− 0·9723
$y(4)$	1·0436	1·0436	0·9123	0·8668
$z(4)$	1·8187	1·8187	1·6365	1·5720
$u(4)$	− 0·6013	− 0·6013	− 0·5405	− 0·5204
$y(5)$	0·6084	0·6084	0·5226	0·4918
$z(5)$	1·0436	1·0436	0·9123	0·8668
$u(5)$	− 0·3255	− 0·3255	− 0·2805	− 0·2668
$y(6)$	0·3776	0·3776	0·3254	0·3044
$z(6)$	0·6084	0·6084	0·5226	0·4918
$u(6)$	− 0·1439	− 0·1439	− 0·1240	− 0·1161
$y(7)$	0·2878	0·2878	0·2479	0·2321
$z(7)$	0·3776	0·3776	0·3254	0·3044
$u(7)$	0·0182	0·0000	0·0000	0·0000
$y(8)$	0·3214	0·3188	0·2746	0·2570
$z(8)$	0·2778	0·2878	0·2479	0·2321

Table 7.4 *Comparison of performance indices of approximate and exact solutions of Example 7.2*

Nature of the solution	Performance index
Degenerate	127·3160
Zeroth order	127·3153
First order	127·0612
Exact	127·0424

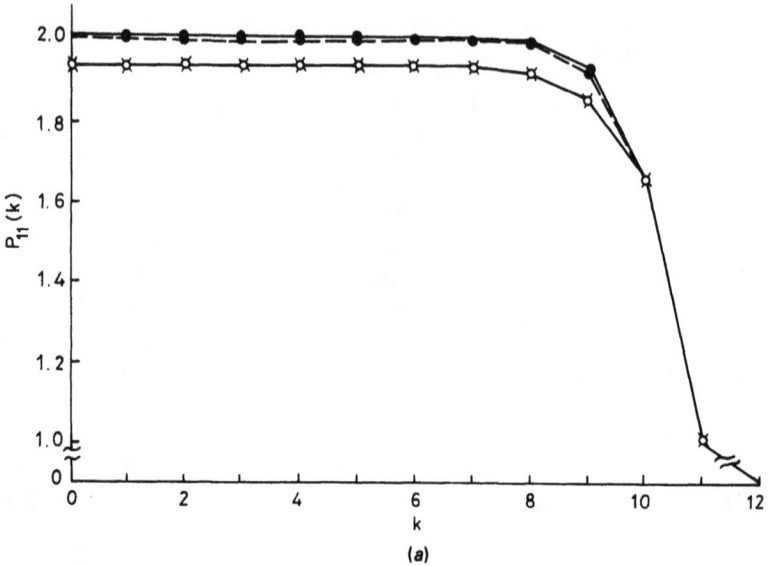

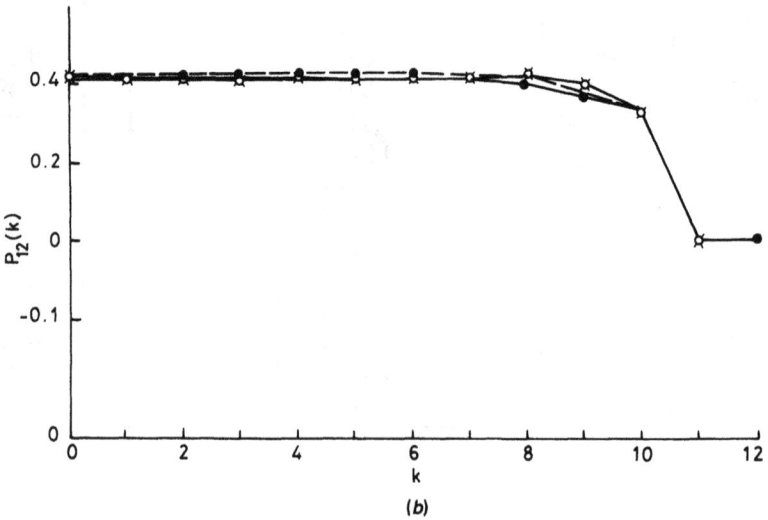

Fig. 7.1a and b   *Exact and approximate solutions of $p_{11}(k)$ and $p_{12}(k)$ of Example 7.3*

●———●———● exact solution
●– – –●– – –● first-order solution
×———×———× zeroth-order solution
○———○———○ degenerate solution
¤———¤———¤ solution common to degenerate and zeroth order

$R = 1 \cdot 0$ and $N = 12$.

The eigenvalues of (7.44) are

$$b_1 = 1 \cdot 1238; \quad b_2 = 0 \cdot 98549; \quad b_3 = -0 \cdot 09298; \quad b_4 = 0 \cdot 084659$$

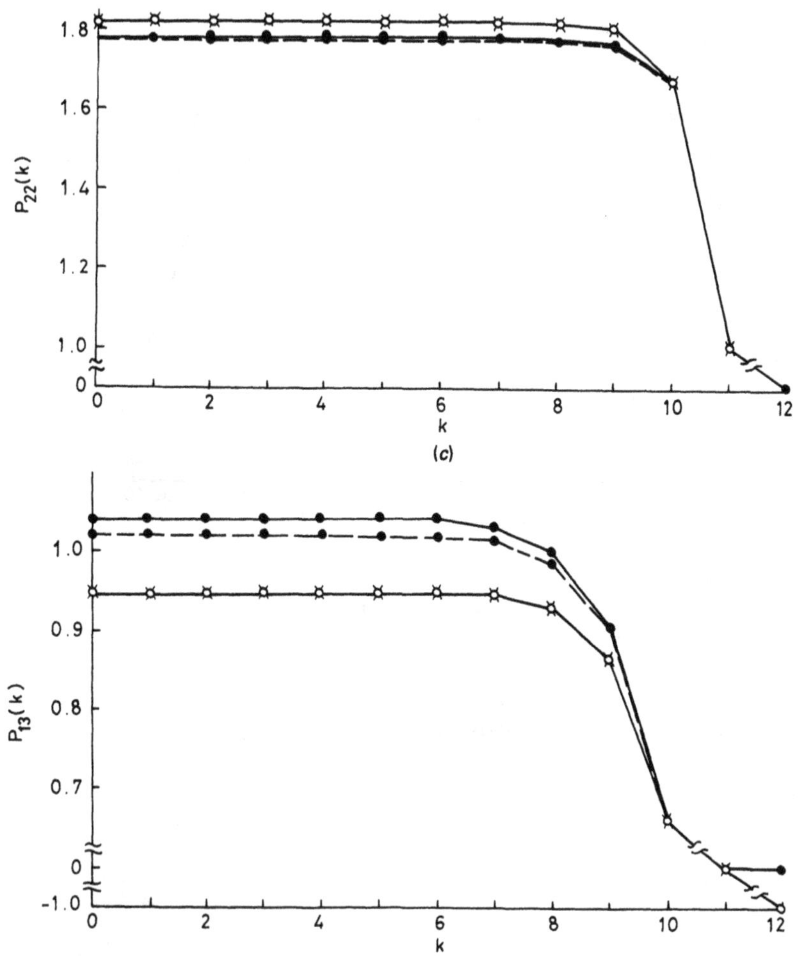

Fig. 7.1c and d *Exact and approximate solutions of $p_{22}(k)$ and $p_{13}(k)$ of Example 7.3*
●——●——● exact solution
●– – –●– – –● first-order solution
×——×——× zeroth-order solution
○——○——○ degenerate solution
¤——¤——¤ solution common to degenerate and zeroth order

Following the method described, the degenerate, zeroth- and first-order series equations of the matrix Riccati equation are solved and the series solutions obtained are compared with the exact solution in Fig. 7.1. Using the series solutions obtained as above, the optimal control and the state trajectories are found (7.16) and (7.14), respectively. The optimal control of the series solutions are compared with the exact optimal control in Table 7.5. The state trajectories are shown in Fig. 7.2. Finally, the performance indices are

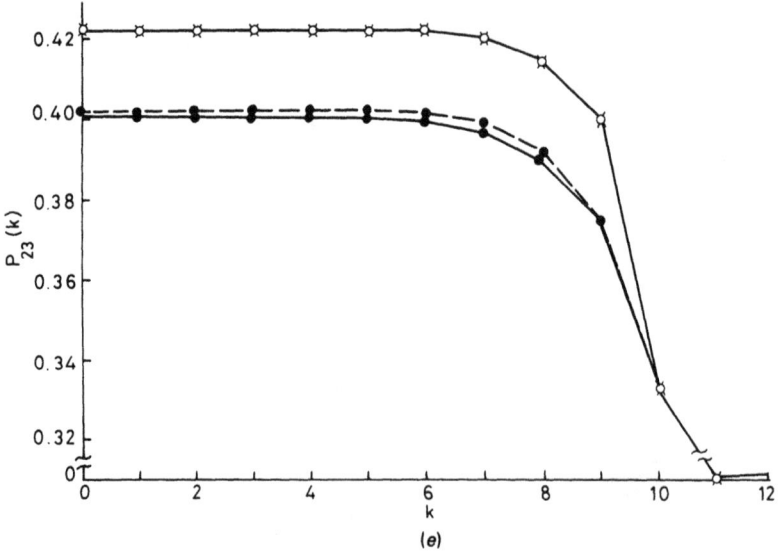

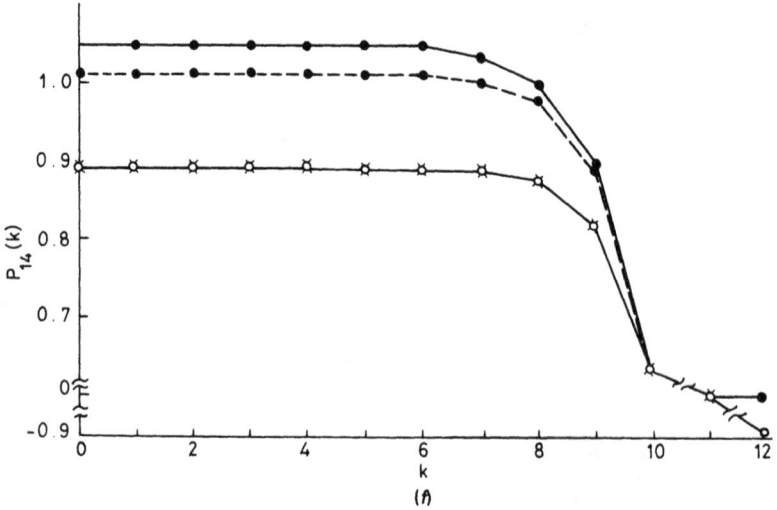

Fig. 7.1e and f *Exact and approximate solutions of $p_{23}(k)$ and $p_{14}(k)$ of Example 7.3*

●———●———● exact solution
●----●----● first-order solution
×———×———× zeroth-order solution
○———○———○ degenerate solution
¤———¤———¤ solution common to degenerate and zeroth order

evaluated for degenerate, zeroth-, and first-order solutions and compared with the exact performance index in Table 7.6.

With the increased order of approximation, the series solutions approach the exact solution very closely.

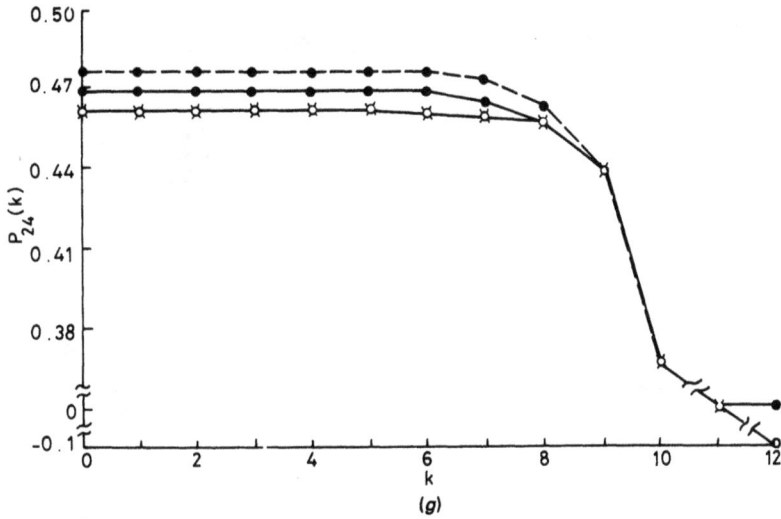

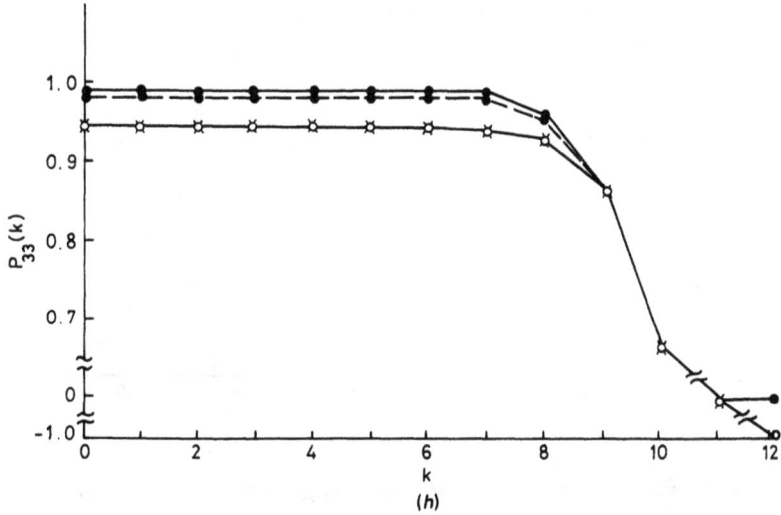

Fig. 7.1g and h   *Exact and approximate solutions of $p_{24}(k)$ and $p_{33}(k)$ of Example 7.3*
- ●———●———● exact solution
- ●– – –●– – –● first-order solution
- ×———×———× zeroth-order solution
- ○———○———○ degenerate solution
- ¤———¤———¤ solution common to degenerate and zeroth order

7.4   Steady-state solution

In many practical situations, the steady-state solution of the matrix Riccati equation (7.17) is of vital importance. The steady-state equation is an algebraic equation which is obtained from (7.18) by writing $P(k + 1) = P(k) = P$.

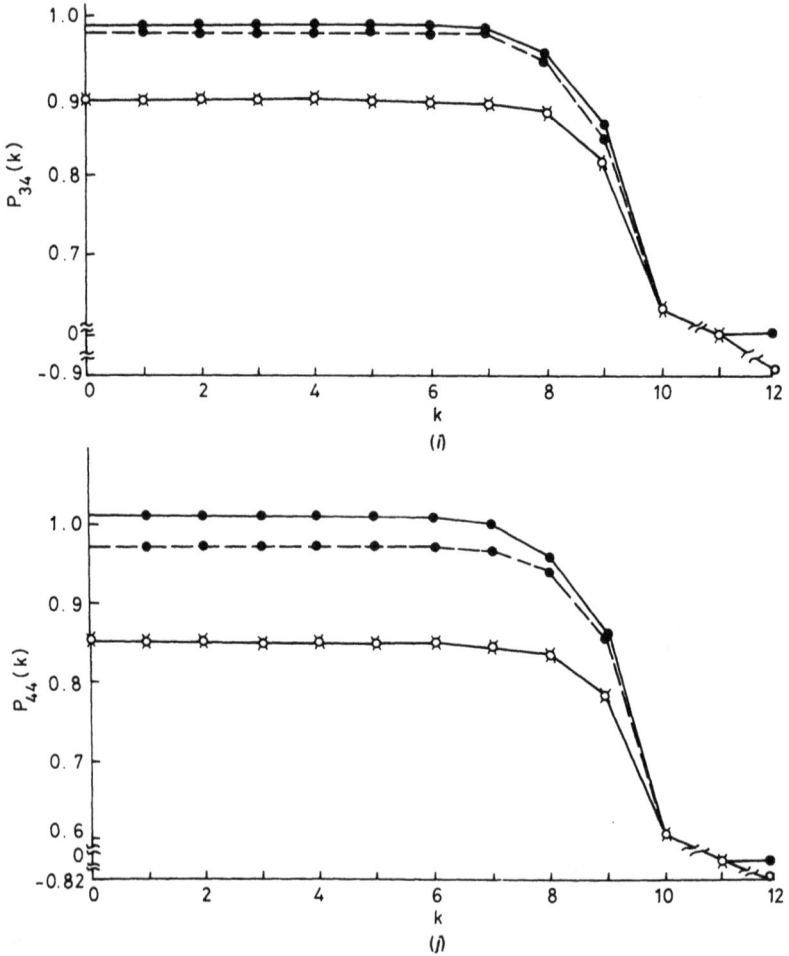

Fig. 7.1i and j Exact and approximate solutions of $p_{34}(k)$ and $p_{44}(k)$ of Example 7.3
●——●——● exact solution
●---●---● first-order solution
×——×——× zeroth-order solution
○——○——○ degenerate solution
¤——¤——¤ solution common to degenerate and zeroth order

That is,

$$\begin{bmatrix} P_1 & P_2 \\ P_2' & P_3 \end{bmatrix} = \begin{bmatrix} Q_1 & Q_2 \\ Q_2' & Q_3 \end{bmatrix} + \begin{bmatrix} A_{11}' & A_{21}' \\ A_{12}' & A_{22}' \end{bmatrix}$$

$$\times \begin{bmatrix} F_1 & \varepsilon F_2 \\ \varepsilon F_2' & \varepsilon^2 F_3 \end{bmatrix} \begin{bmatrix} A_{11} & A_{12} \\ A_{21} & A_{22} \end{bmatrix} \tag{7.46}$$

Table 7.5 *Comparison of approximate and exact optimal control of Example 7.3*

$u(k)$	Degenerate solution	Zeroth-order solution	First-order solution	Exact solution
$u(0)$	$-11{\cdot}2047$	$-11{\cdot}2047$	$-11{\cdot}3912$	$-11{\cdot}4260$
$u(1)$	$-\ 1{\cdot}0440$	$-\ 1{\cdot}0440$	$-\ 0{\cdot}9284$	$-\ 0{\cdot}9049$
$u(2)$	$1{\cdot}1394$	$1{\cdot}1394$	$1{\cdot}2481$	$1{\cdot}2680$
$u(3)$	$1{\cdot}1911$	$1{\cdot}1911$	$1{\cdot}2386$	$1{\cdot}2463$
$u(4)$	$0{\cdot}8172$	$0{\cdot}8172$	$0{\cdot}8203$	$0{\cdot}8199$
$u(5)$	$0{\cdot}4837$	$0{\cdot}4837$	$0{\cdot}4672$	$0{\cdot}4637$
$u(6)$	$0{\cdot}2651$	$0{\cdot}2651$	$0{\cdot}2447$	$0{\cdot}2409$
$u(7)$	$0{\cdot}1386$	$0{\cdot}1386$	$0{\cdot}1213$	$0{\cdot}1182$
$u(8)$	$0{\cdot}0701$	$0{\cdot}0701$	$0{\cdot}0576$	$0{\cdot}0555$
$u(9)$	$0{\cdot}0341$	$0{\cdot}0341$	$0{\cdot}0262$	$0{\cdot}0248$
$u(10)$	$0{\cdot}0147$	$0{\cdot}0147$	$0{\cdot}0102$	$0{\cdot}0095$
$u(11)$	$0{\cdot}0209$	$0{\cdot}0000$	$0{\cdot}0000$	$0{\cdot}0000$

Table 7.6 *Comparison of performance indices of approximate and exact solutions of Example 7.3*

Degenerate solution	$205{\cdot}46780$
Zeroth-order solution	$205{\cdot}46736$
First-order solution	$205{\cdot}24781$
Exact solution	$205{\cdot}21266$

In order to describe a singular perturbation method for (7.46), we write the outer series as

$$P_i = P_i^{(0)} + \varepsilon P_i^{(1)} + \ldots, \quad i = 1, 2, 3 \tag{7.47}$$

Inserting (7.47) in (7.46) and collecting coefficients of like powers of ε on either side, we get for the zeroth-order approximation

$$\begin{bmatrix} P_1^{(0)} & P_2^{(0)} \\ P_2^{(0)}{}' & P_3^{(0)} \end{bmatrix} = \begin{bmatrix} Q_1 & Q_2 \\ Q_2' & Q_3 \end{bmatrix} + \begin{bmatrix} A_{11}' & A_{21}' \\ A_{12}' & A_{22}' \end{bmatrix}$$

$$\times \begin{bmatrix} F_1^{(0)} & 0 \\ 0 & 0 \end{bmatrix} \begin{bmatrix} A_{11} & A_{12} \\ A_{21} & A_{22} \end{bmatrix} \tag{7.48}$$

where $F_1^{(0)} = P_1^{(0)}[I_1 + E_1 P_1^{(0)}]^{-1}$

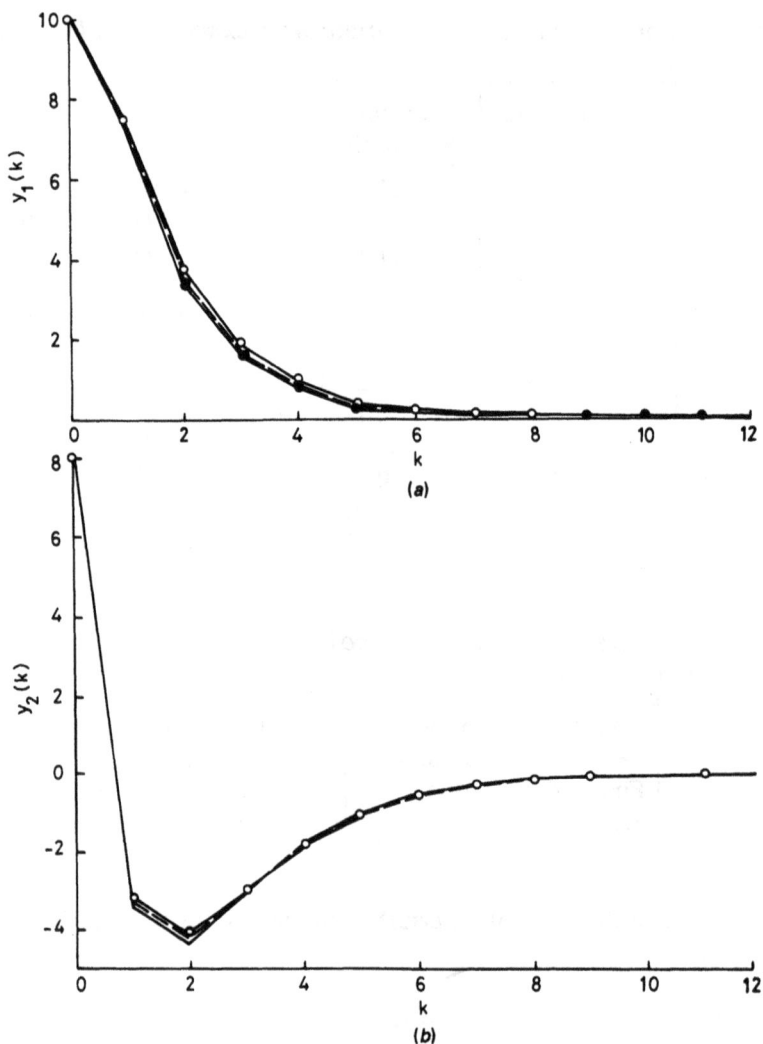

Fig. 7.2a and b   *Exact and approximate solutions of $y_1(k)$ and $y_2(k)$ of Example 7.3*
- ●——●——● exact solution
- ●－－●－－● first-order solution
- ○——○——○ zeroth-order solution

Rearranging (7.48), we get

$$P_1^{(0)} = Q_1 + A_{11}' F_1^{(0)} A_{11} \qquad\qquad (7.49a)$$

$$P_2^{(0)} = Q_2 + A_{11}' F_1^{(0)} A_{12} \qquad\qquad (7.49b)$$

$$P_3^{(0)} = Q_3 + A_{12}' F_1^{(0)} A_{12} \qquad\qquad (7.49c)$$

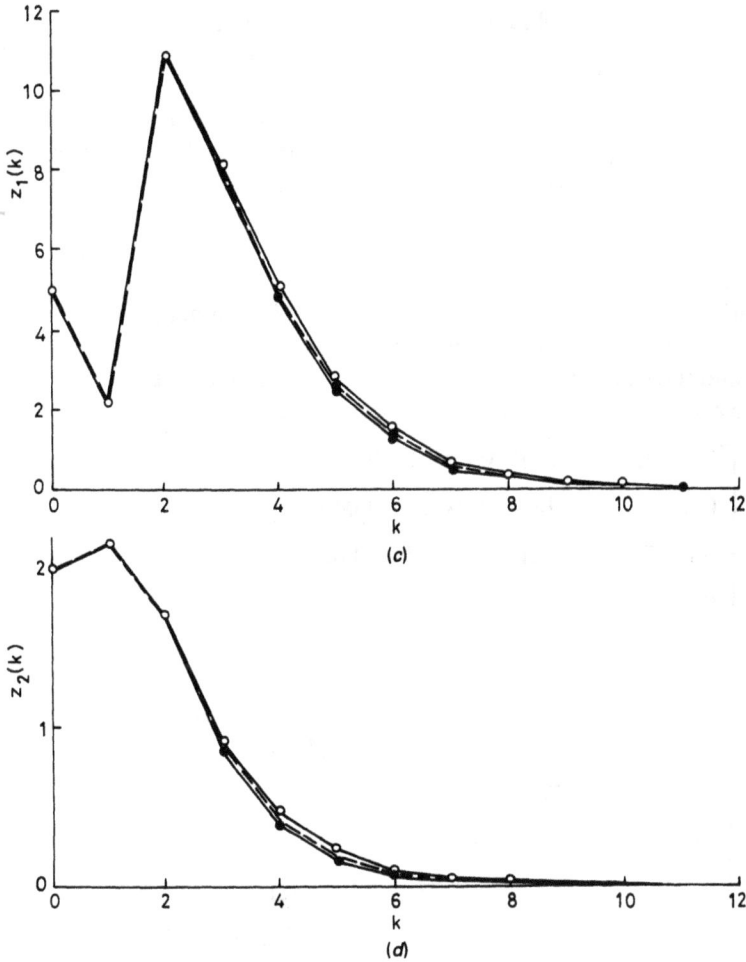

Fig. 7.2c and d   Exact and approximate solutions of $z_1(k)$ and $z_2(k)$ of Example 7.3
● ——— ● ——— ● exact solution
● – – – ● – – – ● first-order solution
○ ——— ○ ——— ○ zeroth-order solution

It is clear from (7.49) that the first equation is an algebraic equation in $P_1^{(0)}$ of order n_1 and decoupled from $P_2^{(0)}$ and $P_3^{(0)}$ which are obtained once $P_1^{(0)}$ is known. On the other hand, the original steady-state equation (7.46) is an algebraic equation of order $n_1 + n_2$ and coupled with reference to P_1, P_2 and P_3.

For the first-order approximation

$$\begin{bmatrix} P_1^{(1)} & P_2^{(1)} \\ P_2^{(1)\prime} & P_3^{(1)} \end{bmatrix} = \begin{bmatrix} A_{11}' & A_{21}' \\ A_{12}' & A_{22}' \end{bmatrix} \begin{bmatrix} F_1^{(1)} & F_2^{(0)} \\ F_2^{(0)\prime} & 0 \end{bmatrix} \begin{bmatrix} A_{11} & A_{12} \\ A_{21} & A_{22} \end{bmatrix} \tag{7.50}$$

Similar equations are obtained for higher-order approximations.

Once again, (7.50) is an algebraic equation in $P_1^{(1)}$, and $P_2^{(1)}$ and $P_3^{(1)}$ are obtained once $P_1^{(1)}$ is known.

In the steady-state case, there is no finite final condition specified for P, and hence there is no need for any final boundary-layer-correction series. The total series solution is given by the outer series itself (7.47).

Example 7.4

Consider the steady-state case of Example 7.3. The matrix Riccati algebraic equation is solved for zeroth- and first-order approximations. The series solutions and the exact steady-state solution are obtained as below.

The zeroth-order solution of the Riccati matrix is:

$$\begin{bmatrix} 1{\cdot}94712 & 0{\cdot}42208 & 0{\cdot}09471 & 0{\cdot}08946 \\ 0{\cdot}42208 & 1{\cdot}82185 & 0{\cdot}04221 & 0{\cdot}04621 \\ 0{\cdot}09471 & 0{\cdot}04221 & 0{\cdot}00947 & 0{\cdot}00895 \\ 0{\cdot}08946 & 0{\cdot}04621 & 0{\cdot}00895 & 0{\cdot}00851 \end{bmatrix}$$

The first-order solution is:

$$\begin{bmatrix} 2{\cdot}05944 & 0{\cdot}42964 & 0{\cdot}10198 & 0{\cdot}10118 \\ 0{\cdot}42964 & 1{\cdot}77564 & 0{\cdot}04028 & 0{\cdot}04767 \\ 0{\cdot}10198 & 0{\cdot}04028 & 0{\cdot}00980 & 0{\cdot}00973 \\ 0{\cdot}10118 & 0{\cdot}04767 & 0{\cdot}00973 & 0{\cdot}00973 \end{bmatrix}$$

The exact solution is:

$$\begin{bmatrix} 2{\cdot}09185 & 0{\cdot}42208 & 0{\cdot}10394 & 0{\cdot}10477 \\ 0{\cdot}42208 & 1{\cdot}78101 & 0{\cdot}04010 & 0{\cdot}04690 \\ 0{\cdot}10394 & 0{\cdot}04010 & 0{\cdot}00995 & 0{\cdot}00995 \\ 0{\cdot}10477 & 0{\cdot}04690 & 0{\cdot}00995 & 0{\cdot}01013 \end{bmatrix}$$

Using the steady-state values of Riccati coefficients shown above, the state trajectories and the optimal control are evaluated for zeroth- and first-order approximations. The results are compared with the exact steady-state solutions in Table 7.7 (optimal control) and Fig. 7.3 (state trajectories). The performance indices are shown in Table 7.8.

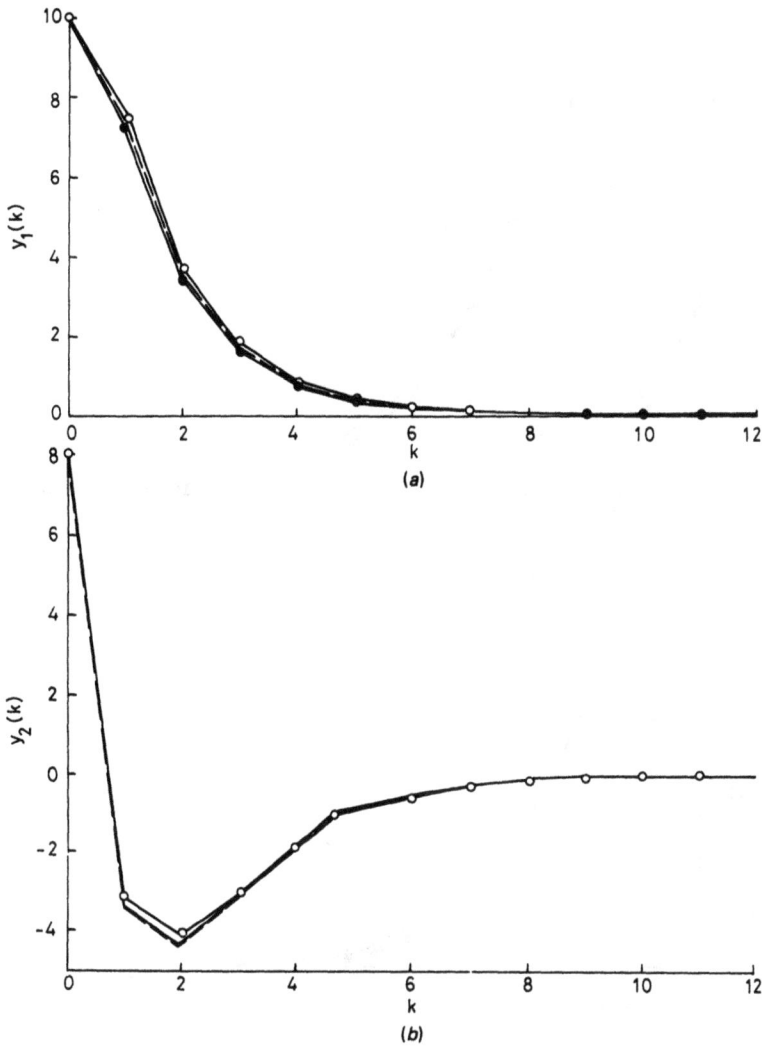

Fig. 7.3a and b   *Exact and approximate solutions of $y_1(k)$ and $y_2(k)$ of Example 7.4*
●——●——● exact solution
●– – –●– – –● first-order solution
○——○——○ zeroth-order solution

7.5   Time-scale analysis of a regulator problem

The spread of reliable and inexpensive micro-computers has recently aroused considerable interest in digital control systems (Mahmoud and Singh, 1984). Digital flight-control systems are known for their flexibility, reliability and easy implementability. The theory of singular perturbations and time scales

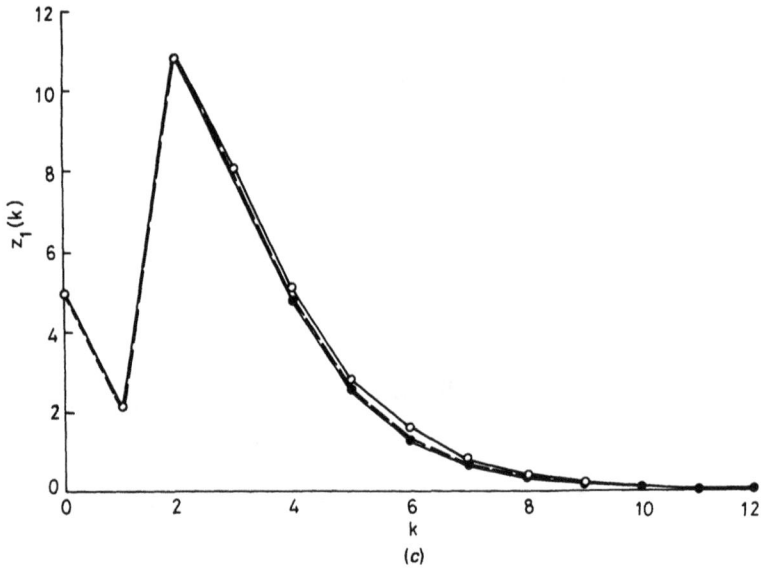

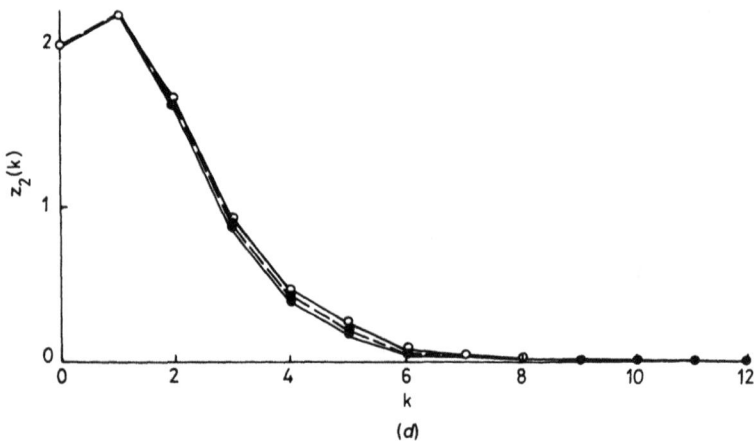

Fig. 7.3c and d   *Exact and approximate solutions of $z_1(k)$ and $z_2(k)$ of Example 7.4*
●———●———● exact solution
●– – –●– – –● first-order solution
○———○———○ zeroth-order solution

with their twin advantages of order reduction and stiffness removal has been successful in the analysis and synthesis of continuous flight-control systems (Ardema, 1977; Mehra *et al.*, 1979). However the digital flight-control system with two-time-scale character has not received due attention.

In this Section, we consider a two-time-scale discrete control system. The closed-loop optimal control of the system using regulator theory involves the

Table 7.7 *Comparison of approximate and exact optimal control of Example 7.4*

$u(k)$	Zeroth-order solution	First-order solution	Exact solution
$u(0)$	−11·2047	−11·3912	−11·4260
$u(1)$	− 1·0440	− 0·9284	− 0·9049
$u(2)$	1·1394	1·2481	1·2680
$u(3)$	1·1911	1·2386	1·2463
$u(4)$	0·8172	0·8203	0·8199
$u(5)$	0·4837	0·4672	0·4636
$u(6)$	0·2651	0·2447	0·2409
$u(7)$	0·1385	0·1212	0·1182
$u(8)$	0·0700	0·0576	0·0555
$u(9)$	0·0345	0·0264	0·0251
$u(10)$	0·0167	0·0118	0·0110
$u(11)$	0·0080	0·0051	0·0047

Table 7.8 *Comparison of performance indices of series and exact solutions of Example 7.4*

Zeroth-order solution	205·5675
First-order solution	205·2479
Exact solution	205·2428

solution of the full-order algebraic matrix Riccati equation. On the other hand, we decompose the original system into slow and fast subsystems (Phillips, 1980a; Kando and Iwazumi, 1983c). The closed-loop optimal control of the slow and fast subsystems leads to the solution of two reduced-order algebraic matrix Riccati equation. A composite, feedback, sub-optimal control is composed out of the slow and fast feedback optimal controls. The method is applied to a fifth-order digital aircraft model possessing three slow modes and two fast modes. The comparison of performance indices with the original optimal control and the composite sub-optimal control are found to be in excellent agreement (Naidu and Price, 1987).

The highlight of the method is that it is computationally much simpler to obtain the composite suboptimal control from the lower-order sub-systems than to obtain the optimal control from the original higher-order system. This aspect of computational simplicity is believed to be a very desirable feature for the onboard real-time computation of adaptive feedback control laws for digital flight systems.

7.5.1   Slow and fast subsystems

Consider a stable, linear, time-invariant discrete control system possessing a two-time-scale (slow and fast) character as

$$\begin{bmatrix} x(k+1) \\ z(k+1) \end{bmatrix} = \begin{bmatrix} A_{11} & A_{12} \\ A_{21} & A_{22} \end{bmatrix} \begin{bmatrix} x \\ z \end{bmatrix} + \begin{bmatrix} B_1 \\ B_2 \end{bmatrix} u(k) \tag{7.51}$$

with initial conditions $x(k=0) = x(0)$; $z(k=0) = z(0)$, where $x(k)$ and $z(k)$ are n- and m-dimensional slow and fast state variables, respectively, $u(k)$ is an r-dimensional control variable, A_{ij}s and B_is are submatrices of appropriate dimensionality.

The two-time-scale character of (7.51) implies that

$$\max |p(A_f)| \ll \min |p(A_s)| \tag{7.52}$$

where the eigenvalues are the disjoint union

$$\left. \begin{aligned} p(A) &= p(A_s) \cup p(A_f) \\ p(A_s) &= \{p_{s1}, p_{s2}, \ldots, p_{sn}\} \\ p(A_f) &= \{p_{f1}, p_{f2}, \ldots, p_{fm}\} \\ |p_{s1}| &\geq \ldots \geq |p_{sn}| \gg |p_{f1}| \geq \ldots \geq |p_{fm}| \end{aligned} \right\} \tag{7.53}$$

The original system of (7.51) is decoupled into slow and fast subsystems of n and m dimensions, respectively. We use here a slight modification of the block diagonalisation given in Section 3.5. The decoupling transformations to be used are given by (Kando and Iwazumi, 1983a,b)

$$\begin{bmatrix} x_s(k) \\ z_f(k) \end{bmatrix} = \begin{bmatrix} I_s + ED & E \\ D & I_f \end{bmatrix} \begin{bmatrix} x(k) \\ z(k) \end{bmatrix} \tag{7.54}$$

and

$$\begin{bmatrix} x(k) \\ z(k) \end{bmatrix} = \begin{bmatrix} I_s & -E \\ -D & I_f + DE \end{bmatrix} \begin{bmatrix} x_s(k) \\ z_f(k) \end{bmatrix} \tag{7.55}$$

where $D(m \times n)$ and $E(n \times m)$ satisfy the algebraic Riccati-type equations

$$A_{22}D - DA_{11} + DA_{12}D - A_{21} = 0 \tag{7.56}$$

$$E(A_{22} + DA_{12}) - (A_{11} - A_{12}D)E + A_{12} = 0 \tag{7.57}$$

Let us note that the interesting feature of the decoupling transformations is that the inverse transformation (7.55) does not require any matrix inverses of D and E.

By using the transformation (7.54) in (7.51), we get the decoupled slow and fast subsystems

$$x_s(k + 1) = A_s x_s(k) + B_s u(k) \qquad (7.58)$$

$$z_f(k + 1) = A_f z_f(k) + B_f u(k) \qquad (7.59)$$

where

$$A_s = A_{11} - A_{12}D; \qquad A_f = A_{22} + DA_{12}$$

$$B_s = (I_s + ED)B_1 + EB_2; \quad B_f = DB_1 + B_2$$

Because of the nonlinear nature of (7.58) and (7.59), one has to obtain an iterative solution as

$$D_{i+1} = [A_{22}D_i + D_i A_{12}D_i - A_{21}]A_{11}^{-1}; \quad D_0 = -A_{21}A_{11}^{-1} \qquad (7.60)$$

$$E_{i+1} = A_{11}^{-1}[E_i(A_{22} + DA_{12}) + A_{12}DE_i + A_{12}]; \quad E_0 = A_{11}^{-1}A_{12} \qquad (7.61)$$

where we assume that A_{11} is nonsingular. We note that the decoupled systems (7.58) and (7.59) have the advantage of mode decoupling.

7.5.2 Original and subsystem regulator problems

The performance index to be minimised with the original system (7.51) is

$$J = \sum_{k=0}^{\infty} [y'(k)Qy(k) + u'(k)Ru(k)] \qquad (7.62)$$

where

$$y'(k) = [x'(k), z'(k)]$$

$$Q = \begin{bmatrix} Q_{11} & 0 \\ 0 & Q_{22} \end{bmatrix}$$

Q is a positive semidefinite symmetric matrix and R is a positive definite symmetric matrix.

The closed-loop optimal control is given by (Sage and White, 1977)

$$u(k) = -R^{-1}B'P[I + BR^{-1}B'P]^{-1}Ay(k) \qquad (7.63)$$

where P, of order $(n + m) \times (n \times m)$, is the positive definite symmetric solution of the algebric matrix Riccati equation

$$P = A'T[I + BR^{-1}B'P]^{-1}A + Q \qquad (7.64)$$

where

$$A = \begin{bmatrix} A_{11} & A_{12} \\ A_{21} & A_{22} \end{bmatrix}; \quad B = \begin{bmatrix} B_1 \\ B_2 \end{bmatrix}$$

The closed-loop optimal system is given by

$$y(k+1) = (A - BF)y(k) \tag{7.65}$$

where

$$F = R^{-1}B'P[I + BR^{-1}B'P]^{-1}A$$

Instead of tackling the original regulator problem (7.51) and (7.62) directly, we decompose it appropriately into two regulator problems for slow and fast subsystems. The slow regulator problem consists of the slow subsystem of (7.58) and a quadratic performance index J_s. The fast regulator problem is composed of the fast subsystem of (7.59) and a quadratic performance measure J_f. We note that the subsystem measures are constructed in such a way that $J = J_s + J_f$.

Consider the performance index (7.62) of the original system. Using the transformation (7.55) in (7.62), it is easily shown that (Othman et al., 1985)

$$J_s = \sum_{k=0}^{\infty} [x_s'(k)Q_s x_s(k) + u_s'(k)Ru_s(k)] \tag{7.66}$$

and

$$J_f = \sum_{k=0}^{\infty} [z_f'(k)Q_f z_f(k) + u_f'(k)Ru_f(k)] \tag{7.67}$$

where the positive semidefinite, symmetric matrices Q_{ss} and Q_{ff} are given by

$$Q_s = Q_{11} + D'Q_{22}D$$

$$Q_f = E'Q_{11}E + (I_f + DE)'Q_{22}(I_f + DE)$$

The slow regulator problem is to minimise J_s for the slow subsystem

$$x_s(k+1) = A_s x_s(k) + B_s u_s(k) \tag{7.68}$$

The optimal feedback control of the slow subsystem is given by

$$u_s(k) = -R^{-1}B_s'P_s[I_s + B_s R^{-1}B_s'P_s]^{-1}A_s x_s(k) \tag{7.69}$$

where P_s, of order $n \times n$, is the positive definite symmetric solution of the reduced-order algebraic matrix Riccati equation

$$P_s = A_s'P_s[I_s + B_s R^{-1}B_s'P_s]^{-1}A_s + Q_s \tag{7.70}$$

Similarly, the fast regulator problem is to minimise J_f for the fast subsystem

$$z_f(k+1) = A_f z_f(k) + B_f u_f(k) \tag{7.71}$$

The optimal feedback control of the fast subsystem becomes

$$u_f(k) = -R^{-1}B_f'P_f[I_f + B_f R^{-1}B_f'P_f]^{-1}A_f z_f(k) \tag{7.72}$$

where P_f, of order $(m \times m)$, is the positive definite symmetric solution of the reduced-order algebraic matrix Riccati equation

$$P_f = A_f' P_f [I_f + B_f R^{-1} B_f' P_f]^{-1} A_f + Q_f \tag{7.73}$$

Rewrite the control laws (7.69) and (7.72) as

$$u_s(k) = -F_s x_s(k) \tag{7.74}$$

$$u_f(k) = -F_f z_f(k) \tag{7.75}$$

where

$$F_s = R^{-1} B_s' P_s [I_s + B_s R^{-1} B_s' P_s]^{-1} A_s$$

$$F_f = R^{-1} B_f' P_f [I_f + B_f R^{-1} B_f' P_f]^{-1} A_f$$

We note that the control laws (7.74) and (7.75) are optimal with respect to the slow and fast subsystems (7.68) and (7.71) only. It is, however, computationally simpler to determine these control laws compared with the optimal control law (7.63) for the original system.

7.5.3 Composite control

The composite control is formulated as the sum of the slow and fast feedback optimal controls (7.74) and (7.75). That is

$$u_c(k) = u_s(k) + u_f(k)$$

$$= -[F_s x_s(k) + F_f z_f(k)] \tag{7.76}$$

Now, using the transformation (7.54) in (7.76), we get

$$u_c(k) = -[F_{sc} x(k) + F_{fc} z(k)]$$

$$= -F_c y(k) \tag{7.77}$$

where

$$F_{sc} = F_s(I_s + ED) + F_f D$$

$$F_{fc} = F_s E + F_f$$

$$F_c = [F_{sc}, F_{fc}]$$

Using the composite control (7.77) in the original system (7.51),

$$y_c(k + 1) = (A - BF_c)y_c(k) \tag{7.78}$$

We know that minimising the original performance index (7.62) with reference to the composite system (7.78) results in the suboptimal performance index

$$J_c = \frac{1}{2} y'(0) P_c y(0) \tag{7.79}$$

where P_c is the positive definite symmetric solution of the discrete Lyapunov equation (Othman *et al.*, 1985)

$$P_c = (A - BF_c)'P_c(A - BF_c) + Q + F_c'RF_c \qquad (7.80)$$

Example 7.5   Aircraft model
The dynamical equations for the aircraft model are obtained based on a rigid-body assumption (Naidu and Price, 1987). Also, the angle of attack is taken to be small. It has been observed that the linearised model possesses a two-time-scale property in the sense that the pitch angle, velocity and altitude are the 'slow' variables and the angle of attack and pitch rate are the 'fast' variables. For digital implementation, we obtain the discrete model using a zero-order hold. For the aircraft under consideration, the discrete model is given by

$$y(k + 1) = Ay(k) + Bu(k) \qquad (7.81)$$

where

$$A = \begin{bmatrix} 0{\cdot}923701 & -0{\cdot}308096 & 0{\cdot}000000 & 0.053043 & -0{\cdot}090367 \\ 0{\cdot}039705 & 0{\cdot}995525 & 0{\cdot}000000 & -0{\cdot}107454 & 0{\cdot}588883 \\ 0{\cdot}087127 & 1{\cdot}899490 & 1{\cdot}000000 & -0{\cdot}635270 & 0{\cdot}394015 \\ -0{\cdot}035537 & 0{\cdot}010123 & 0{\cdot}000000 & 0{\cdot}007748 & 0{\cdot}137407 \\ 0{\cdot}069562 & -0{\cdot}012706 & 0{\cdot}000000 & -0{\cdot}097108 & 0{\cdot}287411 \end{bmatrix}$$

and

$$B = \begin{bmatrix} 0{\cdot}042825 & -0{\cdot}000395 & -0{\cdot}154048 \\ -0{\cdot}484628 & -0{\cdot}515349 & -0{\cdot}002237 \\ -0{\cdot}161525 & -0{\cdot}067522 & -0{\cdot}005257 \\ -0{\cdot}202010 & -0{\cdot}289303 & 0{\cdot}005061 \\ 0{\cdot}806770 & -0{\cdot}852161 & -0{\cdot}006353 \end{bmatrix}$$

The states $x_1(k)$ and $x_3(k)$ are scaled down by a factor of 100 to facilitate easy representation.

The eigenvalues of the discrete model (7.81) are $1{\cdot}0$; $0{\cdot}962103 \pm j0{\cdot}175343$; $0{\cdot}217298$; $0{\cdot}072881$ indicating that there are three $[x_1(k), x_2(k), x_3(k)]$ 'slow' states and two $[z_1(k), z_2(k)]$ 'fast' states with eigenvalue separation ratio of $4{\cdot}5005$.

The various performances measures in (7.62) are taken as

$$Q_{11} = Q_{22} = I; \quad R = r = 1$$

Using the method described in the previous Section, the results are summarised below.

The positive definite matrix P is

$$P = \begin{bmatrix} 4\cdot822583 & 0\cdot427518 & 0\cdot709917 & -0\cdot272287 & 0\cdot059216 \\ 0\cdot427518 & 11\cdot654952 & 3\cdot732418 & -2\cdot665579 & 3\cdot042291 \\ 0\cdot709917 & 3\cdot732418 & 2\cdot690923 & -1\cdot096067 & 0\cdot882337 \\ -0\cdot272287 & -2\cdot665579 & -1\cdot096067 & 1\cdot742674 & -0\cdot689301 \\ 0\cdot059216 & 3\cdot042291 & 0\cdot882337 & -0\cdot689301 & 2\cdot015052 \end{bmatrix}$$

Using P, the optimal control (7.63) is given as

$$F = \begin{bmatrix} -0\cdot021105 & -1\cdot136159 & -0\cdot268959 & 0\cdot260873 & -0\cdot507318 \\ -0\cdot110915 & -0\cdot700095 & -0\cdot101750 & 0\cdot142104 & -0\cdot410844 \\ -0\cdot629305 & -0\cdot025190 & -0\cdot100651 & 0\cdot033007 & 0\cdot003334 \end{bmatrix}$$

The closed-loop optimal system (7.65) has the eigenvalues

$$0\cdot843545; \quad 0\cdot25879 \pm j0\cdot326649; \quad 0\cdot06457 \pm j0\cdot047911$$

For the slow and fast subsystems, the corresponding values are

$$P_s = \begin{bmatrix} 4\cdot621454 & 1\cdot051044 & 0\cdot945863 \\ 1\cdot051044 & 13\cdot073895 & 4\cdot463674 \\ 0\cdot945863 & 4\cdot463674 & 2\cdot710497 \end{bmatrix}$$

and

$$P_f = \begin{bmatrix} 1\cdot204794 & 0\cdot244345 \\ 0\cdot244345 & 1\cdot914426 \end{bmatrix}$$

For the composite control, the feedback matrix F_c is obtained as

$$F_c = \begin{bmatrix} -0\cdot010457 & -1\cdot138702 & -0\cdot270112 & 0\cdot262312 & -0\cdot507373 \\ -0\cdot110238 & -0\cdot699239 & -0\cdot101678 & 0\cdot142112 & -0\cdot410528 \\ -0\cdot654236 & -0\cdot018681 & -0\cdot100062 & 0\cdot031065 & 0\cdot004602 \end{bmatrix}$$

and the positive definite matrix P_c, obtained from the discrete Lyapunov equation (7.80), is

$$P_c = \begin{bmatrix} 5\cdot045733 & 0\cdot338736 & 0\cdot671223 & -0\cdot205383 & -0\cdot013507 \\ 0\cdot338736 & 11\cdot734399 & 3\cdot769142 & -2\cdot716177 & 3\cdot103791 \\ 0\cdot671223 & 3\cdot769142 & 2\cdot714962 & -1\cdot128617 & 0\cdot918364 \\ -0\cdot205383 & -2\cdot716177 & -1\cdot128617 & 1\cdot790493 & -0\cdot744196 \\ -0\cdot013507 & 3\cdot103791 & 0\cdot918364 & -0\cdot744196 & 2\cdot083318 \end{bmatrix}$$

The corresponding closed-loop suboptimal composite system has the eigenvalues

$$0\cdot840132; \quad 0\cdot259351 \pm j0\cdot334107; \quad 0\cdot063358 \pm j0\cdot04656$$

The performance indices of the original optimal system and the composite suboptimal system are obtained as

612·1122   and   616·1734

respectively, with a percentage error of 0·66347.

The responses of the various states for the exact (optimal) system and the composite (suboptimal) system are shown in Fig. 7.4. These responses are

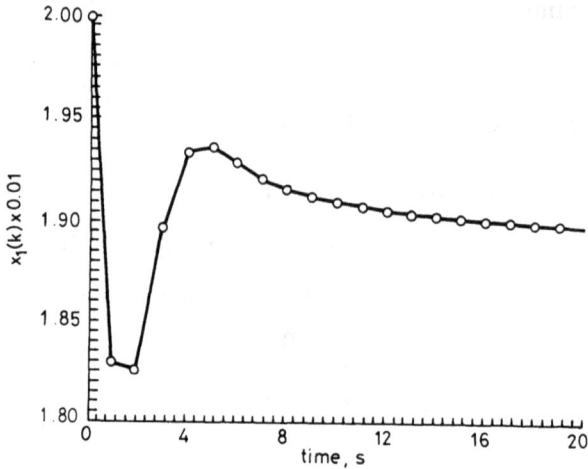

(a)

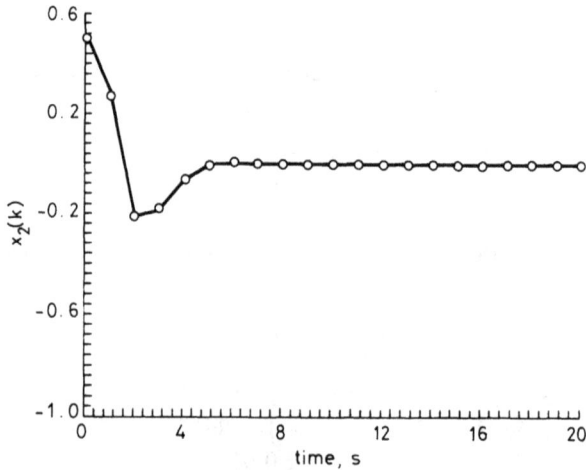

(b)

Fig. 7.4a and b   *Solution of $x_1(k)$ and $x_2(k)$ of Example 7.5*
——— exact system
○ ○ ○ composite system

obtained for the case when the aircraft trajectories are to be regulated to the equilibrium flight conditions of $x_{1e} = 190 \cdot 66 \, \text{ft/sec}$; $x_{2e} = 0$; $x_{3e} = 2000 \, \text{ft}$; $z_{1e} = 0$; $_{2e} = 0$ with equilibrium controls.

The above results clearly show an excellent agreement between the exact system and the composite system. Once again, we note that the composite control is obtained from the low-order slow and fast subsystems with a considerable reduction in the overall computation.

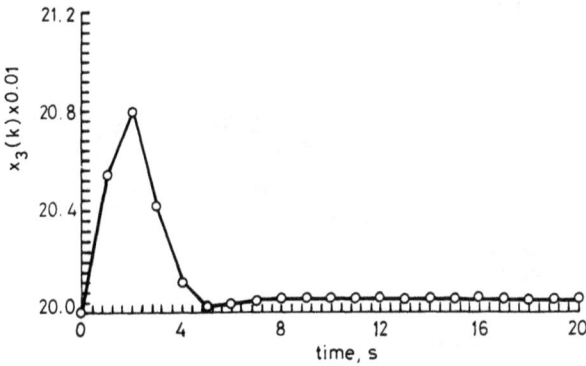

(c)

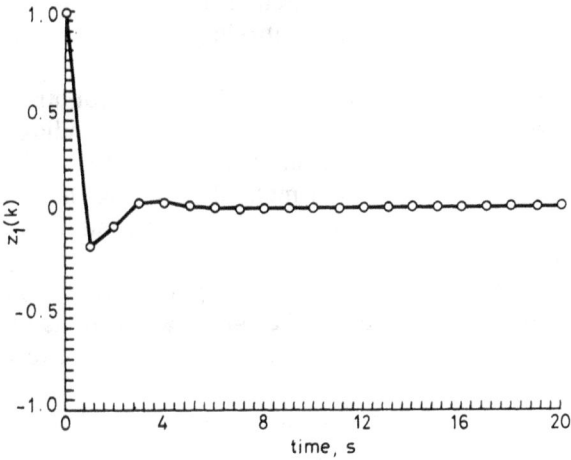

(d)

Fig. 7.4c and d *Solution of $x_3(k)$ and $z_1(k)$ of Example 7.5*
—————— exact system
o   o   o composite system

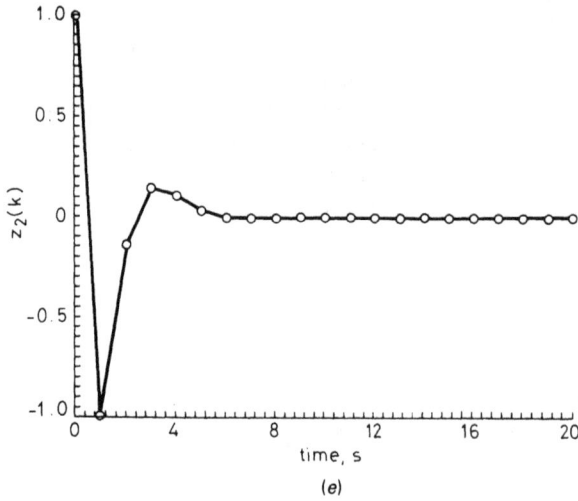

Fig. 7.4e   *Solution of $z_2(k)$ of Example 7.5*
——— exact system
o o o composite system

7.6   Conclusions and discussion

In this Chapter, first, a method has been given to analyse singularly perturbed nonlinear difference equations for initial and boundary-value problems. The approximate solutions have been obtained in terms of the outer series and the correction series. It has been found that considerable care has to be taken in the formulation of the equations controlling the boundary-layer correction series.

Secondly, the closed-loop optimal control problem for a linear shift-invariant discrete system has been formulated. The corresponding matrix Riccati difference equation is cast in the singularly perturbed form. The approximate solutions have been obtained by a method requiring the outer series and the final BLC series to take care of the lost final conditions in the process of degeneration.

A method has been given for the more important case of the steady-state matrix Riccati algebraic equation. The time-scale analysis of the regulator problem has been discussed, and numerical examples have been provided to illustrate the methods.

The special features of this Chapter in describing singular perturbation methods for closed-loop optimal control are:

(i) In solving the high-order matrix Riccati difference equation, order reduction has been effectively used in evaluating the inverse of the matrix in F of order $(n_1 + n_2)$ given by (7.19). That is, the solution of the zeroth-order outer-series (7.33a) requires the inverse of the matrix,

$[I_1 + E_1 P_1^{(0)}(k + 1)]^{-1}$ which is of order n_1 only. In addition, the first-order approximation (7.31*b*) (or high-order approximation) uses the same reduced-order matrix inversion, and hence is solved by simple recursion.

This is also true for the steady-state case of the matrix Riccati equation.

(ii) The correction-series (7.33) need be solved for only a few values of k near the final boundary layer, using the reduced-order matrix inversion which is already evaluated for the solution of the outer series (7.31). As such, the solution of the correction series does not demand any more extra matrix inversions.

These two features are mainly responsible for weighing in favour of the proposed methods, which offer considerable reduction in overall computation.

Other related problems in closed-loop optimal control of singular perturbed discrete systems has been discussed in Glizer and Dmitriev (1979*b*), Blankenship (1981), Rajagopalan and Naidu (1981), Kando and Iwazumi (1983*a*,*b*), Kimura (1983), Litkouhi and Khalil (1984, 1985), Naidu and Rao (1985*a*), Othman *et al.* (1985).

7.7 Appendices

7.7.1 Appendix

In (7.26*a*), consider

$$[I + E\bar{P} + \varepsilon^{N-k}EW]^{-1} = [(I + E\bar{P})(I + \varepsilon^{N-k}(I + E\bar{P})^{-1}EW)]^{-1}$$
$$= [I + \varepsilon^{N-k}(I + E\bar{P})^{-1}EW]^{-1}[I + E\bar{P}]^{-1}$$
$$= [I + \varepsilon^{N-k}CEW]^{-1}C, \quad C = [I + E\bar{P}]^{-1} \quad (7.82)$$

When $k < N$, $\varepsilon^{N-k} \ll 1$, ε being the singular perturbation parameter. Therefore the norm of $\varepsilon^{N-k}CEW$ is much less than unity. Ignoring terms with ε and higher powers of ε, (7.82) becomes

$$[I + E\bar{P} + \varepsilon^{N-k}EW]^{-1} = (I - \varepsilon^{N-k}CEW)C$$
$$= C - \varepsilon^{N-k}CEWC \quad (7.83)$$

Substituting (7.82) in (7.26*a*),

$$F = (\bar{P} + \varepsilon^{N-k}W)(C - \varepsilon^{N-k}CEWC)$$
$$= \bar{P}C + \varepsilon^{N-k}(WC - PCEWC - \varepsilon^{N-k}WCEWC)$$
$$= \bar{F} + \varepsilon^{N-k}G \quad (7.84)$$

7.7.2 Appendix

The matrix $\bar{F}(k + 1)$ can be expressed as a power series in ε as shown below.

Consider the expression for $F(k + 1)$ in (7.26b). (The arguments k and $(k + 1)$ are omitted for the sake of notational simplicity)

$$\bar{F} = \bar{P}(I + E\bar{P})^{-1} \tag{7.85}$$

Consider

$$(I + E\bar{P})^{-1} = \left[\begin{bmatrix} I_1 & 0 \\ 0 & I_2 \end{bmatrix} + \begin{bmatrix} E_1 & E_2 \\ E_2' & E_3 \end{bmatrix} \begin{bmatrix} \bar{P}_1 & \varepsilon\bar{P}_2 \\ \varepsilon\bar{P}_2' & \varepsilon^2\bar{P}_3 \end{bmatrix} \right]^{-1}$$

$$= \left[\underbrace{\begin{bmatrix} I_1 + E_1\bar{P}_1 & 0 \\ E_2'\bar{P}_1 & I_2 \end{bmatrix}}_{L} + \varepsilon \underbrace{\begin{bmatrix} E_2\bar{P}_2' & E_1\bar{P}_2 \\ E_3\bar{P}_2' & E_2\bar{P}_2 \end{bmatrix}}_{M} + \varepsilon^2 \underbrace{\begin{bmatrix} 0 & E_2\bar{P}_3 \\ 0 & E_3\bar{P}_3 \end{bmatrix}}_{N} \right]^{-1}$$

$$= (L + \varepsilon M + \varepsilon^2 N)^{-1}$$

$$= [L(I + \varepsilon L^{-1}M + \varepsilon^2 L^{-1}N)]^{-1}$$

$$= (I + \varepsilon L^{-1}M + \varepsilon^2 L^{-1}N)^{-1}L^{-1} \tag{7.86}$$

For sufficiently small values of ε, the norms of $\varepsilon L^{-1}M$ and $\varepsilon^2 L^{-1}N$ are much less than unity. Therefore (7.86) becomes

$$(I + E\bar{P})^{-1} = L^{-1} - \varepsilon L^{-1}ML^{-1} + 0(\varepsilon^2) \tag{7.87}$$

Since L is a triangular matrix, its inverse is readily written as

$$L^{-1} = \begin{bmatrix} I_1 + E_1\bar{P}_1 & 0 \\ E_2'\bar{P}_1 & I_2 \end{bmatrix}^{-1} = \begin{bmatrix} (I_1 + E_1\bar{P}_1)^{-1} & 0 \\ -E'\bar{P}(I + E\bar{P})^{-1} & I_2 \end{bmatrix} \tag{7.88}$$

Using (7.87) and (7.88) in (7.85)

$$\bar{F} = \begin{bmatrix} \bar{F}_1 & \bar{F}_2 \\ \bar{F}_2' & \bar{F}_3 \end{bmatrix} = \begin{bmatrix} \bar{P}_1 & \varepsilon\bar{P}_2 \\ \varepsilon\bar{P}_2' & \varepsilon^2\bar{P}_3 \end{bmatrix} \left[\begin{Bmatrix} C_1 & 0 \\ -E_2'\bar{P}_1C_1 & 0 \end{Bmatrix} \right.$$

$$\left. - \varepsilon \begin{Bmatrix} C_1 & 0 \\ -E_2'\bar{P}_1C_1 & I_2 \end{Bmatrix} \begin{Bmatrix} E_2\bar{P}_2C_1 - E_1\bar{P}_2E_2'\bar{P}_1C_1 & E_1\bar{P}_2 \\ E_3\bar{P}_2'C_1 - E_2'\bar{P}_2E_2'\bar{P}_1C_1 & E_2'\bar{P}_2 \end{Bmatrix} \right] \tag{7.89}$$

where $C_1 = (I_1 + E_1\bar{P}_1)^{-1}$

Expanding (7.89),

$$\bar{F}_1 = \bar{P}_1C_1 - \varepsilon\bar{P}_1C_1(E_2\bar{P}_2'C_1 - E_1\bar{P}_2E_2'\bar{P}_1C_1) - \varepsilon\bar{P}_2E_2'\bar{P}_1C_1$$
$$\qquad + \varepsilon(I_1 - \bar{P}_1C_1E_1)\bar{P}_2 + 0(\varepsilon^2)$$

$$\bar{F}_2 = \varepsilon(I_1 - \bar{P}_1C_1E_1)\bar{P}_2 + 0(\varepsilon^2)$$

$$\bar{F}_2' = \varepsilon\bar{P}_2'C_1 + 0(\varepsilon^2)$$

$$\bar{F}_3 = 0(\varepsilon^2) \tag{7.90}$$

For the matrix $\bar{F}$ to be symmetric, it is necessary to show that from (7.90,)

$$(\bar{P}'_2 C_1)' = (I_1 - \bar{P}_1 C_1 E_1)\bar{P}_2$$

i.e. $$C_1'\bar{P}_2 = (I_1 - \bar{P}_1 C_1 E_1)\bar{P}_2$$

i.e. $$C_1' = I_1 - \bar{P}_1 C_1 E_1$$

$$[(I_1 + E_1\bar{P}_1)^{-1}]' = I_1 - \bar{P}_1(I_1 + E_1\bar{P}_1)^{-1}E_1 \tag{7.91}$$

Since E_1 and P_1 are symmetric

$$[(I_1 + E_1\bar{P}_1)^{-1}]' = (I_1 + \bar{P}_1 E_1)^{-1}$$

Using the above relationship in (7.91),

$$(I_1 + \bar{P}_1 E_1)^{-1} = I_1 - \bar{P}_1(I_1 + E_1\bar{P}_1)^{-1}E_1$$

Multiplying both sides by $(I_1 + \bar{P}_1 E_1)$

$$\begin{aligned} I_1 &= [I_1 - \bar{P}_1(I_1 + E_1\bar{P}_1)^{-1}E_1](I_1 + \bar{P}_1 E_1) \\ &= I_1 - \bar{P}_1(I_1 + E_1\bar{P}_1)^{-1}E_1 + \bar{P}_1 E_1 - \bar{P}(I_1 + E_1\bar{P}_1)^{-1}E_1\bar{P}_1 E_1 \\ &= (I_1 + \bar{P}_1 E_1) - \bar{P}_1(I_1 + E_1\bar{P}_1)^{-1}(I_1 + E_1\bar{P}_1)E_1 \\ &= I_1 \end{aligned}$$

Thus (7.91) is established.
   Using (7.91) in (7.90),

$$\begin{aligned} \bar{F}_1 &= \bar{P}_1 C_1 - \varepsilon(\bar{P}_1 C_1 E_2 \bar{P}_2 C_1 + C_1 \bar{P}_2 E_2 \bar{P}_1 C_1) + 0(\varepsilon^2) \\ F_2 &= \varepsilon C_1 \bar{P}_2 + 0(\varepsilon^2) \\ \bar{F}_3 &= 0(\varepsilon^2) \end{aligned} \tag{7.92}$$

Similarly, it is shown that G can be expressed as a power series in ε

$$G = WC - \bar{P}CEWC - \varepsilon^{N-k}WCEWC$$

$$= [I - (\bar{P} + \varepsilon^{N-k}W)CE]WC \tag{7.93}$$

where $C = (I + E\bar{P})^{-1}$
   It is shown in (7.87) that C is expressible as a power series in ε. Thus it follows from (7.93) that G is also expressible as a power series in ε.
   From (7.93), it is clear that the terms associated with ε^{N-k} are of first order for $k = N - 1$, second order for $k = N - 2$ etc. When writing the zeroth- and higher-order terms for this equation, this aspect should be kept in mind.

For the zeroth-order approximation.

$$G^{(0)} = \begin{bmatrix} G_1^{(0)} & G_2^{(0)} \\ G_2^{(0)'} & G_3^{(0)} \end{bmatrix} = \left\{ \begin{bmatrix} I_1 & 0 \\ 0 & I_2 \end{bmatrix} - \begin{bmatrix} \bar{P}_1 & 0 \\ 0 & 0 \end{bmatrix} \right.$$

$$\times \left. \begin{bmatrix} C_1 & 0 \\ -E_2'P_1C_1 & I_2 \end{bmatrix} \begin{bmatrix} E_1 & E_2 \\ E_2' & E_3 \end{bmatrix} \right\} \begin{bmatrix} W_1 & W_2 \\ W_2' & 0 \end{bmatrix} \begin{bmatrix} C_1 & 0 \\ -E_2'P_1C_1 & 0 \end{bmatrix} \qquad (7.94)$$

or

$$G_1^{(0)} = C_1(W_1C_1 - W_2E_2\bar{P}_1C_1) - \bar{P}_1C_1E_2W_2C_1$$
$$G_2^{(0)} = C_1'W_2 \qquad\qquad\qquad\qquad\qquad\qquad\qquad\qquad (7.95)$$
$$G_3^{(0)} = 0$$

The first- and higher-order terms of G can be evaluated on similar lines.

References

ABED, E. H. (1985*a*): 'Singularly perturbed Hopf bifurcation', *IEEE Trans.*, **CAS-32**, pp. 1270–1280

ABED, E. H. (1985*b*): 'Multiparameter singular perturbation problems: interactive expansions and asymptotic stability', *Syst. & Control Lett.*, **5**, pp. 279–282

ABED, E. H. (1985*c*): 'A new parameter estimate in singular perturbations', *Syst. & Control Lett.*, **6**, pp. 193–198

ABED, E. H. (1986*a*): 'Strong D-stability', *Syst. & Control Lett.*, **7**, pp. 207–212

ABED, E, H. (1986*b*): 'Decomposition and stability of multiparameter singular perturbation problems', *IEEE Trans.*, **AC-31**, pp. 925–934

ABED, E. H. and TITS, A. L. (1986): 'On the stability of multiple time-scale systems', *Int. J. Control*, **44**, pp. 211–218

ABRAHAM-SHRAUNER, B. (1974): 'Perturbation expansions for the potential of a small radius charged dielectric sphere in 1-1 electrolytes', *SIAM J. Appl. Math.*, **27**, pp. 656–665

ABRAHAMSON, L. R., KELLER, H. B. and KREISS, H. D. (1974): 'Difference approximations for singular perturbations of systems of ordinary differential equations', *Num. Math.*, **22**, pp. 367–391

ACKERBERG, R. C. and O'MALLEY, R. E., Jun. (1970): 'Boundary layer problems exhibiting resonance', *Studies in Applied Math.*, **49**, pp. 277–295

AKULENKO, L. D. (1975): 'The asymptotic solutions of some time-optimal control problems', *Appl. Math. Mech.*, **39**, pp. 565–578

ALDEN, R. T. H. and NOLAN, P. J. (1976): 'Evaluating alternate models for power systems dynamic stability studies', *IEEE Trans.*, **PAS-95**, pp. 433–440

ALLEMONG, J. J. and KOKOTOVIC, P. V. (1980): 'Eigen-sensitivities in reduced order modelling', *IEEE Trans.*, **AC-25**, pp. 821–822

ALTSHULER, D. and HADDAD, A. H. (1978): 'Near optimal smoothing for singularly perturbed linear systems', *Automatica*, **14**, pp. 81–88

ANDERSON, B. D. O. and JOHNSON, C. R., Jun. (1982): 'On reduced-order adaptive output error identification and adaptive IIR filtering', *IEEE Trans.*, **AC-27**, pp. 927–933

ANDERSON, B. D. O. and MOORE, J. B. (1971): 'Linear Optimal Control' (Prentice Hall, Englewood Cliffs)

ANDERSON, L. R. (1982): 'Pole placement and order reduction in two-time scale control systems through Riccati iteration', *Mathematical Modelling*, **3**, pp. 93–101

ANDERSON, L. R., BREWER, D. W. and BAYKAN, A. R. (1983): 'Numerical solution of the symmetric Riccati equation through Riccati iteration', *Optimal Control: Appl. & Methods*, **4**, pp. 239–251

ANDERSON, L. R. and HALLAUER, W. L., Jun. (1981): 'A method of order reduction for structural dynamics based on Riccati iteration', *AIAA J.*, **19**, pp. 796–800

ANDREWS, J. G. and ATTHEY, D. R. (1975): 'On the motion of an intensely heated evaporating boundary', *J. Int. Math. Appl.*, **15**, pp. 59–72

ANDREWS, J. G. and ATTHEY, D. R. (1976): 'Hydrodynamic limit to penetration of a material by a high power beam', *J. Physics, D: Appl. Physics*, **9**, pp. 2181–2194

AOKI, M. (1978): 'Some approximation methods for estimation and control of large scale systems', *IEEE Trans.*, **AC-23**, pp. 173–182

ARAPOSTATHIS, A., SASTRY, S. S. and VARIAYA, P. (1982): 'Global analysis of swing dynamics', *IEEE Trans.*, **CAS-29**, pp. 673–679

ARDEMA, M. D. (1976):'Solution of the minimum time-to-climb problem by matched asymptotic expansions', *AIAA J.*, **14**, pp. 843–850

ARDEMA, M. D. (1977): 'Singular perturbations in flight mechanics'. NASA Technical Memorandum TM X-62, 330, 2nd Revision, Ames Research Center, Moffett Field, CA

ARDEMA, M. D. (1979): 'Linearization of the boundary-layer equations of the minimum time-to-climb problem', *AIAA J. Guidance and Control*, **2**, pp. 434–436

ARDEMA, M. D. (1980): 'Nonlinear singularly perturbed optimal control problems with singular arcs', *Automatica*, **16**, pp. 99–104

ARDEMA, M. D. (Ed.) (1983a): 'Singular Perturbations in Systems and Control'. CISM Courses and Lectures 280 (Springer-Verlag, Wien)

ARDEMA, M. D. (1983b): 'Solution algorithms for nonlinear singularly perturbed optimal control problems', *Optimal Control: Applications and Methods*, **4**, pp. 283–302

ARDEMA, M. D. and RAJAN, N. (1985a): 'Separation of time-scales in aircraft trajectory optimization', *J. Guidance, Control and Dynamics*, **8**, pp. 275–278

ARDEMA, M. D. and RAJAN, N. (1985b): 'Slow and fast state variables for three-dimensional flight dynamics', *J. Guidance, Control and Dynamics*, **8**, pp. 532–535

ARGEMI, J., GOLA., M. and CHAGNEUX, H. (1979): 'Qualitative analysis of a model generating long potential waves in BA-treated nerve cells-I: reduced systems', *Bull. Math. Biology*, **41**, pp. 665–686

ARIS, R. (1968): 'A note on mechanism and memory in the kinetics of biochemical reactions', *Math. Biosci.*, **3**, pp. 177–202

ARIS, R. (1975): 'The Mathematical Theory of Diffusions and Reactions in Permeable Catalysts' Vol. I (Clarendon Press, Oxford)

ARIS, R. (1979): 'Method in modeling in chemical engineering systems' *in* LEONDES, C. T. (Ed.) 'Control and Dynamic Systems' (Academic Press, New York) Vol. **15**, pp. 41–98

ASATANI, K. (1974): 'Suboptimal control of fixed-end-point minimum energy problem via singular perturbation theory', *J. Math. Anal. Appl.*, **45**, pp. 684–697

ASATANI, K. (1976): 'Near optimum control of distributed parameter systems via singular perturbation theory', *J. Math. Anal. Appl.*, **54**, pp. 799–819

ASATANI, K., IWAZUMI, T. and HATTORI, Y. (1977a): 'Error estimation of prompt jump approximation by singular perturbation theory', *J. Nucl. Sci. Technology*, **8**, pp. 653–656

ASATANI, K., SHIOTANI, M. and HATTORI, Y. (1977b): 'Suboptimal control of nuclear reactors with distributed parameters using singular perturbation theory', *Nuclear Sci. & Enging.*, **62**, pp. 9–19

ASCHER, U. and WEISS, R. (1983): 'Collocation for singular perturbation problems—I: first order systems with constant coefficients', *SIAM J. Num. Anal.*, **20**, pp. 537–557

ASCHER, U. and WEISS, R. (1984): 'Collocation for singular perturbation problems—II: Linear first order systems without turning points', *Math. Comp.*, **43**, pp. 157–187

ASFAR, O. R. and NAYFEH, A. H. (1983): 'The application of the method of multiple scales to wave propagation in periodic structures', *SIAM Rev.*, **25**, pp. 455–480

ATHANS, M. and FALB, P. L. (1966): 'Optimal control: An introduction to the theory and applications' (McGraw-Hill, New York)

ATLURI, R. and KAO, Y. K. (1981): 'Sampled-data control of systems with widely varying time constants', *Int. J. Control*, **33**, pp. 555–564

AUCHMUTY, J. F. G. and NICHOLIS, G. (1976): 'Bifurcation analysis of nonlinear reaction diffusion equations-III, chemical oscillations', *Bull. Math. Biol.*, **38**, pp. 325–350

AVRAMOVIC, B., KOKOTOVIC, P. V., WINKLEMAN, J. R. and CHOW, J. H. (1980): 'Area decomposition for electromechanical models of power systems', *Automatica*, **16**, pp. 637–648

BALACHANDRA, M. (1975): 'Periodic solution of singularly perturbed equations arising from gyroscopic systems', *J. Math. Anal. Appl.*, **49**, pp. 302–316

BALACHANDRA, M. (1976): 'An averaging theorem for two-point boundary value problems with applications to optimal control', *J. Math. Anal. Appl.*, **55**, pp. 46–60

BALAS, M. J. (1978): 'Observer stabilization of singularly perturbed systems', *J. Guidance & Control*, **1**, pp. 93–95

BALAS, M. J. (1982a): 'Reduced-order feedback control of distributed parameter system via singular perturbation theory', *J. Math. Anal. Appl.*, **87**, pp. 281–294

BALAS, M. J. (1982b): 'Towards a more practical control theory for distributed parameter systems' *in* LEONDES, C. T. (Ed), 'Control and dynamic systems' (Academic Press, New York) Vol. **18**, pp. 362–421

BALAS, M. J. (1984): 'Stability of distributed parameter systems with finite-dimensional controller-compensators using singular perturbations', *J. Math. Anal. Appl.*, **99**, pp. 90–108

BARCILON, V., COLE, J. D. and EISENBERG, R. S. (1971): 'A singular perturbation analysis of induced electric fields in nerve cells', *SIAM J. Appl. Math.*, **21**, pp. 339–354

BARKER, J. W. (1984): 'Interaction of fast and slow waves in problems with two-time scales', *SIAM J. Math. Anal.*, **15**, pp. 500–513

BAVNICK, H. and GRASMAN, J. (1974): 'The method of matched asymptotic expansions for the periodic solution of the Van der Pol equations', *Int. J. Nonlinear Mech.*, **9**, pp. 421–434

BELL, D. J. (Ed.) (1973): 'Recent developments in control' (Academic Press, New York)

BELL, D. J. and COOK, L. P. (1978): 'On the solutions of a nerve conduction equation', *SIAM J. Appl. Math.*, **35**, pp. 678–688

BELL, D. J. and JACOBSON, D. H. (1970): 'Singular optimal control problems' (Academic Press, New York)

BENSOUSSAN, A. (1981): 'Singular perturbation results for a class of stochastic control problems', *IEEE Trans.*, **AC-26**, pp. 1071–1080

BENSOUSSAN, A., BLANKENSHIP, G. L. and KOKOTOVIC, P. V. (Eds.) (1984): 'Singular perturbations and optimal control'. Lecture Notes (Springer–Verlag, Berlin)

BENSOUSSAN, A., LIONS, J. L. and PAPANICOLAOU, G. (1981): 'Asymptotic analysis for periodic structures' (North Holland, Amsterdam)

BENSOUSSAN, A. and TITLI, A. (1982): 'Interconnected dynamical systems: stability, decomposition and decentralization' (North Holland, Amsterdam)

BENTWITCH, M. (1971): 'Singular perturbation solution on time dependent mass-transfer with nonlinear chemical reaction', *J. Inst. Math. Appl.*, **7,** pp. 228–240

BERKEY, D. D. and FREEDMAN, M. I. (1976): 'A perturbation problem from control exhibiting on and off the boundary behavior', *SIAM J. Control & Opt.*, **14,** pp. 857–876

BERKEY, D. D. and FREEDMAN, M. I. (1979a): 'A perturbation method for control systems with multiple switch points', *IEEE Trans.*, **AC-24,** pp. 79–88

BERKEY, D. D. and FREEDMAN, M. I. (1979b): 'Buckled elastic at contact: a perturbation analysis', *SIAM J. Appl. Math.*, **37,** pp. 55–68

BERKEY, D. D., FREEDMAN, M. I. and GRANOFF, B. (1978): 'Bifurcation analysis in systems with switch-type coefficients', *SIAM J. Appl. Math.*, **35,** pp. 83–96

BINDING, P. (1976): 'Singularly perturbed optimal control problems—I: convergence', *SIAM J. Control & Opt.*, **14,** pp. 591–622

BIRKOFF, G. (1966): 'Numerical solution of the reactor kinetics equations' *in* GREENSPAN, D. (Ed.) 'Numerical solution of non-linear differential equations' (Wiley, New York)

BIRONT, D., GOUSSENS, M., COUSENS, A. and MESTAL, L. (1982): 'A singular perturbation approach to the effect of a weak magnetic field on stellar oscillations', *R. Astronom. Soc.*, Monthly Notices, **201,** pp. 619–633

BJUREL, G., DAHLQUIST, G., LINDBERG, B., LINDE, S. and ODEN, L. (1970): 'Survey of stiff ordinary differential equations'. Royal Inst. of Tech. Report NA 70.11, Stockholm, Sweden

BLANKENSHIP, G. L. (1979): 'Asymptotic analysis in mathematical physics and control theory: some problems with common features', *in* BROCKETT, R. W. (Ed.) 'Special issue on system theory and physics', *Researche de Automatica*, **10,** pp. 265–315

BLANKENSHIP, G. L. (1981): 'Singularly perturbed difference equations in optimal contol problems', *IEEE Trans.*, **AC-26,** pp. 911–917

BOBISUD, L. E. (1967): 'On the behavior of the solution of the Telegraphis's equation for large velocities', *Pacific J. Math.*, **22,** pp. 213–219

BOBISUD, L. E. (1979): 'A nonlinear perturbation problem arising in chemical engineering', *J. Nonlinear Anal.: Theory, Methods & Appl.*, **3,** pp. 337–345

BOBISUD, L. E. and CHRISTENSON, C. I. (1980): 'A singularly perturbed system of nonlinear parabolic equation from chemical kinetics', *J. Math. Anal. Appl.*, **74,** pp. 293–310

BOBROVSKY, B. Z. and SCHUSS, Z. (1982): 'A singular perturbation method for the computation of the mean first passage time in a nonlinear filter', *SIAM J. Appl. Math.*, **42,** pp. 174–187

BOGLAEV, J. P. (1979): 'The generalized Frrobenius formula in singularly perturbed linear equations', *Soviet Math. Dokl.*, **20,** pp. 731–734

BRADSHAW, A., MAK, K. L. and PORTER, B. (1982): 'Singular perturbation methods in the synthesis of control policies for production-inventory systems', *Int. J. Systems Sci.*, **13,** pp. 589–600

BRADSHAW, A. and PORTER, B. (1979): 'Design of linear multivariable continuous time tracking systems incorporating high gain decentralized error-actuated controllers', *Int. J. Systems Sci.*, **10,** pp. 961–970

BRADSHAW, A. and PORTER, B. (1981): 'Singular perturbation methods in the design of tracking systems incorporating fast sampling error actuated controllers', *Int. J. Systems Sci.*, **12,** pp. 1181–1191

BRAUNER, C. M., GAY, G. and MATHIEU, J. (Eds.) (1977) 'Singular perturbations and boundary layer theory'. Lecture Notes in Math. 594 (Springer–Verlag, New York)

BRAYTON, R. K., GUSTAVSON, F. G. and LINIGER, W. (1966): 'A numerical analysis of the transient behavior of a transistor circuit', *IBM J. Res. Develop.*, **10**, pp. 292–299

BREAKWELL, J. V., SHINAR, J. and VISSER, H. G. (1985): 'Uniformly valid feedback expansions for optimal control of singularly perturbed dynamic systems', *J. Opt. Theory & Appl.*, **46**, pp. 441–453

BURGHARDT, A. and ZALESKI, T. (1968): 'Longitudinal dispersion at small and large Peclet numbers in chemical flow reactors', *Chem. Engg. Sci.*, **23**, pp. 575–591

BUSH, W. B. (1971): 'On the Lagerstrom mathematical model for viscous flow at low Reynolds number', *SIAM J. Appl. Math.*, **20**, pp. 279–287

BUTI, G. (1965): 'The singular perturbation theory of differential equations in control theory', *Periodica Polytechnica*, **10**

BUTUZOV, V. F. (1978): 'The asymptotics of solutions of some model problems on chemical kinetics, taking diffusion into account', *Soviet Math. Dokl.*, **19**, pp. 1079–1083

BUTUZOV, V. F. and VASILEVA, A. B. (1970): 'Differential and difference systems of equations with a small parameter in the sense when the unperturbed (degenerate) system is situated on the spectrum', *Diff. Urav.*, **6**, pp. 650–664 (Russian)

BUTUZOV, V. F., VASILEVA, A. B. and FEDORUK, M. V. (1970): 'Asymptotic methods in the theory of ordinary differential equations' *in* GAMKRELIDGE, R. V. (Ed.) 'Progress in Mathematics, (Plenum Publishing, New York) Vol. **8**, pp. 1–82

CALISE, A. J. (1976): 'Singular perturbation methods for variational problems in aircraft flight', *IEEE Trans.*, **AC-21**, pp. 345–352

CALISE, A. J. (1977): 'Extended energy management methods for flight performance optimization', *AIAA J.*, **115**, pp. 314–321

CALISE, A. J. (1978): 'A new boundary layer matching procedure for singularly perturbed systems', *IEEE Trans.*, **AC-23**, pp. 434–438

CALISE, A. J. (1979): 'A singular perturbation analysis of optimal aerodynamic and thrust magnitude control', *IEEE Trans.*, **AC-24**, pp. 720–729

CALISE, A. J. (1980): 'A singular perturbation analysis of optimal thrust control with proportional navigation guidance', *AIAA J. Guidance & Control*, **3**, pp. 312–318

CALISE, A. J. (1981): 'Singular perturbation techniques for on-line optimal flight path control', *AIAA J. Guidance & Control*, **4**, pp. 398–405

CALISE, A. J. (1984): 'Optimization of aircraft altitude and flight path angle dynamics', *AIAA J. Guidance & Control*, **7**, pp. 123–125

CALISE, A. J. and MOERDER, D. D. (1985): 'Optimal output feedback design of systems with ill-conditioned dynamics', *Automatica*, **21**, pp. 271–276

CALLEGARI, A. J. and TING, L. (1978): 'Motion of a curved vortex filament with decaying vertical core and axial velocity', *SIAM J. Appl. Math.*, **35**, pp. 148–175

CALLIER, F. M., CHAN, W. S. and DESOER, C. A. (1978): 'Input-output stability of interconnected system using decompositions: an improved formulation', *IEEE Trans.*, **AC-23**, pp. 150–163

CAMPBELL, S. L. (1980): 'Singular systems of differential equations' (Pitman, London)

CAMPBELL, S. L. (1981): 'A procedure for analyzing a class of nonlinear semistate equations that arise in circuits and control problems', *IEEE Trans.*, **CAS-28**, pp. 256–261

CAMPBELL, S. L. (1982): 'Singular systems of differential equations—II' (Pitman, London)

CAMPBELL, S. L. and ROSE, N. J. (1982): 'A second order singular linear system arising in electric power system analysis', *Int. J. Syst. Sci.*, **13**, pp. 101–108

CAPRIZ, G. and CIMMATTI, G. (1978): 'On some singular perturbation problems in the theory of lubrication', *Appl. Math. & Opt.*, **4**, pp. 287–297

CARPENTER, G. A. (1977): 'A geometric approach to singular perturbation problems with applications to nerve impulse equations', *J. Diff. Eqn.*, **23**, pp. 335–367

CARRIER, G. F. (1953): 'Boundary layer problems in applied mechanics' *in* 'Advances in Applied Mechanics'. Vol. **3**, pp. 1–19 (Academic Press, New York)

CARRIER, G. F. (1970): 'Singular perturbation theory and geophysics', *SIAM Rev.*, **12**, pp. 175–193

CASTEN, R. G., COHEN, H. and LAGERSTROM, P. A. (1975): 'Perturbation analysis of an approximation to the Hodgkin–Huxley theory', *Qrt. Appl. Math.*, **32**, pp. 365–402

CHAKRAVARTY, A. (1985): 'Four-dimensional fuel-optimal guidance in the presence of winds', *J. Guidance, Control & Dynamics*, **8**, pp. 16–22

CHANG, K. W. (1974): 'Diagonalisation method for a vector boundary problem of singular perturbation type', *J. Math. Anal. Appl.*, **48**, pp. 652–665

CHANG, K. W. and HOWES, F. A. (1984): 'Nonlinear singular perturbation phenomena: Theory and application' (Springer–Verlag, New York)

CHEMOUIL, P. and WAHDAN, A. M. (1980): 'Output feedback control of systems with slow and fast modes', *Large Scale Systems*, **1**, pp. 257–264

CHEN, J. and O'MALLEY, R. E., Jun. (1972): 'Multiple solutions of a singularly perturbed boundary value problem arising in chemical reactor theory'. Lecture Notes in Mathematics 280 (Springer–Verlag, Berlin) pp. 214–319

CHEN, J. and O'MALLEY, R. E., Jun. (1974): 'On the asymptotic solution of a two-parameter boundary value problem of chemical reactor theory', *SIAM J. Appl. Math.*, **26**, pp. 717–729

CHERNOUSKO, F. L. (1974): 'Some problems of optimal control with a small parameter', *J. Appl. Math.*, **26**, pp. 717–729

CHERNOUSKO, F. L. (1983): 'Dynamics of systems with elastic elements of large stiffness', *Mech. of Solids*, **18**, pp. 99–112

CHERNOUSKO, F. L. and SHAMAEV, A. S. (1983): 'Asymptotic behaviour of singular perturbations in the problem of dynamics of a rigid body with elastic joints and dissipative elements', *Mech. of Solids*, **18**, pp. 31–41

CHEW, W. C. and KONG, J. A. (1982): 'Microstrip capacitance for circular disc through matched asymptotic expansions', *SIAM J. Appl. Math.*, **42**, pp. 301–317

CHIDAMBARA, M. R. and SCHAINKER, R. B. (1971): 'Lower order generalized aggregated model and suboptimal control', *IEEE Trans.*, **AC-16**, pp. 175–180

CHOW, J. H. (1977): 'Preservation of controllability in linear time-invariant perturbed systems', *Int. J. Control*, **25**, pp. 697–704

CHOW, J. H. (1978a): 'Pole-placement design of multiple controller systems via weak and strong controllability', *Int. J. Systems Sci.*, **9**, pp. 129–135

CHOW, J. H. (1978b): 'Asymptotic stability of a class of nonlinear, singularly perturbed systems', *J. Franklin Inst.*, **305**, pp. 275–281

CHOW, J. H. (1979): 'A class of singularly perturbed, nonlinear, fixed-end-point control problem', *J. Opt. Theory & Appl.*, **29**, pp. 231–252

CHOW, J. H. (1982): 'Time-scale modelling of dynamic networks with applications to power systems. Lecture Notes in Control & Inf. Sciences 46 (Springer–Verlag, Berlin)

CHOW, J. H., ALLEMONG, J. J. and KOKOTOVIC, P. V. (1978): 'Singular perturbations analysis of systems with sustained high frequency oscillations', *Automatica*, **14**, pp. 271–279

CHOW, J. H. and KOKOTOVIC, P. V. (1976): 'A decomposition of near optimum regulators for systems with slow and fast modes', *IEEE Trans.*, **AC-21**, pp. 701–705

CHOW, J. H. and KOKOTOVIC, P. V. (1978*a*): 'Two-time-scale feedback design of a class of nonlinear systems', *IEEE Trans.*, **AC-23**, pp. 438–443

CHOW, J. H. and KOKOTOVIC, P. V. (1978*b*): 'Near optimal feedback stabilization of a class of nonlinear singularly perturbed systems', *SIAM J. Control & Opt.*, **16**, pp. 756–770

CHOW, J. H. and KOKOTOVIC, P. V. (1981) 'A two-stage Lyapunov–Bellman feedback design of a class of nonlinear systems', *IEEE Trans.*, **AC-26**, pp. 656–663

CHOW, J. H. and KOKOTOVIC, P. V. (1985): 'Time scale modelling of sparse dynamic networks', *IEEE Trans.*, **AC-30**, pp. 714–722

CHRISTENSEN, H. N. (1969): 'Some kinetic problems of transport', *in* 'Advances in Enzymology' (Wiley, New York) pp. 1–20

CLIFF, E. M., KELLEY, H. J. and LEFTON, L. (1982): 'Thrust-vectored energy turns', *Automatica*, **18**, pp. 559–564

COBB, D. J. (1981): 'Feedback and pole placement in descriptor variable systems', *Int. J. Control*, **33**, pp. 1135–1146

COBB, D. J. (1983): 'Descriptor variable systems and optimal state regulation', *IEEE Trans.*, **AC-28**, pp. 601–611

COBB, D. J. (1986*a*): 'Global analyticity of a geometric decomposition for linear singularly perturbed systems', *Circuits, Systems & Signal Processing*, **5**, pp. 139–152

COBB, D. J. (1986*b*): 'Fundamental properties of the manifield of singular and regular linear systems', *J. Math. Anal. Appl.*, **120**, pp. 328–353

CODERCH, M., WILLSKY, A. S., SASTRY, S. S. and CASTONON, D. A. (1983*a*): 'Hierarchical aggregation of linear systems with multiple time scales', *IEEE Trans.*, **AC-28**, pp. 1017–1030

CODERCH, M., WILLSKY, A. S., SASTRY, S. S. and CASTONON, D. A. (1983*b*): 'Hierarchical aggregation of singularly perturbed finite state Morkov processes', *Stochastics*, **8**, pp. 259–289

COHEN, D. S. (1971): 'Multiple stable solutions of nonlinear boundary value problems arising in chemical reactor theory', *SIAM J. Appl. Math.*, **20**, 1-13

COHEN, D. S. (1973): 'Multiple solutions and periodic oscillations in non-linear diffusion processes', *SIAM J. Appl. Math.*, **25**, pp. 640–654

COHEN, D. S. (Ed.) (1974): 'Mathematical aspects of chemical and biochemical problems and quantum chemistry'. SIAM-AMS Proceedings, American Mathematical Society, Providence

COHEN, D. S. (1976): Instabilities in chemically reacting mixtures', *Rocky Mountain J. Math.*, **6**, pp. 551–559

COHEN, D. S., FOLKS, A. and LAGERSTROM, P. A. (1978): 'Proof of some asymptotic results for a model equation for low Reynolds number flow', *SIAM J. Appl. Math.*, **35**, pp. 187–207

COHEN, D. S., HOPPENSTEADT, F. C. and MIURA. R. M. (1977): 'Slowly modulated oscillations in nonlinear diffusion processes', *SIAM J. Appl. Math.*, **33**, pp. 217–229

COLE, J. D. (1968): 'Perturbation methods in applied mathematics'. Blaisdell, Waltham, MA

COLLINGE, I. R. and OCKENDON, J. R. (1979): 'Transition through resonance of a Duffing oscillator', *SIAM J. Appl. Math.*, **37**, pp. 350–357

COLLINS, W. D. (1973): 'Singular perturbation in linear time-optimal control' *in* BELL, D. J. (Ed.): 'Recent developments in control' (Academic Press, New York) pp. 123–136

COME, G. M. (1977): 'Radical reaction mechanisms: mathematical theory', *J. Phys. Chem.*, **81**, pp. 2560–2563

COMSTOCK, C. (1975): 'Singular perturbation of the wave equation', *Appl. Anal.*, **5**, pp. 117–123

COMSTOCK, C. and HSIAO, G. C. (1976): 'Singular perturbations for difference equations', *Rocky Mountain J. Math.*, **6**, pp. 561–567

COOK, L. P. and COLE, J. D. (1978): 'Lifting line theory for transonic flow', *SIAM J. Appl. Math.*, **35**, pp. 209–228

COOK, L. P. and LUDFORD, G. S. (1971): 'The behaviour as $\varepsilon - 0$ of solutions to $\varepsilon \nabla^2 \omega = \delta\omega/\delta y$ in $|y| \leqslant 1$ for discontinous data', *SIAM J. Math. Anal.*, **2**, pp. 567–594

COOK, L. P., LUDFORD, G. S. and WALKIER, J. S. (1972): 'Corner regions in the asymptotic solution of $\varepsilon \nabla^2 u = \delta u/\delta y$ with reference to MHD duct flow', *Proc. Camb. Phil. Soc.*, **72**, pp. 117–122

COOPER, L. Y. (1971): 'Constant temperature at the surface of an initially uniform temperature variable conductivity half surfaces', *J. Heat Transfer*, **93**, pp. 55–60

COOPER, L. Y. (1975): 'A singular perturbation solution to a problem of extreme temperature imposed at the surface of a variable-conductivity half-space: small surface conductivity', *Quart. Appl. Math.*, **32**, pp. 427–444

COPE, D. (1980): 'Stability of limit cycle solutions of reaction-diffusion equations', *SIAM J. Appl. Math.*, **38**, pp. 457–479

COPPEL, W. A. (1967): 'Dichotomies and reducibility', *J. Diff. Eqn.*, **3**, pp. 500–521

CORI, R. and MAFFEZONI, C. (1984): 'Practical optimal control of a drum boiler power plant', *Automatica*, **20**, pp. 163–173

CRESPO DA SILVA, M. R. M. and DAVIS, R. T. (1977): 'A study of optimal coordinate theory with applications to several physical problems', *Int. J. Engg. Sci.*, **15**, pp. 455–464

CRONIN, J. (1977): 'Some mathematics of biological oscillations', *SIAM Rev.*, **19**, pp. 100–138

CRUZ, J. B., Jun. (Ed.) (1972): 'Feedback Systems' (McGraw-Hill, New York)

DAUPHIN-TANGUY, G., BORNE, P. and LEBRAUN, M. (1985): 'Order reduction and multitime scale systems using bond graphs, the reciprocal systems and the singular perturbation method', *J. Franklin Inst.*, **319**, pp. 157–271

DAVIS, J. L. and REISS, E. L. (1970): 'An asymptotic analysis of the damped wave equation', *Trans. Soc. Rheology*, **14: 21**, pp. 239–251

DAVIS, R. T. and ALFRIEND, K. T. (1967): 'Solution of van der Pol's equation using a perturbation method', *Int. J. Non-Linear Mech.*, **2,**, pp. 153–162

DAVISON, E. J. (1966): 'A method for simplifying linear dynamic systems', *IEEE Trans.*, **AC-11**, pp. 93–101

DAVISON, E. J. (1976): 'The robust control of a servomechanism problem for linear time invariant multivariable systems', *IEEE Trans.*, **AC-21**, pp. 25–34

DELEBECQUE, F. (1983): 'A reduction process for perturbed Markov chains', *SIAM J. Appl. Math.*, **43**, 324–350

DELEBECQUE, F. and QUADRAT, J. P. (1978): 'Contributions of stochastic control singular perturbation averaging and team theories to an example of large scale system: management of hydropower production', *IEEE Trans.*, **AC-23**, pp. 209–222

DELEBECQUE, F. and QUADRAT, J. P. (1981): 'Optimal control of Markov chains admitting strong and weak interactions', *Automatica*, **17**, pp. 281–296

DELEBECQUE, F., QUADRAT, J. P. and KOKOTOVIC, P. V. (1984): 'A unified view of aggregation and coherency in networks and Morkov chains', *Int. J. Control*, **40**, pp. 939–952

DESOER, C. A. (1970): 'Singular perturbation and bounded-input bounded-state stability', *Electron. Lett.*, **6**, pp. 496–497

DESOER, C. A. (1977): 'Distributed networks with small parasitic elements: input-output stability', *IEEE Trans.*, **CAS-24**, pp. 1–8

DESOER, C. A. and SHENSA, M. J. (1970): 'Networks with very small and very large parasites: natural frequencies and stability', *Proc. IEEE*, **58**, pp. 1933–1938

DESOER, C. A. and VIDYASAGAR, M. (1975): 'Feedback systems: Input-output properties' (Academic Press, New York)

DICAPRIO, U. (1982): 'Emergency control', *Elect. Power & Energy Syst.*, **4**, pp. 19–28

DICKER, D. and BABU, D. K. (1974*a*): 'Two dimensional seepage in layered soil-destabilizing effects of flow with an unsteady free water surface', *Water Resources Research*, **10**, pp. 801–809

DICKER, D., and BABU, D. K. (1974*b*): 'A singular perturbation problem in unsteady ground water flows with a free surface', *Int. J. Engg. Sci.*, **12**, pp. 967–980

DIEKMANN, O., HILHORST, D. and PELETIER, L. A. (1980): 'A singular boundary value problem arising in a pre-breakdown gas discharge', *SIAM J. Appl. Math.*, **39**, pp. 48–66

DILLON, T. S. (1982): 'Dynamical modelling and control of large-scale systems', *Elect. Power & Energy Syst.*, **4**, pp. 29–36

DIPRIMA, R. C. (1968): 'Asymptotic methods for an infinitely long slider squeeze-film bearing', *J. Lubric Technol.*, **90**, pp. 173–183

DMITRIEV, M. G. (1978): 'A class of singularly perturbed optimal control problems', *J. Appl. Math. & Mech.*, **42**, pp. 238–242

DMITRIEV, M. G. and KLISHEVIC, A. M. (1984): 'Iterative solutions of optimal control problems with fast and slow motions', *Systems & Control Lett.*, **4**, pp. 223–226

DO, D. D. and GREENFIELD, P. F. (1981): 'A finite integral transform technique for solving the diffusion-reaction equation with Michaelis–Menten kinetics', *Math. Biology*, **54**, pp. 31–47

DONTCHEV, A. L. (1977): 'Linear model simplification for singularly perturbed optimal control problems', *Compt. Rend. Acad. Bul. Sci.*, **30**, pp. 449–502

DONTCHEV, A. L. (1983): 'Perturbations, approximations and sensitivity analysis of optimal control systems'. Lecture Notes in Control and Inf. Sciences, Vol. 52 (Springer–Verlag, Berlin)

DONTCHEV, A. L. and GICEV, T. R. (1979): 'Convex singularly perturbed optimal control problems with fixed final state controllability and convergence', *Math. Operations for Sch. Statist. Ser. Optimization*, **10**, pp. 345–355

DONTCHEV, A. L. and VELIOV, V. M. (1982): 'A singularly perturbed optimal control problem with fixed final state and constrained control', *Control & Cybernetics*, **11**, pp. 1–2

DONTCHEV, A. L. and VELIOV, V. M. (1983): 'Singular perturbations in Meyer's problem for linear systems', *SIAM J. Control & Opt.*, **21**, pp. 566–581

DONTCHEV, A. L. and VELIOV, V. M. (1985): 'Singular perturbations in linear control systems with weakly coupled stable and unstable fast subsystems', *J. Math. Anal. Appl.*, **110**, pp. 1–30

DORAISWAMI, R. (1982): 'A reduced order stabilizer design ensuring closed loop stability', *Proc. IEEE*, **70**, pp. 92–93

DORATO, P., HIESH, C. M. and ROBINSON, P. M. (1967): 'Optimal bang-bang control of linear stochastic systems with a small noise parameter', *IEEE Trans.*, **AC-12**, pp. 682–689

DORODNITSYN, A. A. (1947): 'The asymptotic solution of the van der Pol equation', *Prikl. Matem. i Mekhan.*, **11**, pp. 313–328 (Russian)

DORR, F. W., PARTER, S. V. and SHAMPINE, L. F. (1973): 'Application of the maximum principle to singular perturbation problem', *SIAM Rev.*, **15**, pp. 43–88

DORSEY, J. F. and SCHLUETER, R. A. (1984): 'Structural archetypes for coherency: a framework for comparing power system equivalents', *Automatica*, **20**, pp. 349–452

DRAGAN, V. (1985): 'Observers with several time scales for systems with several time scales', *Int. J. Control*, **42**, pp. 149–154

DRAGAN, V. and HALANAY, A. (1982): 'Suboptimal stabilization of linear systems with several time scales', *Int. J. Control*, **36**, pp. 109–126

DUC, G., MICHAILESCO, G., SIRET, J. M. and BERTRAND, P. (1983): 'Design of large scale system control laws by aggregation and perturbation methods', *Large Scale Systems*, **41**, pp. 91–99

DYGAS, M. M., MATKOWSKY, B. J. and SCHUSS, Z. (1986): 'A singular perturbation approach to non-Markovian escape rate problems', *SIAM J. Appl. Math.*, **46**, pp. 265–298

ECKHAUS, W. (1973): 'Matched asymptotic expansions and singular perturbations', (North Holland, Amsterdam)

ECKHAUS, W. (1979): 'Asymptotic analysis of singular perturbations' (North Holland, Amsterdam)

ECKHAUS, W., HARTEN, A. V. and PERADZYNSKI, Z. (1985): 'A singularly perturbed free boundary value problem describing a laser sustained plasma', *SIAM J. Appl. Math.*, **45**, pp. 1–31

ECKHAUS, W. and DE JAGER, E. M. (Eds.) (1978): 'Differential equations and applications'. Mathematical Studies 31 (North Holland, Amsterdam)

ECKHAUS, W. and DE JAGER, E. M. (Eds.) (1982): 'Theory and applications of singular perturbations'. Lecture Notes in Mathematics Vol. 942 (Springer–Verlag, Berlin)

EINAUDI, F. (1969): 'Singular perturbation analysis of acoustic-gravity waves', *Phy. Fluids*, **12**, pp. 752–756

EITELBERG, E. (1982): 'Model reduction and perturbation structures', *Int. J. Control*, **35**, pp. 1029–1050

EITELBERG, E. (1983): 'State space for nonlinear algebraically constrained dynamical systems', *Int. J. Control*, **37**, pp. 1433–1447

EITELBERG, E. (1985): 'A transformation of nonlinear dynamical systems with a single singular singularly perturbed differential equation', *Int. J. Control*, **41**, pp. 1301–1316

EITELBERG, E. and RODE, M. (1982): 'On model reduction and iterative calculation of optimal output feedback gains', *Int. J. Control*, **35**, pp. 561–571

EL-ENSARY, M. and KHALIL, H. K. (1986): 'On the interplay of singular perturbations and wide-band stochastic fluctuations', *SIAM J. Control & Optimiz.*, **24**, pp. 83–94

ELRAZAZ, Z. and SINHA, N. K. (1981): 'A review of some model reduction techniques', *Can. Elect. Engg. J.*, **6**, pp. 34–40

ERDELYI, A. (1956): 'Asymptotic expansions' (Dover Publications, New York)

FAIKOVITZ, M. S. and SEGAL, L. A. (1983): 'Spatially inhomegeneous polymerization in unstirred bulk', *SIAM J. Appl. Math.*, **43**, pp. 386–416

FARBER, N. and SHINAR, J. (1980): 'Approximate solution of singularly perturbed nonlinear pursuit-evasion games', *J. Opt. Theory & Appl.*, **32**, pp. 39–75

FATTORINI, H. O. (1985): 'On the Schrodinger singular perturbation problem', *SIAM J. Math. Anal.*, **16**, pp. 1000–1019

FENICHEL, N. (1979): 'Geometric singular perturbation theory for ordinary differential equations', *J. Diff. Eqn.*, **31**, pp. 53–98

FENICHEL, N. (1983a): 'Oscillatory bifurcation in singular perturbation theory—I: Slow oscillations', *SIAM J. Math. Anal.*, **14**, pp. 861–867

FENICHEL, N. (1983b): 'Oscillatory bifurcation in singular perturbation theory—II: Fast oscillations', *SIAM J. Math. Anal.*, **14**, pp. 868–874

FERGUSON. W. E., Jun. (1986): 'Analysis of a singularly perturbed linear two-point boundary-value problem', *SIAM J. Num. Anal.*, **23**, pp. 940–947

FERNANDO, K. V. and NICHOLSON, H. (1982a): 'Singular perturbation model reduction of balanced systems', *IEEE Trans.*, **AC-27**, pp. 466–468

FERNANDO, K. V. and NICHOLSON, H. (1982b): 'Singular perturbation model reduction in the frequency domain', *IEEE Trans.*, **AC-27**, pp. 969–970

FERNANDO, K. V. and NICHOLSON, H. (1983a): 'Reciprocal transformations in balanced model order reduction', *IEE Proc. Pt.D.*, **130**, pp. 359–362

FERNANDO, K. V. and NICHOLSON, H. (1983b): 'Singular perturbational approximations for discrete-time balanced systems', *IEEE Trans.*, **AC-28**, pp. 240–242

FIFE, P. C. (1976): 'Pattern formation in reacting and diffusing systems', *J. Chem. Phys.*, **64**, pp. 554–564

FIFE, P. C. (1979): 'The mathematics of reacting and diffusing systems'. Lecture Notes in Biomathematics 28 (Springer–Verlag, New York)

FIFE, P. C. and NICHOLES, K. R. K. (1975): 'Dispersion in flow through small tubes', *Proc. Royal Soc. London A*, **344**, pp. 131–145

FISHER, T. M., HSIAO G. C. and WENDLAND, W. L. (1985): 'Singular perturbations for the exterior three-dimensional slow viscous problem', *J. Math. Anal. Appl.*, **110**, pp. 583–603

FLAHERTY, J. E. and MATHON, W. (1980): 'Collocation with polynomial and tension splines for singularly-perturbed boundary value problems', *SIAM J. Stat. Comput.*, **1**, pp. 260–289

FLAHERTY, J. E. and O'MALLEY, R. E., Jun. (1977): 'On the computation of singular arcs', *IEEE Trans.*, **AC-22**, pp. 640–648

FLAHERTY, J. E. and O'MALLEY, R. E., Jun. (1982): 'Singularly perturbed boundary value problems for nonlinear systems including a challenging problem for a nonlinear beam' *in* ECKHAUS, W. and DE JAGER, E. M. (Eds.) 'Theory and applications of singular perturbations'. Lecture Notes in Mathematics 942 (Springer–Verlag, Berlin) pp. 170–191

FLATTO, L. and LEVINSON, N. (1955): 'Periodic solutions of singularly perturbed systems', *J. Rational Mech. Anal.*, **4**, pp. 943–950

FLEMING, W. H. (1971): 'Stochastic control for small noise intensities', *SIAM J. Control*, **9**, pp. 473–517

FLEMING, W. H. (1974): 'Stochastically perturbed dynamical systems', *Rocky Mountain J. Math.*, **4**, pp. 407–433

FLETCHER, J. E. (1980): 'On facilitated oxygen diffusion in muscle tissues', *Biophys. J.*, **29**, pp. 437–458

FOSSARD, A. J., BERTHELOT, M. and MAGNI, J. F. (1983): 'On coherency-based decomposition algorithm', *Automatica*, **19**, pp. 247–253

FOSSARD, A. J. and MAGNI, J. F. (1980): 'A frequential analysis of singularly perturbed systems with state or output control', *Large Scale Syst.: Theory & Appl.*, **1**, pp. 223–228

FOWLER, A. C. (1977): 'Convective diffusion of an enzyme reaction', *SIAM J. Appl. Math.*, **33**, pp. 289–297

FRAENKEL, L. E. (1969a): 'On the method of matched asymptotic expansions—I: A matching principle', *Proc. Cambridge Phil. Soc.*, **65**, pp. 209–232

FRAENKEL, L. E. (1969b): 'On the method of matched asymptotic expansions—II: Some applications of the composite series', *Proc. Cambridge Phil. Soc.*, **65**, pp. 233–261

FRAENKEL, L. E. (1969c): 'On the method of matched asymptotic expansions—III: Two boundary value problems', *Proc. Cambridge Phil. Soc.*, **65**, pp. 263–284

FRANCIS, B. A. (1979): 'The optimal linear quadratic time-invariant regulator with cheap regulator', *IEEE Trans.*, **AC-24**, pp. 616–620

FRANCIS, B. A. (1981): 'Convergence in the boundary layer for singularly perturbed equations', *Automatica*, **18**, pp. 57–62

FRANCIS, B. A. and GLOVER, K. (1978): 'Bounded peaking in the optimal linear regulator with cheap control', *IEEE Trans.*, **AC-23**, pp. 608–617

FRANK, L. S. and WENDT, W. D. (1982): 'On singular perturbation in the kinetic theory of enzymes', *in* ECKHAUS, W. and DE JAGER, E. M. (Eds.): 'Theory and applications of singular perturbations'. Lecture Notes in Mathematics 942 (Springer–Verlag, Berlin) pp. 346–363

FREEDMAN, M. I. (1976): 'On the possibility of carrying out a perturbation analysis for control problems involves switch points that coalesce' *in* 'Asymptotic methods and singular perturbations. SIAM–AMS Proceedings, Vol. 10 (American Mathematical Soc., Providence, USA) p. 149

FREEDMAN, M. I. and GRANOFF, B. (1976): 'Formal asymptotic solution of a singularly perturbed nonlinear optimal control problem', *J. Opt. Theory & Appl.*, **19**, pp. 301–325

FREEDMAN, M. I. and KAPLAN, J. L. (1975): 'Use of a nonlinear clock in the perturbation analysis of time optimal control problems', *J. Comp Appl. Math.*, **1**, pp. 219–228

FREEDMAN, M. I. and KAPLAN, J. L. (1976): 'Singular perturbations of two-point boundary value problems arising in optimal control', *SIAM J. Control & Opt.*, **14**, pp. 189–215

FREEDMAN, M. I. and KAPLAN, J. L. (1977): 'Perturbation analysis of an optimal control problem involving bang-bang controls', *J. Diff. Eqn.*, **25**, pp. 11–29

FREEMAN, L. B. and HOUGHTON, G. (1966): 'Singular perturbations of nonlinear boundary value problems arising in chemical flow reactors', *Chem. Engrg. Soc.*, **21**, pp. 1011–1024

FREILING, G. (1984): 'Singular perturbations of multi-point eigenvalue problems', *Arch. Rat. Mech. Anal.*, **86**, pp. 197–210

FRIEDRICHS, K. O. (1955): 'Asymptotic phenomena in mathematical physics', *Bull. Amer. Math. Soc.*, **61**, pp. 485–504

FRIEDRICHS, K. O. and WASOW, W. (1946): 'Singular perturbations of nonlinear oscillations, *Duke Math. J.*, **13**, pp. 367–381

FU, S. J. and SAWAN, M. E. (1986): 'On the reduced order model for control system design and analysis', *Int, J. Syst. Sci.*, **17**, pp. 319–324

GAITSOGORI, V. G. and PERVOZVANSKII, A. A. (1975): 'Aggregation of states in a Markov chain with weak interactions', *Kybernatica*, pp. 91–98

GAITSOGORI, V. G. and PERVOZVANSKII, A. A. (1979): 'Perturbation method in optimal control problems', *J. Syst. Sci.*, **5**, pp. 91–102

GAITSOGORI, V. G. and PERVOZVANSKII, A. A. (1980): 'On the optimization of weakly controlled stochastic system', *Sov. Math. Dokl.*, **21**, pp. 408–410

GAJIC, Z. (1986): 'Numerical fixed-point solution for near-optimum regulators of linear quadratic gaussian control problems for singularly perturbed systems', *Int. J. Control*, **43**, pp. 373–387

GAJIC, Z. and KHALIL, H. K. (1986): 'Multimodel strategies under random disturbances and imperfect partial observations', *Automatica*, **22**, pp. 121–125

GARDNER, B. F., Jun. and CRUZ, J. B., Jun. (1978): 'Well-posedness of singularly perturbed Nash games', *J. Franklin Inst.*, **306**, pp. 355–374

GARDNER, B. F., Jun. and CRUZ, J. B., Jun. (1980): 'Lower order control of systems with fast and slow modes', *Automatica*, **16**, pp. 211–215

GARDNER, R. and SMOLLER, J. (1983): 'The existence of periodic travelling waves for singularly perturbed predator-prey equations via the Couley index', *J. Diff. Eqn.*, **47**, pp. 133–161

GENESIO, R. and MILANESE, M. (1976): 'A note on the derivation and the use of reduced-order models', *IEEE Trans.*, **AC-21**, pp. 118–121

GICHEV, T. R. (1984): 'Singular perturbations in a class of problems of optimal control with integral convex criterion', *J. Appl. Math. Mech.*, **48**, pp. 654–658

GICHEV, T. R. and DONTCHEV, A. L. (1979): 'Convergence of solutions to the linear singular-perturbed time optimal control problem', *Appl. Math. Mech.*, **43**, pp. 466–474

GLIZER, V. J. and DMITRIEV, M. G. (1975a): 'Singular perturbation in a linear optimal control problem with a quadratic functional', *Soviet Math. Dokl.*, **16**, pp. 1555–1558

GLIZER, V. J. and DMITRIEV, M. G. (1975b): 'Singular perturbations in a linear control problem with a quadratic functional', *Diff. Eqns.*, **11**, pp. 1427–1432

GLIZER, V. J. and DMITRIEV, M. G. (1976): 'The connection between singular perturbations and penalty functions', *Soviet Math.*, **17**, pp. 1503–1505

GLIZER, V. J. and DMITRIEV, M. G. (1977a): 'On continuity of the solutions of the problem of a regulator analytic construction in singular perturbations', *Appl. Math. Mech. (PMM)*, **41**, pp. 594–598

GLIZER, V. J. and DMITRIEV, M. G. (1977b): 'About continuity of the regulator problem by singular perturbations', *Appl. Math. Mech. (PMM)*, **41**, pp. 573–576

GLIZER, V. J. and DMITRIEV, M. G. (1978): 'Asymptotic solution of a singularly perturbed Cauchy problem in optimal control', *Differential Eqns.*, **14**, pp. 601–612

GLIZER, V. J. and DMITRIEV, M. G. (1979a): 'Singular perturbations and generalized functions', *Soviet Math.*, **20**, pp. 1360–1364

GLIZER, V. J. and DMITRIEV, M. G. (1979b): 'Asymptotic solution of some discrete optimal control problems with small step size', *Differential Eqns.*, **15**, pp. 1681–1686

GRABMULLER, H. (1978): 'Singular perturbation techniques applied to intego-differential equations' (Pitman, London)

GRACEY, C., CLIFF, E. M., LUTZE, F. H. and KELLEY, H. J. (1982): 'Fixed-trim re-entry guidance analysis', *J. Guidance, Control and Dynamics*, **5**, pp. 558–563

GRASMAN, J. (1979a): 'Non-uniqueness in singular optimal control', Int. Symposium on Math. Theory of Networks and Systems, Vol. 3, pp. 415–420

GRASMAN, J. (1979b): 'Small random perturbations of dynamical systems with applications to population genetics', *in* VERHULST, F. (Ed.): 'Asymptotic analysis'. Lecture Notes in Mathematics 711 (Springer–Verlag, Berlin) pp. 158–175

GRASMAN, J. (1980): 'Relaxation oscillations of a Van der Pol equation with large critical forcing terms', *Q. Appl. Math.*, **308**, pp. 9–16

GRASMAN, J. (1982): 'On a class of optimal control problems with an almost cost-free solutions', *IEEE Trans.*, **AC-27**, pp. 441–445

GRASMAN, J. (1984): 'The mathematical modelling of entrained biological oscillations', *Bull. Math. Biol.*, **46**, pp. 407–422

GRASMAN, J. and JANSEN, M. J. W. (1979): 'Mutually synchronized relaxation oscillations as prototypes of oscillating systems in biology', *J. Math. Biol.*, **7**, pp. 171–197

GRASMAN, J. and VELING, J. M. (1973): 'An asymptotic formula for the period of a Volterra-Lokta system', *Math. Biosci.*, **18**, pp. 185–189

GRASMAN, J., VELING, J. M. and WILLEMS, G. M. (1976): 'Relaxation oscillations governed by Van der Pol equation with periodic forcing term', *SIAM J. Appl. Math.*, **13**, pp. 667–676

GREENBERG, J. M. (1976): 'Periodic solutions to wave equations', *Rocky Mountain J. Math.*, **6**, pp. 755–756

GRIMBLE, M. J. (1978): 'A simple method for the design of single input optimum regulators and servomechanism', *Proc. IEE*, **125**, p. 537

GRUJIC, L. T. (1979a): 'Singular perturbations, large scale systems and asymptotic stability of invariant sets', *Int. J. Syst. Sci.*, **10**, pp. 1323–1341

GRUJIC, L. T. (1979b): 'Singular perturbations and large-scale systems', *Int. J. Control*, **29**, pp. 159–169

GRUJIC, L. T. (1981): 'Uniform asymptotic stability of nonlinear singularly perturbed general and large scale systems', *Int. J. Control*, **33**, pp. 481–504

GUARDABASSI, G. and LOCATELLI, A. (1975): 'Periodic control of singularly perturbed systems' *in* Lecture Notes in Economics and Mathematical Systems Vol. 11, pp. 101–105

GUCKENHEIMER, J. (1980*a*): 'Symbolic dynamics and relaxation oscillations', *J. Phys. D*, **1**, pp. 227–235

GUCKENHEIMER, J. (1980*b*): 'Dynamics of the Van der Pol equation', *IEEE Trans.*, **CAS-27**, pp. 983–989

GUINZY, N. J. and SAGE, A. P. (1973): 'Identification and modelling of large scale systems using sensitivity analysis,' *Int. J. Control*, **17**, pp. 1073–1087

GUSTAFSSON, B. (1980): 'Numerical solution of hyberbolic systems with different time scales using asymptotic expansions', *J. Computational Phys.*, **36**, pp. 209–235

GUTOWSKI, T. (1972): 'The properties of a solution of the equation of motion of a mechanical system subject to irregular (singular) perturbation', *Arch. Mech.*, **24**, pp. 531–538

HABETS, P. (1983): 'Singular perturbations in nonlinear systems and optimal control' *in* ARDEMA, M. D. (Ed.): 'Singular perturbations in systems and control', CISM Courses and Lectures 280 (Springer–Verlag, Wien) pp. 103–142

HADDAD, A. H. (1976): 'Linear filtering of singularly perturbed systems', *IEEE Trans.*, **AC-21**, pp. 515–519

HADDAD, A. H. and KOKOTOVIC, P. V. (1971): 'A note on singular perturbation of linear state regulators', *IEEE Trans.*, **AC-16**, pp. 279–281

HADDAD, A. H. and KOKOTOVIC, P. V. (1977): 'Stochastic control of linear singularly perturbed systems', *IEEE Trans.*, **AC-22**, pp. 815–821

HADLOCK, C. R. (1973): 'Existence and dependence on a parameter of solutions of nonlinear two point boundary value problems', *J. Diff. Eqns.*, **14**, pp. 498–517

HADLOCK, C. R. (1977): 'Asymptotic performance of near-optimum controls obtained by regular and singular perturbations', *J. Math. Anal. Appl.*, **61**, p. 292–301

HALANAY, A. (1966): 'Differential equations: Stability, oscillations, and time lags (Academic Press, New York)

HALANAY, A. and MIRICA, S. (1979): 'The time optimal feedback control for singularly perturbed linear systems', *Rev. Roum. Math. Pures Appl.*, **24**, pp. 585–596

HALLER, M. A. and IUNG, C. (1983): 'A multiple-scale approach in the study of communication phenomena: application to a chopper supplying a step motor', *Elect. Machines & Power Syst.*, **8**, pp. 113–122

HANMANDLU, M., SURYANARAYANA, N. V. and SINHA, A. K. (1986): 'Multi-time-scale decompostion of a high-gain feedback systems', *Int. J. Control*, **44**, pp. 17–41

HARAUX, A. and MURAT, F. (1985): 'Influence of a singular perturbation on the infimum of some functionals', *J. Diff, Eqns.*, **58**, pp. 43–75

HARRIS, W. A., Jun. (1960): 'Singular perturbations of two-point boundary value problems for systems of ordinary differential equations', *Arch. Rational Mech. Anal.*, **5**, pp. 212–225

HARRIS, W. A., Jun. (1973): 'Singularly perturbed boundary value problems revisited' *in* Lecture Notes in Mathematics 312 (Springer–Verlag, Berlin) pp. 54–84

HARTEN, A. VAN (1979*a*): 'Feedback control of singularly perturbed heating problems' *in* VERHULST, F. (Ed.): 'Asymptotic analysis'. Lecture Notes in Mathematics 711 (Springer–Verlag, Berlin) pp. 33–62

HARTEN, A. VAN (1979*b*): 'Asymptotic approximations in magneto-hydrodynamic singular perturbation problems' *in* VERHULST, F. (Ed.): 'Asymptotic analysis'. Lecture Notes in Mathematics 711 (Springer–Verlag, Berlin) pp. 94–124

HARTEN, A. VAN (1982): 'Application of singular perturbation technique to combustion theory', *in* ECKHAUS, W. and DE JAGER, E. M. (Eds.): 'Theory and applications of singular perturbations'. Lecture Notes in Mathematics 942 (Springer–Verlag, Berlin) pp. 295–308

HARTEN, A. VAN (1984): 'Singularly perturbed systems of diffusion type and feedback control', *Automatica*, **20**, pp. 79–91

HARTWIG, R. E. (1979): 'Theory and applications of matrix perturbations with respect to Hankel matrices and power formulated', *Int. J. Syst. Sci.*, **10**, pp. 437–460

HEINEKEN, F. G., TSUCHIYA, H. M. and ARIS, R. (1967): 'On the mathematical status of the pseudo-steady state hypothesis of biochemical kinetics', *Math. Biosci.*, **1**, pp. 95–113

HEKIMOVA, M. A. and BAINOV, D. D. (1985): 'Periodic solutions of singularly perturbated systems of differential equations with impulse effect', *J. Appl. Math. Phys.*, **36**, pp. 520–537

HEMKER, P. W. and MILLER, J. J. H. (Eds.) (1979): 'Numerical analysis of singular perturbation problems' (Academic Press, New York)

HENDRY, W. L. (1970): 'Application of the method of matched asymptotic expansions to a problem in linear transport theory', *J. Math. Phys.*, **11**, pp. 1743–1749

HENDRY, W. L. (1971): 'Solutions to the linear time-dependent neutron transport equation with time-dependent source and cross section', *Nuclear Sci. & Engg.*, **45**, pp. 1–6

HENDRY, W. L. and BELL, G. I. (1969): 'An analysis of the time-dependent neutron transport equation with delayed neutron by the method of matched asymptotic expansions', *Nuclear Sci. & Engg.*, **35**, pp. 240–252

HERMANS, A. J. (1982): 'Wave pattern of a ship sailing at low speeds' *in* ECKHAUS, W. and DE JAGER, E. M. (Eds.): 'Theory and applications of singular perturbations'. Lecture Notes in Mathematics 942 (Springer–Verlag, Berlin) pp. 281–294

HICKIN, J. D. (1980): 'Approximate aggregation for linear multivariable systems', *Electron. Lett.*, **16**, pp. 518–519

HICKIN, J. D. and SINHA, N. K. (1980): 'Model reduction for linear multivariable systems', *IEEE Trans.*, **AC-25**, pp. 1121–1127

HILDEBRAND, F. B. (1968): 'Finite difference equations and simulations' (Prentice Hall, Englewood Cliffs)

HILHORST, D. (1982): 'A perturbed free boundary value problem arising in the physics of ionized gases', *in* ECKHAUS, W. and DE JAGER, E. M. (Eds.): 'Theory and applications of singular perturbations'. Lecture Notes in Mathematics 942 (Springer–Verlag, Berlin) pp. 309–317

HIRSCHEFELDER, J. O., CURTIS, C. F. and CAMPBELL, D. F. (1953): 'The theory of flame propagation', *J. Phys. Chem.*, **34**, pp. 324–330

HO, T. C. and HSIAO, G. C. (1977): 'Estimation of the effectiveness factor for a cylindrical catalyst support: a singular perturbation approach', *Chem. Engg. Sci.*, **32**, pp. 63–73

HOLDEN, A. V. and WINLOW, W. (1983): 'Neuronal activity as the behavior of a differential system', *IEEE Trans.*, **SMC-13**, pp. 711–719

HOLLAND, C. J. (1976): 'Singular perturbation problems using probabilistic methods', *Rocky Mountain J. Math.*, **6**, pp. 585–590

HOLLAND, C. J. (1981): 'An approximate technique for small noise open-loop control problem', *Optimal Control: Appl. & Methods*, **2**, pp. 89–94

HOPKINS, W. E., Jun. and BLANKENSHIP, G. L. (1981): 'Perturbation analysis of a system of quasi-variational inequalities for optimal scheduling', *IEEE Trans.*, **AC-26**, pp. 1054–1070

HOPPENSTEADT, F. C. (1971): 'Properties of solutions of ordinary differential equations with small parameters', *Comm. Pure Appl. Math.*, **24**, pp. 807–840

HOPPENSTEADT, F. C. (1975): 'Mathematical theories of demographics, genetics and epidemics, Vol. 20 (SIAM Publications, Philadelphia)

HOPPENSTEADT, F. C. (1983): 'An algorithm for approximate solutions to weakly coupled filtered synchronous control systems and nonlinear processes', *SIAM J. Appl. Math.*, **43**, pp. 834–843

HOPPENSTEADT, F. C. and MIRANKER, W. L. (1977): 'Multitime methods for systems of difference equations', *Studies in Appl. Math.*, **56**, pp. 273–289

HOULIHAN, S. G., CLIFF, E. M. and KELLEY, H. J. (1982): 'Study of chattering cruise', *J. Aircraft*, **19**, pp. 119–124

HOWARD, L. N. and KOPPEL, N. (1977): 'Slowly varying waves and shock structures in reaction-diffusion equations', *Studies in Appl. Math.*, **56**, pp. 95–145

HOWES, F. A. (1977): 'Singular perturbation analysis of a class boundary value problems arising in catalytic reaction theory', *Arch. Rat. Mech. Anal.*, **66**, pp. 237–265

HOWES, F. A. (1982): 'The asymptotic solution of singularly perturbed Divichlet problem with applications to the study of incompressible flows at high Reynolds number', *in* ECKAUS, W. and DE JAGER, E. M. (Eds.): 'Theory and applications of singular perturbations'. Lecture Notes in Mathematics 942 (Springer–Verlag, Berlin) pp. 245–257

HOWES, F. A. (1983): 'The asymptotic solution of a class of third order boundary value problems arising in the theory of thin film flows', *SIAM J. Appl. Math.*, **43**, pp. 993–1004

HSIAO, G. C. and MACCAMY, R. C. (1982): 'Singular perturbations for the two-dimensional viscous flow problem' *in* ECKHAUS, W. and DE JAGER, E. M. (Eds.): 'Theory and applications of singular perturbations'. Lecture Notes in Mathematics 942 (Springer–Verlag, Berlin) pp. 229–244

HUANG, T. C. and DAS, A. (1980): 'Singular perturbation equation for flexible satellites', *Int. J. Non-Linear Mech.*, **15**, pp. 355–365

HUNT, F. (1980): 'On the persistence of spatially homogeneous solutions of a population genetics model with slow selection', *Math. Biosci.*, **52**, pp. 185–206

HUTTON, M. F. and FRIEDLAND, B. (1975): 'Routh approximations for reducing the order of linear systems', *IEEE Trans.*, **AC-20**, pp. 329–337

HWANG, C., SHIP, Y. P. and HWANG, R. Y. (1981): 'A combined time and frequency domain method for model reduction of discrete systems', *J. Franklin Inst.*, **311**, pp. 391–402

IGNETIK, R. and DEAKIN, M. A. B. (1981): 'Asymptotic analysis of the Michaelis-Menten reaction equation', *Bull, Math. Biol.*, **43**, pp. 375–388

IOANNOU, P. A. (1981): 'Robustness of absolute stability', *Int. J. Control*, **33**, pp. 1027–1033

IOANNOU, P. (1986): 'Robust adaptive controller with zero residual tracking', *IEEE Trans.*, **AC-31**, pp. 773–776

IOANNOU, P. A. and JOHNSON, C. R., Jun. (1983): 'Reduced order performance of parallel and series parallel identifiers with weakly observable parasitics,' *Automatica*, **19**, pp. 75–88

IOANNOU, P. A. and KOKOTOVIC, P. V. (1982): 'An asymptotic error analysis of identifiers and adaptive observers in the presence of parasitics', *IEEE Trans.*, **AC-27**, pp. 921–927

IOANNOU, P. A. and KOKOTOVIC, P. V. (1983): 'Adaptive systems with reduced models'. Lecture Notes in Control & Inf. Sciences, Vol. 47 (Springer–Verlag, Berlin)

IOANNOU, P. A. and KOKOTOVIC, P. V. (1984a): 'Instability analysis and improvement of robustness of adaptive control', *Automatica*, **20**, pp. 583–594

IOANNOU, P. A. and KOKOTOVIC, P. V. (1984b): 'Robust redesign of adaptive control', *IEEE Trans.*, **AC-29**, pp. 202–211

IOANNOU, P. A. and KOKOTOVIC, P. V. (1985): 'Decentralized adaptive control of interconnected systems with reduced order models', *Automatica*, **21**, pp. 401–412

IOANNOU, P. A. and TSAKALIS, K. S. (1986): 'A robust direct adaptive controller', *IEEE Trans.*, **AC-31**, pp. 1033–1043

JAMES, E. C. (1975): 'Lifting-line theory for an unsteady wing as a singularly perturbation problem', *J. Fluid Mech.*, **70**, pp. 753–771

JAMESON, A. and O'MALLEY, R. E., Jun. (1975): 'Cheap control of the time invariant regulator', *Appl. Math. Optimiz.*, **1**, pp. 337–355

JAMSHIDI, M. (1972): 'A near optimum design controller for cold rolling-mills', *Int. J. Control*, **16**, pp. 1137–1154

JAMSHIDI, M. (1974): 'Three-stage near-optimum design of nonlinear control processors', *Proc. IEE*, **121**, pp. 886—892

JAMSHIDI, M. (1976): 'Applications of three-time-scale near-optimum design to control systems', *Automatic Control Theory & Appl.*, **4**, pp. 7–13

JAMSHIDI, M. (1980): 'An overview on the solutions of algebraic matrix Riccati equation and related problems', *Large Scale Systems*, **1**, pp. 167–192

JAMSHIDI, M. (1983): 'Large scale systems: Modeling and control' (Elsevier, New York)

JAVID, S. H. (1978a): 'The time-optimal control of a class of nonlinear singularly perturbed systems', *Int. J. Control*, **27**, pp. 831–836

JAVID, S. H. (1978b): 'Uniform asymptotic stability of linear time-varying singularly perturbed systems', *J. Franklin Inst.*, **305**, pp. 27–37

JAVID, S. H. (1980): 'Observing the slow states of singularly perturbed system', *IEEE Trans.*, **AC-25**, pp. 277–280

JAVID, S. H. (1982): 'Stabilization of singularly perturbed system by observer based slow state feedback', *IEEE Trans.*, **AC-27**, pp. 702–704

JAVID, S. H. and KOKOTOVIC, P. V. (1977): 'A decomposition of time-scales for iterative computation of time-optimal controls', *J. Opt. Theory Appl.*, **21**, pp. 459–468

JEFFREY, A. and KAWAHARA, T. (1982): 'Asymptotic methods in nonlinear wave theory' (Pitman, London)

KAMEL, A. and DUHAMEL, T. (1984): 'A second order solution of the main problem of artificial satellites using multiple scales', *J. Guidance & Control*, **8**, pp. 125–133

KANDO, H. and IWAZUMI, T. (1978): 'Asymptotic series solution of singularly perturbed fixed-end-point problem of nuclear reactor', *J. Nuclear Sci. Tech.*, **15**, pp. 466–468

KANDO, H. and IWAZUMI, T. (1981): 'Sub-optimal feedback control of large-scale systems using the boundary layer method', *Int. J. Syst. Sci.*, **12**, pp. 1083–1109

KANDO, H. and IWAZUMI, T. (1983a): 'Suboptimal control of discrete regulator problems via time-scale decomposition', *Int. J. Control*, **37**, pp. 1323–1347

KANDO, H. and IWAZUMI, T. (1983b): 'Initial value problems of singularly perturbed discrete systems via time scale decomposition', *Int. J. Syst. Sci.*, **14**, 555-570

KANDO, H. and IWAZUMI, T. (1983c): 'Suboptimal control of large scale linear quadratic problems using time-scale decomposition', *Large Scale Syst.*, **4**, pp. 1–25

KANDO, H. and IWAZUMI, T. (1984): 'Stabilizing feedback controllers for singularly perturbed discrete systems', *IEEE Trans.*, **SMC-14**, pp. 903–911

KANDO, H. and IWAZUMI, T. (1985): 'Design of observers and stabilizing feedback controllers for singularly perturbed discrete systems', *IEE Proc. Pt.D.*, **132**, pp. 1–10

KAO, Y. K. and BANKOFF, S. G. (1974): 'The singular perturbation analysis of free-time optimal control problem', *Int. J. Syst. Sci.*, **5**, pp. 335–350

KAPLUN, S. (1967): 'Fluid mechanics and singular perturbations' (Academic Press, New York)

KAPLUN, S. and LAGERSTROM, P. A. (1957): 'Asymptotic expansions of Navier-Stokes solutions for small Reynolds numbers', *J. Math. Mech.*, **6**, pp. 585–593

KARPINSKAYA, N. N., MARKECHKO, M. I. and RYBASHOV, M. V. (1985): 'Design of control systems using singularly perturbed differential equations', *Automat. & Remote Control*, **46**, pp. 434–441

KASSOY, D. R. (1976): 'Extremely rapid transient phenomena in combustion, ignition and explosion' *in* 'Asymptotic methods and singular perturbations'. SIAM-AMS Proc. (Amer. Math. Soc., Providence) p. 10

KASSOY, D. R. (1982): 'A note on asymptotic methods for jump phenomena', *SIAM J. Appl. Math.*, **42**, pp. 926–932

KATH, W. L. and COHEN, D. S. (1982): 'Waiting-time behavior in a nonlinear diffusion equation', *Studies in Appl. Math.*, **67**, pp. 79–105

KEENER, J. P. (1983): 'Analog circuitry for the Van der Pol and Fitz Hugh-Nagumo equations', *IEEE Trans.*, **SMC-13**, pp. 1010–1014

KELLER, H. B. (1973): 'Stability theory for multiple equilibrium states of a nonlinear diffusion process: a singularly perturbed eigenvalue problem', *SIAM J. Math. Anal.*, **4**, pp. 134–140

KELLER, H. B. (1974): 'Tubular chemical reactors with recycle'. SIAM-AMS Proc., Amer. Math. Soc., Providence, **8**, pp. 85–98

KELLEY, H. J. (1964): 'An optimal guidance approximation theory', *IEEE Trans.*, **AC-9**, pp. 375–380

KELLEY, H. J. (1970*a*): 'Singular perturbations for a Meyer variational problem', *AIAA J.*, **8**, pp. 1177–1178

KELLEY, H. J. (1970*b*): 'Boundary layer approximations to powered-flight altitude transients', *J. Spacecraft & Rockets*, **7**, p. 879

KELLEY, H. J. (1971*a*): 'Flight path optimization with multiple time scales', *J. Aircraft*, **8**, pp. 238–240

KELLEY, H. J. (1971*b*): 'Reduced-order modeling in aircraft mission analysis', *AIAA J.*, **9**, pp. 349–350

KELLEY, H. J. (1972): 'State variable selection and singular perturbations' *in* 'Singular perturbations: Order reduction in control system design'. American Society of Mechanical Engineers, New York, pp. 37–43

KELLEY, H. J. (1973): 'Aircraft maneuver optimization by reduced order approxima-tion' *in* LEONDES, C. T. (Ed.): 'Control and dynamic systems', Vol. 10 (Academic Press, New York) pp. 132–175

KELLEY, H. J. (1978): 'Comment on a new boundary layer matching procedure for singularly perturbed systems', *IEEE Trans.*, **AC-23**, pp. 510–511

KELLEY, H. J. and EDELBAUM, T. N. (1970): 'Energy climbs, energy turns, and asymptotic expansions', *J. Aircraft*, **7**, pp. 93–95

KELLEY, W. G. (1984): 'Boundary and interior layer phenomena for singularly per-turbed systems', *SIAM J. Math. Anal.*, **15**, pp. 635–641

KEVORKIAN, J. and COLE, J. D. (1981): 'Perturbation methods in mathematics' (Springer–Verlag, New York)

KHALIL, H. K. (1978): 'Control of linear singularly perturbed systems with colored noise disturbance', *Automatica*, **14**, pp. 153–156

KHALIL, H. K. (1979): 'Stablilzation of multiparameter singularly perturbed systems', *IEEE Trans.*, **AC-24**, pp. 790–791

KHALIL, H. K. (1980*a*): 'Multimodel design of a Nash strategy', *J. Opt. Theory Appl.*, **31**, pp. 553–564

KHALIL, H. K. (1980b): 'Approximation of Nash strategies', *IEEE Trans. Control*, **AC-25**, pp. 247–250

KHALIL, H. K. (1981a): 'Asymptotic stability of nonlinear multiparameter singularly perturbed systems', *Automatica*, **17**, pp. 791–804

KHALIL, H. K. (1981b): 'On the robustness of output feedback control methods to modelling errors', *IEEE Trans.*, **AC-26**, pp. 524–526

KHALIL, H. K. (1984): 'Time scale decomposition of linear implicit singularly perturbed systems', *IEEE Trans.*, **AC-29**, pp. 1054–1056

KHALIL, H. K. and GAJIC, Z. (1984): 'Near optimum regulators for stochastic linear singularly perturbed systems', *IEEE Trans.*, **AC-29**, pp. 531–541

KHALIL, H. K. and KOKOTOVIC, P. V. (1978): 'Control strategies for decision makers using different models of the same system', *IEEE Trans.*, **AC-23**, pp. 289–298

KHALIL, H. K, and KOKOTOVIC, P. V. (1979a): 'Control of linear systems with multiparameter singular perturbations', *Automatica*, **15**, pp. 197–207

KHALIL, H. K. and KOKOTOVIC, P. V. (1979b): 'D-stability and multiparameter singular perturbation', *SIAM J. Control Opt.*, **17**, pp. 56–65

KHALIL, H. K. and KOKOTOVIC, P. V. (1979c): 'Feedback and well-posedness of singularly perturbed Nash games', *IEEE Trans.*, **AC-24**, pp. 699–678

KHALIL, H. K. and KOKOTOVIC, P. V. (1980): 'Decentralized stabilization of systems with slow and fast modes', *Large Scale Syst.*, **1**, pp. 141–148

KHALIL, H. K. and MEDANIC, J. V. (1980): 'Closed-loop Stackelberg strategies for singularly perturbed linear quadratic problems', *IEEE Trans.*, **AC-25**, pp. 66–71

KHALIL, H. K. and SABERI, A. (1982): 'Decentralized stabilization of nonlinear interconnected systems using high-gain geedback', *IEEE Trans.*, **AC-27**, pp. 265–268

KHORASANI, K. and KOKOTOVIC, P. V. (1985): 'Feedback linearization of a flexible manipulator near its rigid body manifold', *Syst. & Controls Lett.*, **6**, pp. 187–192

KHORASANI, K. and KOKOTOVIC, P. V. (1986): 'A corrective feedback design for nonlinear systems with fast actuators', *IEEE Trans.*, **AC-31**, pp. 67–69

KHORASANI, K. and PAI, M. A. (1985a): 'Asymptotic stability of nonlinear singularly perturbed systems using high order corrections', *Automatica*, **21**, pp. 717–727

KHORASANI, K. and PAI, M. A. (1985b): 'Asymptotic stability improvements of multiparameter nonlinear singularly perturbed systems', *IEEE Trans.*, **AC-30**, pp. 802–804

KIMURA, H. (1981): 'A new approach to the perfect regulation and the bounded peaking in linear multivariable control systems', *IEEE Trans.*, **AC-26**, pp. 253–270

KIMURA, H. (1982): 'Perfect and subperfect regulation in linear multivariable control systems', *Automatica*, **18**, pp. 125–145

KIMURA, H. (1983): 'On the matrix Riccati equation for a singularly perturbed linear discrete control system', *Int. J. Control*, **38**, pp. 959–975

KLIMUSHEV, A. I. and KRASOVSKII, N. N. (1961): 'Uniform asymptotic stability of systems of differential equations with small parameter in the derivative terms', *J. Appl. Math. Mech., (PMM)*, **25**, pp. 1011–1025

KNOWLES, J. K. and MESSICK, R. E. (1964): 'On a class of singular perturbation problem', *J. Math. Anal. Appl.*, **9**, pp. 42–58

KODA, M. (1982): 'Sensitivity analysis of singularly perturbed systems', *Int. J. Systems Sci.*, **13**, pp. 909–919

KOKOTOVIC, P. V. (1975): 'A Riccati equation for block-diagonalization of ill-conditioned systems', *IEEE Trans.*, **AC-20**, pp. 812–814

KOKOTOVIC, P. V. (1976): 'Singular perturbation in optimal control', *Rocky Mountain J. Math.*, **6**, pp. 767–773

KOKOTOVIC, P. V. (1981): 'Subsystems, time-scales, and multi-modeling', *Automatica*, **17**, pp. 789–795

KOKOTOVIC, P. V. (1984): 'Applications of singular perturbation techniques to control problems', *SIAM Rev.*, **26**, pp. 501–550

KOKOTOVIC, P. V. (1985): 'Recent trends in feedback design: an overview', *Automatica*, **21**, pp. 225–236

KOKTOVIC, P. V., ALLEMONG, J. J., WINKELMAN, J. R. and CHOW, J. H. (1980): 'Singular perturbations and iterative separation of time scales', *Automatica*, **16**, pp. 23–33

KOKOTOVIC, P. V., AVRAMOVIC, B., CHOW, J. H. and WINKELMAN, J. R. (1982): 'Coherency based decomposition and aggregation', *Automatica*, **18**, 47–56

KOKOTOVIC, P. V., CRUZ, J. B., Jun., HELLER, J. E. and SANNUTI, P. (1968): 'Synthesis of optimally sensitive systems', *Proc. IEEE*, **56**, pp. 1318–1324

KOKOTOVIC, P. V. and HADDAD, A. H. (1975a): 'Controllability and time-optimal control of systems with slow and fast modes', *IEEE Trans.*, **AC-20**, pp. 111–113

KOKOTOVIC, P. V. and HADDAD, A. H. (1975b): 'Singular perturbation of a class of time-optimal controls', *IEEE Trans.*, **AC-20**, pp. 163–164

KOKOTOVIC, P. V. and KHALIL, H. K. (Eds.) (1986): 'Singular perturbations in systems and control' (IEEE Press, New York)

KOKOTOVIC, P. V., KHALIL, H. K. and O'REILLY, J. (1986): 'Singular perturbation methods in control: Analysis and design' (Academic Press, London)

KOKOTOVIC, P. V., O'MALLEY, R. E., Jun. and SANNUTI, P. (1976): 'Singular perturbations and order reduction in control theory-an overview', *Automatica*, **12**, pp. 123–132

KOKOTOVIC, P. V. and PERKINS, W. R. (Eds.) (1972): 'Singular perturbations: Order reduction in control systems design' (American Society of Mechanical Engineers, New York)

KOKOTOVIC, P. V., PERKINS, W. R., CRUZ, J. B., Jun. and D'ANS, G. (1969): 'ε-coupling method for near optimum design of large scale system', *Proc. IEEE*, **116**, pp. 889–892

KOKOTOVIC, P. V. and SANNUTI, P. (1968): 'Singular perturbation method for reducing model order in optimal control design', *IEEE Trans.*, **AC-13**, pp. 377–384

KOKOTOVIC, P. V. and YACKEL, R. A. (1972): 'Singular perturbation of linear regulators: basic theorems', *IEEE Trans.*, **AC-17**, pp. 29–37

KOPPEL, N. (1979): 'A geometric approach to boundary layer problems exhibiting resonance', *SIAM J. Appl. Math.*, **37**, pp. 436–458

KORTUM, W. (1979): 'Computational techniques in optimal state estimation-a tutorial review', *ASME J. Dynamic Syst., Measurement & Control*, **101**, pp. 99–107

KREISS, H. O. (1978): 'Difference methods for stiff differential equations', *SIAM J. Num. Anal.*, **15**, pp. 21–58

KREISS, H. O., NICHOLS, N. K. and BROWN, D. R. (1986): 'Numerical methods for two-point boundary value problems', *SIAM J. Num. Anal.*, **23**, pp. 325–368

KRIEGSMANN, G. A. and REISS, E. L. (1983): 'Low frequency scattering by local inhomogeneities', *SIAM J. Appl. Math.*, **43**, pp. 923–934

KRIKORIAN, K. V. and LEONDES, C. T. (1982a): 'Dynamic programming using singular perturbations', *J. Opt. Theory Appl.*, **38**, pp. 221–240

KRIKORIAN, K. V. and LEONDES, C. T. (1982b): 'Application of singular perturbations to optimal control' *in* LEONDES, C. T. (Ed.): 'Control and dynamic systems. Vol. 18' (Academic Press, New York) pp. 131–160

KRTOLICA, R. (1984): 'A singular perturbation model or reliability in system control', *Automatica*, **20**, pp. 51–57

KUNG, C. F. (1976): 'Singular perturbation of an infinite interval linear state regulator problem in optimal control', *J. Math. Anal. Appl.*, **55**, pp. 365–374

KUPPURAJULU, A. and ELANGOVAN, S. A. (1970): 'System analysis by simplified models', *IEEE Trans.*, **AC-15**, pp. 234–237

KURINA, G. A. (1977*a*): 'On a degenerate optimal control problem and singular perturbation', *Soviet Math.*, **18**, pp. 1452–1456

KURINA, G. A. (1977*b*): 'Asymptotic solution of a classical singularly perturbed optimal control problems', *Soviet Math.*, **18**, pp. 722–726

KURINA, G. A. (1983*a*): 'On a classical singularly perturbed optimal control problem', *Diff. Eqn.*, **19**, pp. 710–712

KURINA, G. A. (1983*b*): 'An asymptotic solution to a class of singularly perturbed problems of optimal control', *Prikl. Mat. Mekh.*, **47**, pp. 363–371

KURUOGLU, N., CLOUGH, N. DE amd RAMIREZ, W. F. (1981): 'Distributed parameter estimation for systems with fast and slow dynamics', *Chem. Engg. Sci.*, **36**, pp. 1357–1363

KUSHNER, H. J. (1965): 'Near optimal control in the presence of small stochastic perturbations', *Trans. ASME: Series D, J. Basic Engg.*, pp. 103–108

KUSHNER, H. J. (1967): 'Approximations to optimal nonlinear filters', *IEEE Trans.*, **AC-12**, pp. 546–556

KUSHNER, H. J. (1984): 'Approximation and weak convergence methods for random processes, with applications to stochastic systems theory' (MIT Press, Cambridge, USA)

LADDE, G. S. and RAJYALAKSHMI, S. G. (1985): 'Diagonalization and stability of multi-time-scale singularly perturbed linear systems', *Applied Math. & Computation*, **16**, pp. 115–140

LADDE, G. S. and SIRISAENGATAKSIN, O. (1986): 'Multi-time-scale singularly perturbed linear stochastic systems', *Stochastic Anal. & Appl.*, **4**, pp. 213–238

LADDE, G. S. and SILJAK, D. D. (1983): 'Multiparameter singular perturbations of linear systems with multiple time scales', *Automatica*, **19**, pp. 385–394

LAGERSTROM, P. A. (1975): 'Solution of the Navier–Stokes equation at large Reynolds numbers', *SIAM J. Appl. Math.*, **28**, pp. 202–214

LAGERSTROM, P. A. and CASTEN, R. G. (1972): 'Basic concepts underlying singular perturbation techniques', *SIAM Rev.*, **14**, pp. 63–120

LAGERSTROM, P. A. and KEVORKIAN, J. (1963): 'Earth-to-moon trajectories in the restricted three body problem', *J. de Mecanique*, **II**, pp. 189–218

LAGERSTROM, P. A. and REINELT, D. A. (1984): 'Note on logarithmic switchback trms in regular and singular perturbation expansions', *SIAM J. Appl. Math.*, **44**, pp. 451–462

LAKIN, W. D. and DRIESSCHE, VAN DEN P. (1977): 'Time scales in population biology', *SIAM J. Appl. Math.*, **32**, pp. 694–705

LAMBA, S. S. and RAO, S. V. (1974): 'On suboptimal control via the simplified model of Davison', *IEEE Trans.*, **AC-19**, pp. 448–450

LANGE, C. G. and MIURA, R. M. (1985): 'Singular perturbation analysis of boundary value problems for differential-difference equations—II: Rapid oscillations and resonance', *SIAM J. Appl. Math.*, **45**, pp. 687–707

LAPIDUS, L., AIKEN, R. C. and LIU, Y. A. (1974): 'The occurrence and numerical solution of physical and chemical systems having widely varying time constants' *in* WILLOUGHY, W. A. (Ed.): 'Stiff differential systems' (Plenum Press, New York) pp. 187–200

LASTMAN, G. J., SINHA, N. K. and ROZSA, P. (1984): 'On the selection of states to be retained in a reduced order model', *IEE Proc. Pt.D.*, **131**, pp. 15–22

LEHMAN, S. L. and STARK, L. W. (1983): 'Perturbation analysis applied to eye, head, and arm movement models', *IEEE Trans.*, **SMC-13**, pp. 972–979

LEVIN, J. J. (1956): 'Singular perturbation of nonlinear systems of differential equations related to conditional stability', *Duke Math. J.*, **23**, pp. 609–620

LEVIN, J. J. and LEVINSON, N. (1954): 'Singular perturbations of nonlinear systems of differential equations and an associated boundary layer equation', *J. Rat. Mech. Anal.*, **3**, pp. 247–270

LI, W. H. (1972): 'Differential equations of hydraulic transients, dispersion, and ground water flow' (Prentice Hall, Englewood Cliffs)

LIN, C. C. and SEGAL, L. A. (1974): 'Mathematics applied to deterministic problems of natural sciences' (Macmillan, New York)

LIONS, J. L. (1973*a*): 'Perturbations singularies dans les problems aux limites et au control optimal'. Lecture Notes in Mathematics, Vol. 323 (Springer–Verlag, New York)

LIONS, J. L. (1973*b*): 'The optimal control of distributed parameter systems', *Russian Math. Surveys*, **8**, pp. 13–46

LIONS, J. L. (1978): 'Asymptotic methods in the optimal control of distributed systems', *Automatica*, **14**, pp. 199–211

LIONS, J. L. (1981): 'Some methods in mathematical analysis of systems and their control' (Gordon & Breach, New York)

LITKOUHI, B. and KHALIL, H. K. (1984): 'Infinite time regulators for singularly perturbed difference equations', *Int. J. Control*, **39**, pp. 567–598

LITKOUHI, B. and KHALIL, H. K. (1985): 'Multirate and composite control of two-time-scale discrete-time systems', *IEEE Trans.*, **AC-30**, pp. 645–651

LITZ, L. and ROTH, H. (1981): 'State decomposition for singular perturbation order reduction: A model approach', *Int. J. Control*, **34**, pp. 937–954

LOCATELLI, A. and SCHIAVONI, N. (1976): 'Two-time scale discrete systems'. 1st Int. Conference on Information Science and Systems, Patras, Greece

LOGAN, J. A., WOLLKIND, D. J., HOYT, S. C. and TANIGOSHI, L. K. (1976): 'An analytic model for the description of temperature dependent rate phenomena in arthropods', *Environmental Entomology*, **5**, pp. 1130–1140

LOMOV, S. A. (1981): 'Introduction to the general theory of singular perturbations' (Nauka, Moscow)

LONGCHAMP, R. (1983): 'Singular perturbation analysis of a receding horizon controller', *Automatica*, **19**, pp. 303–308

LUDWIG, D. (1976): 'A singular perturbation problem in the theory of population extinction' *in* 'Asymptotic methods and singular perturbations'. SIAM-AMS Proc., Amer. Math. Soc., Providence, Vol. 10, pp. 87–104

LUSE, D. W. (1986): 'Frequency domain results for systems with multiple time scales', *IEEE Trans.*, **AC-31**, pp. 918–924

LUSE, D. W. and KHALIL, H. K. (1985): 'Frequency domain results for systems with slow and fast dynamics', *IEEE Trans.*, **AC-30**, pp. 1171–1179

MacGILLIVRAY, A. D. (1983): 'On the leading term of the inner asymptotic expansion of Van der Pol's equation', *SIAM J. Appl. Math.*, **43**, pp. 594–612

MAGINU, K. (1980): 'Existence and stability of periodic travelling wave solutions to Nagumo's nerve equation', *J. Math. Biol.*, **10**, pp. 133–153

MAGINU, K. (1981): 'Stability of periodic travelling solutions with large spatial periods in reaction-diffusion systems', *J. Diff. Eqn.*, **39**, pp. 73–99

MAGNAN, J. F. and GOLDSTIEN, R. A. (1980): 'Perturbed bifurcation of stationary striations in a contaminated nonuniform plasma', *SIAM J. Appl. Math.*, **43**, pp. 16–30

MAHMOUD, M. S. (1982*a*): 'Design of observer-based controllers for a class of discrete systems', *Automatica*, **18**, pp. 323–328

MAHMOUD, M. S. (1982*b*): 'Order reduction and control of discrete system', *IEE Proc. Pt.D.*, **129**, pp. 129–135

MAHMOUD, M. S. (1982*c*): 'Structural properties of discrete systems with slow and fast modes', *Large Scale Syst.: Theory and Appl.*, **3**, pp. 227–236

MAHMOUD, M. S. (1986): 'Stabilization of discrete systems with multiple-time scales', *IEEE Trans.*, **AC-31**, pp. 159–162

MAHMOUD, M. S., CHEN, Y. and SINGH, M. G. (1985a): 'On eigenvalues assignment in discrete systems with fast and slow modes', *Int. J. Control*, **16**, pp. 61–70

MAHMOUD, M. S., CHEN, Y. and SINGH, M.G. (1986): 'Discrete two-time scale systems', *Int. J. Syst. Sci.*, **17**, pp. 1187–1207

MAHMOUD, M. S., HASSAN, M. F. and DARWISH, M. G. (1985): 'Large scale systems: Theories and techniques' (Marcel Dekker, New York)

MAHMOUD, M. S., HASSAN, M. F. and SINGH, M. G. (1982): 'Approximate feedback design for a class of singularly perturbed systems', *IEE Proc. Pt.D.*, **129**, pp. 49–56

MAHMOUD, M. S., and SINGH, M. G. (1981a): 'Large-scale systems modelling' (Pergamon Press, Oxford)

MAHMOUD, M. S. and SINGH, M. G. (1981b): 'Decentralized state reconstruction of interconnected discrete systems', *Large Scale Syst. Theory & Appl.*, **2**, pp. 151–158

MAHMOUD, M. S. and SINGH, M. G. (1984): 'Discrete systems: Analysis, control, and optimization' (Springer–Verlag, Berlin)

MAHMOUD, M. S. and SINGH, M. G. (1985): 'On the use of reduced order models in output feedback design of discrete systems', *Automatica*, **21**, pp. 485–489

MAJDA, G. (1984): 'Filtering techniques for systems of differential equations—I', *SIAM J. Num. Anal.*, **21**, pp. 535–556

MALEK-ZAVAREI, M. (1980): 'Near-optimum design of nonstationary linear systems with state and control delays', *J. Opt. Theory & Appl.*, **30**, pp. 73–88

MARGOLIS, S. B. and MATKOWSKY, B. J. (1982a): 'Flame propagation with multiple fuels', *SIAM J. Appl. Math.*, **42**, pp. 982–1003

MARGOLIS, S. B. and MATKOWSKY, B. J. (1982b): 'Flame propagation with a sequential reaction mechanism', *SIAM J. Appl. Math.*, **42**, pp. 1175–1182

MARINO, R. (1985): 'High-gain feedback in nonlinear control systems', *Int. J. Control*, **42**, pp. 1369–1385

MARINO, R. and NICOSIA, S. (1984): 'On the feedback control of industrial robots with elastic joints: A singular perturbation approach'. Report R-84.01, University of Rome, Rome, Italy

MARKOWICH, P. A. (1986): 'The stationary semiconductor device equations' (Springer–Verlag, New York)

MARKOWICH, P. A. and RINGHOFFER, C. A. (1984): 'A singularly perturbed boundary value problem modelling a semiconductor device', *SIAM J. Appl. Math.*, **44**, pp. 231–256

MATHUNA, D. O. (1971): 'The differential equation of enzyme reaction kinetics – the regularisation of a singular perturbation', *Proc. Royal Irish Acad.*, **A71**, pp. 27–51

MATKOWSKY, B. J. (1975): 'On boundary layer problems exhibiting resonance', *SIAM Rev.*, **17**, pp. 82–100

MATKOWSKY, B. J. and REISS, E. L. (1977): 'Singular perturbations of bifurcations', *SIAM J. Appl. Math.*, **33**, pp. 230–255

MATKWOSKY, B. J. and SIEGMANN, W. L. (1976): 'The flow between counter-rotating disks at high Reynolds number', *SIAM J. Appl. Math.*, **30**, pp. 720–727

MATSUMOTO, T., CHUA, L. O., KAWAAKAMI, H. and ICHIRAKU, S. (1981): 'Geometric properties of dynamic nonlinear networks: transversality, local solubility, and eventual passivity', *IEEE Trans.*, **CAS-28**, pp. 406–428

McKELVEY, R. and BOHAC, R. (1976): 'Ackerberg–O'Malley resonance revisited', *Rocky Mountain J. Math.*, **6**, pp. 637–650

McLEOD, J. B. and PARTER, S. V. (1974): 'On the flow between two counter-rotating infinite plane disks', *Arch. Rat. Mech. Anal.*, **54**, pp. 301–327

MEHRA, R. K., WASHBURN, R. B., SAJAN, S. and CARROL, J. V. (1979): 'A study of the application of singular perturbation theory'. NASA Contractor Report 3167 (Scientific System, Cambridge, MA, USA)

MEISKE, W. (1978): 'An approximate solution of the Michalis–Menton mechanism for quasi-steady state and quasi-equilibrium', *Math. Biosci.*, **42**, pp. 63–71

MEYER, R. E. and PARTER, S. V. (Eds.) (1980): 'Singular perturbations and asymptotics' (Academic Press, New York)

MIKA, J. (1976): 'Asymptotic expansion method in reactor physics calculation', *Nukleonka*, **21**, pp. 573–587

MILUSHEVA, S. D. and BAINOV, D. D. (1985): 'Justification of the averaging method for a system of singularly perturbed differential equation with impulses', *J. Appl. Math. Phys.*, **36**, pp. 293–308

MIRANKER, W. L. (1962): 'Singular perturbation analysis of the differential equations of a tunnel diode circuit', *Quart. Appl. Math.*, **20**, pp. 279–299

MIRANKER, W. L. (1980): 'Numerical methods for stiff equations and singular perturbation problems' (Reidel Publ., Dordrecht, Holland)

MIRANKER, W. L. and HOPPENSTEADT, F. C. (1974): 'Numerical methods for stiff systems of differential equations related with transistors, tunnel diodes, etc.'. Lecture Notes in Computer Science Vol. 10 (Springer–Verlag, New York)

MISHCHENKO, E. F. and ROZOV, N. K. (1980): 'Differential equations with small parameters and relaxation oscillations' (Plenum, New York)

MISHRA, R. N. (1980): 'Design of low-order schemes using reduction techniques', *Int. J. Control*, **32**, pp. 899–906

MIYAZAKI, F. and ARIMOTO, S. (1979): 'Singular perturbation for the analysis of biped locomotion system with many degrees of freedom', *Trans. Soc. Inst. Control Engg.*, **14**, pp. 428–433

MIYAZAKI, F. and ARIMOTO, S. (1980): 'A control theoretic study on dynamical biped locomotion', *Trans. ASME, J, Dyn. Systems, Meas. Control*, **102**, pp. 233–239

MOERDER, D. D. and CALISE, A. J. (1985): 'Two-time-scale stabilization of systems with output feedback', *J. Guidance, Control and Dynamics*, **8**, pp. 731–736

MOISEEV, M. N. and CHERNOUSKO, F. L. (1981): 'Asymptotic methods in the theory of optimal control', *IEEE Trans.*, **AC-26**, pp. 993–1000

MOORE, B. C. (1976): 'On the flexibility offered by state feedback multivariable systems beyond closed-loop eigenvalue assignment', *IEEE Trans.*, **AC-21**, pp. 689–692

MOTTANI, P. DE and ROTHE, F. (1980): 'A singular perturbation analysis for a reaction diffusion system describing pattern formation', *Studies in Appl. Math.*, **63**, pp. 227–247

MUFTI, I. H. (1970): 'Computational methods in optimal control problems'. Lecture Notes in Operational Research and Mathematical Systems, Vol. 27 (Springer–Verlag, Berlin)

MURRAY, J. D. (1977): 'Lectures in nonlinear differential equation models in biology' (Oxford Univ. Press, Oxford)

MURRAY, J. D. (1984): 'Asymptotic analysis' (Springer–Verlag, Berlin)

MURTHY, D. N. P. (1978): 'Solution of linear regulator problem via two-parameter singular perturbation', *Int. J. Systems Sci.*, **9**, pp. 1113–1132

NAIDU, D. S. (1977): 'Applications of singular perturbation technique to problems in control systems'. Ph.D. Thesis, Indian Inst. Tech., Kharagpur, India

NAIDU, D. S. and PRICE, D. B. (1987): 'Time scale analysis of a closed-loop discrete optimal control system'. *J. Guidance, Control and Dynamics*, **10**

NAIDU, D. S. and RAJAGOPALAN, P. K. (1979): 'Application of Vasileva's singular perturbation method to a problem in ecology', *Int. J. Syst. Sci.*, **10**, pp. 761–774

NAIDU, D. S. and RAJAGOPALAN, P. K. (1980): ' Singular perturbation method for a closed-loop optimal control problem', *IEE Proc. Pt.D.*, **127**, pp. 1–6

NAIDU, D. S.and RAO, A. K. (1981): 'A singular perturbation method for initial value problems with inputs in discrete control systems', *Int. J. Control*, **33**, pp. 953–965

NAIDU, D. S. and RAO, A.K. (1982): 'Singular perturbation methods for a class of initial and boundary value problems', *Int. J. Control*, **36**, pp. 77–94

NAIDU, D. S. and RAO, A. K. (1984): 'Singular perturbation analysis of the closed-loop discrete optimal control problem', *Opt. Control: Appl. & Methods*, **5**, pp. 19–38

NAIDU, D. S. and RAO, A. K. (1985a): 'Singular perturbation analysis of discrete control systems'. Lecture Notes in Mathematics, Vol. 1154 (Springer–Verlag, Berlin)

NAIDU, D. S. and RAO, A. K. (1985b): 'Application of singular perturbation method to a steam power system', *Elect. Power Syst. Res.*, **8**, pp. 219–226

NAIDU, D. S. and RAVINDER, R. (1985): 'On three-time scale analysis'. 24th IEEE Conference on Decision and Control, Fort Lauderdale, USA

NAIDU, D. S. and SEN, S. (1982): 'Singular perturbation method for the transient analysis of a transformer', *Elect. Power Syst. Res.*, **5**, pp. 307–313

NAYFEH, A. H. (1965): 'A comparison of three perturbation methods for earth-moon-spaceship problem', *AIAA J.*, **3**, pp. 1682–1687

NAYFEH, A. H. (1973): 'Perturbation methods' (Wiley–Interscience, New York)

NAYFEH, A. H. (1981): 'Introduction to perturbation techniques' (Wiley, New York)

NAYFEH, A. (1985): 'Problems in perturbation' (Wiley, New York)

NAYFEH, A. H. and MOOK, D. T. (1979): 'Nonlinear oscillations' (Wiley, New York)

NEU, J. C. (1979): 'Coupled chemical oscillations', *SIAM J. Appl. Math.*, **37**, pp. 307–315

NEU, J. C. (1980): 'Chemical waves and the diffusive coupling of limit cycle oscillators', *SIAM J. Appl. Math.*, **36**, pp. 509–515

NEWCOMB, R. W. (1981): 'The semi-state description on nonlinear time variable circuits', *IEEE Trans.*, **CAS-28**, pp. 62–71

NIPP, K. (1983): 'An extension of Tikhonov's theorem in singular perturbations for the planar case', *J. Appl. Math. Phys.*, **34**, pp. 277–290

NIPP, K. (1985): 'Invariant manifolds of singularly perturbed ordinary differential equations', *J. Appl. Math. Phys.*, **36**, pp. 309–320

NIPP, K. (1986): 'Breakdown of stability in singularly perturbed autonomous systems I: orbit equations', *SIAM J. Math. Anal.*, **17**, pp. 512–523

OLVER, F. W. J. (1973): 'Asymptotics and special functions' (Academic Press, New York)

OLVER, F. W. J. (1978): 'Sufficient conditions for Ackerberg-O'Malley resonance', *SIAM J. Math. Anal.*, **9**, pp. 328–355

O'MALLEY, R. E., Jun. (1968): 'Topics in singular perturbations' *in* 'Advances in mathematics. Vol. 2' (Academic Press, New York) pp. 356–470

O'MALLEY, R. E., Jun. (1970): 'A nonlinear singular perturbation problem arising in the study of chemical flow reactors', *J. Inst. Math. Appl.*, **6**, pp. 12–21

O'MALLEY, R. E., Jun. (1971): 'Boundary layer methods for nonlinear initial value problems', *SIAM Rev.*, **13**, pp. 425–434

O'MALLEY, R. E., Jun. (1972*a*): 'Singular perturbation of the time-invariant linear state regulator problem', *J. Diff. Eqns.*, **12**, pp. 117–128

O'MALLEY, R. E., Jun. (1972*b*): 'The singularly perturbed linear state regulator problem', *SIAM J. Control*, **10**, pp. 399–413

O'MALLEY, R. E., Jun. (1972*c*): 'On multiple solutions of a singular perturbation problem', *Arch. Rat. Mech. Anal.*, **49**, pp. 89–98

O'MALLEY, R. E., Jun. (1974*a*): 'Introduction to singular perturbations' (Academic Press, New York)

O'MALLEY, R. E., Jun. (1974*b*): 'Boundary layer methods for certain nonlinear singularly perturbed optimal control problems', *J. Math. Anal. Appl.*, **45**, pp. 468–484

O'MALLEY, R. E., Jun. (1975): 'On two methods of solution for a singularly perturbed linear state regulator problem', *SIAM Rev.*, **17**, pp. 16–37

O'MALLEY, R. E., Jun. (1976*a*): 'A more direct solution of the nearly singular linear regulator problem', *SIAM J. Control & Optimization*, **14**, pp. 1063–1077

O'MALLEY, R. E., Jun. (1976*b*): 'Phase plane solution to some singular perturbation problems', *J. Math. Anal. Appl.*, **54**, pp. 449–466

O'MALLEY, R. E., Jun. and ANDERSON, L. R. (1982): 'Time-scale decoupling and order reduction for linear time-varying systems', *Opt. Control: Appl. & Methods*, **3**, pp. 133–153

O'MALLEY, R. E., Jun. and KUNG, C. F. (1974): 'On the matrix Riccati approach to a singularly perturbed regulator problem', *J. Diff. Eqns.*, **16**, pp. 413–427

O'MALLEY, R. E., Jun. and KUNG, C. F. (1975): 'The singularly perturbed linear state regulator problem—II', *SIAM J. Control*, **13**, pp. 299–309

O'REILLY, J. (1979*a*): 'Full-order observers for a class of singularly perturbed linear time-varying systems', *Int. J. Control*, **30**, pp. 745–756

O'REILLY, J. (1979*b*): 'Two time-scale feedback stabilization of linear time-varying singularly perturbed systems', *J. Franklin Inst.*, **308**, pp. 465–474

O'REILLY, J. (1980): 'Dynamical feedback control for a class of singularly perturbed systems using a full-order observer', *Int. J. Control*, **31**, pp. 1–10

O'REILLY, J. (1983*a*): 'Observers for linear systems' (Academic Press, London)

O'REILLY, J. (1983*b*): 'Partial cheap control of time-invariant regulator', *Int. J. Control*, **37**, pp. 909–927

O'REILLY, J. (1986): 'Robustness of linear feedback control systems to unmodelled high-frequency dynamics', *Int. J. Control*, **44**, pp. 1077–1088

OTHMAN, H. A., KHRAISHI, N. M. and MAHMOUD, M. S. (1985): 'Discrete regulators with time-scale separation', *IEEE Trans.*, **AC-30**, pp. 293–297

OZGUNER, U. (1979): 'Near-optimal control of composite systems: the multi-time-scale approach', *IEEE Trans.*, **AC-24**, pp. 652–655

OZGUNER, U. (1982): 'Near-Nash feedback control of a composite system with time-scale hierarchy', *IEEE Trans.*, **SMC-12**, pp. 62–66

PAI, M. A. and ADGAONKAR, R. P. (1983): 'Singular perturbation analysis of transients in multi-machine power systems', *Elect. Machines & Power Syst.*, **8**

PANDOLFI, L. (1981): 'On the regulator problem for linear degenerate control problem', *J. Opt. Theory & Appl.*, **33**, pp. 241–254

PARASKEVOPOULOS, P. N. (1986): 'Techniques in model reduction for large scale systems' *in* LEONDES, C. T. (Ed.): 'Control and dynamic systems' Vol. 23 (Academic Press, New York) pp. 165–193

PARASKEVOPOULOS, P. N. and CHRISTOODOULOU, M. A. (1984): 'On the computation of the transfer function matrix of singular systems', *J. Franklin Inst.*, **317**, pp. 403–412

PARTER, S. V., STEIN, M. L. and STEIN, P. R. (1975): 'On the multiplicity of a differential equation arising in a chemical reactor theory', *Studies in Appl. Math.*, **54**, pp. 293–315

PEARSON, C. E. (1968): 'A numerical method for ordinary differential equation of boundary layer type', *J. Math. Phys.*, **47**, pp. 134–154

PERKINS, W. R. and KOKOTOVIC, P. V. (1971): 'Deterministic parameter estimation for near-optimum feedback control', *Automatica*, **7**, pp. 439–444

PEPONIDES, G. M. and KOKOTOVIC, P. V. (1983): 'Weak connections, time-scales, and aggregation of nonlinear systems', *IEEE Trans. Aut. Control*, **AC-28**, pp. 729–734

PEPONIDES, G. M., KOKOTOVIC, P. V. and CHOW, J. H. (1982): 'Singular perturbations and time scales in nonlinear models of power systems', *IEEE Trans.*, **CAS-29**, pp. 758–767

PERELSON, A. S. (1980): 'Receptor clustering on cell surface—II: Theory of receptor cross-linking by ligands bearing two chemically distinct functional groups', *Math. Biosci.*, **49**, pp. 87–110

PERELSON, A. S. and DELISI, C. (1980): 'Receptor clustering on a cell surface—I: Theory or receptor cross-linking by ligands bearing two chemically identical functional groups', *Math. Biosci.*, **48**, pp. 71–110

PERVOZVANSKII, A. A. and GAITSGORI, V. G. (1979): 'Decomposition, aggregation and suboptimization (Nauka, Mascow) (Russian)

PHILLIPS, R. G. (1980a): 'Reduced order modelling and control of two-time scale discrete systems', *Int. J. Control*, **31**, pp. 765–780

PHILLIPS, R. G. (1980b): 'A two-stage design of linear feedback controls', *IEEE Trans.*, **AC-25**, pp. 1220–1223

PHILLIPS, R. G. (1983): 'The equivalence of time-scale decomposition techniques used in the analysis and design of linear system', *Int. J. Control*, **37**, pp. 1239–1257

PHILLIPS, R. G. and KOKOTOVIC, P. V. (1981): 'A singular perturbation approach to modeling and control of Markov chains', *IEEE Trans.*, **AC-26**, pp. 1087–1094

POLK, J. F. (1976): 'Asymptotic approximations to the solution of the heat equation', *Rocky Mountain J. Math.*, **6**, 697–708

PONTRYAGIN, L. S. (1957): 'Asymptotic behaviour of solutions of systems of differential equations with a small parameter in the derivatives of highest order', *Isz. Akad. Nauk SSSR, Ser. Mat.*, **21**, pp. 605–626

PONZA, P. J. and WAX, N. (1972a): 'Stability, singular perturbations and the vector lineard equation', *IEEE Trans.*, **AC-17**, pp. 563–565

PONZA, P. J. and WAX, N. (1972b): 'Relaxation oscillations, parasities and singular perturbations', *IEEE Trans.*, **AC-19**, pp. 623–625

PORTER, B. (1974): 'Singular perturbation methods in the design of stabilizing feedback controllers for multivariable systems', *Int. J. Control*, **20**, pp. 689–692

PORTER, B. (1977): 'Singular perturbation methods in the design of full-order observers for multivariable linear systems', *Int. J. Control*, **26**, pp. 589–594

PORTER, B. (1982): 'Singular perturbation methods in the design of tracking systems incorporating error-actuated controllers for plants with explicit actuator dynamics', *Int. J. Control*, **35**, pp. 383–389

PORTER, B. and BRADSHAW, A. (1981): 'Singular perturbation methods in the design of tracking systems incorporating high-gain error actuated conrollers', *Int. J. Syst. Sci.*, **12**, pp. 1169–1179

PORTER, B. and SHENTON, A. T. (1975a): 'Singular perturbation methods of asymptotic eigenvalue assignment in multivariable systems', *Int. J. Syst. Sci.*, **6**, pp. 33–37

PORTER, B. and SHENTON, A. T. (1975b): 'Singular perturbation analysis of the transfer function matrices of a class of multivariable systems', *Int. J. Control*, **21**, pp. 655–660

PRANDTL, L. (1905): 'Uber Flussigkeits-bewegung bei kleiner Reibung'. Verh. 3rd Int. Math. Kongr. (Tuebner, Leipzig) pp. 484–491

PRICE, B. (1976): 'Singular perturbation methods in the synthesis of feedback-control laws for eigenvalue assignment in multivariable systems', *Int. J. Systems Sci.*, **7**, pp. 597–601

PRICE, D. B. (1979): Comments on 'Linear filtering of singularly perturbed systems', *IEEE Trans.*, **AC-24**, pp. 675–677

PRICE, D. B., CALISE, A. J. and MOERDER, D. D. (1984): 'Piloted simulation of an onboard trajectory optimization algorithm', *J. Guidance, Control & Dynamics*, **7**, pp. 355–360

PRIEL, B. and SHAKED, U. (1984): 'Cheap optimal control of discrete single input single output systems', *Int. J. Control*, **38**, pp. 1087–1113

RAJAGOPALAN, P. K. and NAIDU, D. S. (1980*a*): 'A singular perturbation method for discrete control systems', *Int. J. Control*, **32**, pp. 925–936

RAJAGOPALAN, P. K. and NAIDU, D. S. (1980*b*): 'Singular perturbation analysis of a closed-loop fixed-end-point optimal control problem', *IEE Proc. Pt.D.*, **127**, pp. 194–203

RAJAGOPALAN, P. K. and NAIDU, D. S. (1981): 'Singular perturbation method for discrete models of continuous systems in optimal control', *IEE Proc. Pt.D.*, **128**, pp. 142–148

RAJAGOPALAN, P. K. and NAIDU, D. S. (1983): Authors reply to 'Singular perturbation method for discrete models of continuous systems in optimal control', *IEE Proc. Pt.D.*, **130**, p. 136

RAJAN, N. and ARDEMA, M. D. (1984): 'Barriers and dispersal surfaces in minimum time interception', *J. Opt. Theory & Appl.*, **42**, pp. 201–228

RAJAN, N. and ARDEMA, M. D. (1985): 'Interception in three dimensions: an energy formulation', *J. Guidance, Control & Dynamics*, **8**, pp. 23–30

RAO, K. R., JENKINS, S. L., PARTHASARATHI, K. and BALASUBRAMA-NIAM, R. (1985): 'Multitime scale analysis applied to long time simulation of power systems', *Int. J. Elect. Power & Energy Syst.*, **7**, pp. 7–12

RAO, A. K. and NAIDU, D. S. (1981): 'Singularly perturbed boundary value problems in discrete systems', *Int. J. Control*, **34**, pp. 1163–1173

RAO, A. K. and NAIDU, D. S. (1982): 'Singular perturbation method applied to open-loop discrete optimal control problem', *Optimal Control: Appl. & Methods*, **3**, pp. 121–131

RAO, A. K. and NAIDU, D. S. (1984): 'Singular perturbation method for Kalman filter in discrete systems', *IEE Proc. Pt.D.*, **131**, pp. 39–46

RAO, S. V. and LAMBA, S. S. (1974): 'Suboptimal control of linear systems via simplified models of Chidambara', *Proc. IEE*, **121**, pp. 879–882

REDDY, P. B. and SANNUTI, P. (1975): 'Optimal control of a coupled core nuclear reactor by a singular perturbation method', *IEEE Trans.*, **AC-20**, pp. 776–779

REDDY, P. B. and SANNUTI, P. (1976): 'Asumptotic approximation of optimal control applied to a power system problem', *Proc. IEE*, **123**, pp. 371–376

REIDLE, B. D. and KOKOTOVIC, P. V. (1985): 'Stability analysis of an adaptive system with unmodelled dynamics', *Int. J. Control*, **41**, pp. 389–402

REIDEL, B. D. and KOKOTOVIC, P. V. (1986): 'Integral manifolds of slow adaptation', *IEEE Trans.*, **AC-31**, pp. 316–324

REINHARDT, H. J. (1979): 'On asymptotic expansions in nonlinear singularly perturbed difference equations', *Num. Funct. Anal. Opt.*, **1**, pp. 565–587

REISS, E. L. (1961): 'On the quasi-static theory of viscoelasticity', *Arch. Rat. Mech. Anal.*, **7**, pp. 402–411

REISS, E. L. (1980): 'A new asymptotic method for jump phenomena', *SIAM J. Appl. Math.*, **39**, pp. 440–455

RENARDY, M. (1984): 'Singularly perturbed hyperbolic evolution problems with infinite delay and an application to polymer rheology', *SIAM J. Math. Anal.*, **15**, pp. 333–349

RINZEL, J. and TERMAN, D. (1982): 'Propagation phenomena in a bistable reaction diffusion systems', *SIAM J. Appl. Math.*, **42**, pp. 1111–1137

ROBERTS, S. M. (1986): 'An approach to singular perturbation problems insoluble by asymptotic methods', *J. Opt. Theory & Appl.*, **48**, pp. 325–340

ROBERTS S. M. and SHIPMAN, J. S. (1972): 'Two-point boundary value problems: Shooting methods' (Elsevier, New York)

ROHRS, C. E., VALAVANI, L., ATHANS, M. and STEIN, G. (1985): 'Robustness of continuous-time adaptive control algorithms in the presence of unmodelled dynamics', *IEEE Trans.*, **AC-30**, pp. 881–889

ROZOV, N. H. and GICHEV, T. R. (1983): 'Singularly perturbed problems with minimal pulse', *Diff. Eqns.*, **19**, pp. 269–266

RUBINOW, S. I. (1975): 'Introduction to mathematical biology' (Wiley, New York)

RUBLEIN, G. (1979): '2nd order estimates in singular perturbation of output regulator problems', *IEEE Trans.*, **AC-24**, pp. 350–352

RUDRAIAH, N., CHANDRASEKHARA, B. C. and JANAKAMMA, C. (1974): 'A singular perturbation problem of non-Newtonian flow between porous disks', *Int. J. Engg. Sci.*, **12**, pp. 31–44

RUDRAIAH, N. and MUSUOKA, T. (1982): 'Asymptotic analysis of natural convection through horizontal porous layer', *Int. J. Engg. Sci.*, **30**, pp. 27–39

RUIJTER, W. P. M. DE (1979): 'Boundary layers in large scale ocean circulation' *in* VERHULST, F.(Ed.): 'Asymptotic analysis Vol. 711' (Springer–Verlag, Berlin) pp. 125–145

SABERI, A. and KHALIL, H. K. (1982): 'Decentralized stabilization of a class of nonlinear interconnected systems', *Int. J. Control*, **36**, pp. 803–818

SABERI, A. and KHALIL, H. K. (1984*a*): 'Quadratic-type Lyapunov functions for singularly perturbed system', *IEEE Trans.*, **AC-29**, pp. 542–550

SABERI, A. and KHALIL, H. K. (1984*b*): 'An initial value theorem for nonlinear singularly perturbed systems', *Systems & Control Lett.*, **4**, pp. 301–305

SABERI, A. AND KHALIL, H. K. (1985): 'Stabilization and regulation of nonlinear singularly perturbed systems-composite control', *IEEE Trans.*, **AC-30**, pp. 739–747

SAGE, A. P. and WHITE, C. C. (1977): 'Optimum systems control, $2/E$' (Prentice Hall, Englewood Cliffs, USA)

SAKSENA, V. R. and BASAR, T. (1982): 'A multimodel approach to stochastic team problems', *Automatica*, **18**, pp. 713–720

SAKSENA, V. R. and BASAR, T. (1986): 'Multimodelling, singular perturbations and stochastic decision problems' *in* LEONDES, C. T. (Eds.): 'Control and dynamic systems Vol. 23' (Academic Press, New York) pp. 1—58

SAKSENA, V. R. and CRUZ, J. B., Jun. (1981*a*): 'Stabilization of singularly perturbed linear time-invariant systems using low-order observers', *IEEE Trans.*, **AC-26**, pp. 510–513

SAKSENA, V. R. and CRUZ, J. B., Jun. (1981*b*): 'Nash strategies in decentralized control of multiparameter singularly perturbed large scale systems', *Large Scale Syst.*, **2**, pp. 219–234

SAKSENA, V. R. and CRUZ, J. B., Jun. (1982): 'A multimodel approach to stochastic Nash games', *Automatica*, **18**, pp. 295–305

SAKSENA, V. R. and CRUZ, J. B., Jun. (1984): 'Robust Nash strategies for a class of nonlinear singularly perturbed problems', *Int. J. Control*, **39**, pp. 293–310

SAKSENA, V. R. and CRUZ, J. B., Jun. (1985a): 'Optimal and near optimal incentive strategies in the hierarchical control of Markov chains', *Automatica*, **21**, pp. 181–191

SAKSENA, V. R. and CRUZ, J. B., Jun. (1985b): 'A unified approach to reduced order modelling and control of large scale systems with multiple decision makers', *Optimal Control: Appl. & Methods*, **4**, pp. 403–420

SAKSENA, V. R. and CRUZ, J. B., Jun., PERKINS, W. R. and BASAR, T. (1983): 'Information induced multimodel solutions in multiple decision maker problems', *IEEE Trans.*, **AC-28**, pp. 716–728

SAKSENA, V. R. and KOKOTOVIC, P. V. (1981): 'Singular perturbation of the Popov–Kalman–Yakubrovich terms', *Syst. & Control Lett.*, **1**, pp. 65–68

SAKSENA, V. R., O'REILLY, J. and KOKOTOVIC, P. V. (1984): 'Singular perturbations and time-scale methods in control theory: survey 1976–1983', *Automatica*, **20**, pp. 273–293

SALATHE, E. P., WANG, T. C. and GROSS, J. F. (1980): 'Mathematical analysis of oxygen transport tissue', *Math. Biosci.*, **51**, pp. 89–115

SALMAN, M. A. and CRUZ, J. B., Jun. (1979): 'Well-posedness of linear closed-loop Stackelberg strategies for singularly perturbed systems', *J. Franklin Inst.*, **308**, pp. 25–37

SALMAN, M. A. and CRUZ, J. B., Jun. (1983): 'Optimal co-ordination of multimodel interconnected systems with slow and fast modes', *Large Scale Syst.*, **5**, pp. 207–219

SANDELL, N. R., Jun. (1979) 'Robust stability of systems with applications to singular perturbations', *Automatica*, **15**, pp. 467–470

SANDELL, N. R., Jun., VARAIYA, P., ATHANS, M. and SAFANOV, M. G. (1978): 'Survey of decentralized control methods for large scale systems', *IEEE Trans.*, **AC-23**, pp. 108–128

SANNUTI, P. (1968): 'Singular perturbation method in the theory of optimal control'. Ph.D. Thesis, University of Illinois, Urbana, USA

SANNUTI, P. (1969): 'Continuity and differentiability properties of optimal control with respect to singular perturbations', *IEEE Trans.*, **AC-14**, pp. 762–763

SANNUTI, P. (1974a): 'A note on obtaining reduced order optimal control problems by singular perturbations', *IEEE Trans.*, **AC-19**, pp. 256–257

SANNUTI, P. (1974b): 'Asymptotic series solution of singularly perturbed optimal control problems', *Automatica*, **10**, pp. 183–194

SANNUTI, P. (1975): 'Asymptotic expansions of singularly perturbed quasilinear optimal systems', *SIAM J. Control*, **13**, pp. 572–592

SANNUTI, P. (1976): 'Use of singular perturbation methods to formulate electric networks equations', *Rocky Mountain J. Math.*, **6**, pp. 709–710

SANNUTI, P. (1977): 'On the controllability of singularly perturbed systems', *IEEE Trans.*, **AC-22**, pp. 622–624

SANNUTI, P. (1978): 'On the controllability of some singularly perturbed nonlinear systems', *J. Math. Anal. Appl.*, **64**, pp. 579–591

SANNUTI, P. (1981): 'Singular perturbations in the state space approach of electric networks', *Circuit Theory and Appl.*, **9**, pp. 47–57

SANNUTI, P. (1983): 'A direct singular perturbation analysis of high-gain and cheap control problems', *Automatica*, **19**, pp. 41–51

SANNUTI, P. and KOKOTOVIC, P. V. (1969a): 'Near optimum design of linear systems by singular perturbation method', *IEEE Trans.*, **AC-14**, pp. 15–22

SANNUTI, P. and KOKOTOVIC, P. V. (1969b): 'Singular perturbation method for near optimum design of high-order nonlinear systems', *Automatica*, **5**, pp. 773–779

SANNUTI, P. and REDDY, P. B. (1973): 'Asymptotic series solution of optimal systems with small time delay', *IEEE Trans.*, **AC-18**, pp. 250–259

SANNUTI, P. and WASON, H. S. (1983): 'A singular perturbation canonical form of invertible systems: determination of multivariable root loci', *Int. J. Control*, **37**, pp. 1259–1286

SANNUTI, P. and WASON, H. S. (1985): 'Multiple time-scale decompostion in cheap control problems-singular control', *IEEE Trans.*, **AC-30**, pp. 633–644

SARLET, W. (1978): 'On a common derivation of the averaging method and the two-time-scale method', *Celestial Mech.*, **17**, pp. 299–312

SASTRY, S. S. (1983): 'The effects of small noise on implicitly defined nonlinear dynamical systems', *IEEE Trans.*, **CAS-30**, pp. 651–663

SASTRY, S. S. and DESOER, C. A. (1981): 'Jump behaviour of circuits and systems', *IEEE Trans.*, **CAS-28**, pp. 1109–1124

SASTRY, S. S. and DESOER, C. A. (1983): 'Asymptotic unbounded root loci-formulas and computation', *IEEE Trans.*, **AC-28**, pp. 557–568

SASTRY, S. S. and HIJAB, O. (1981): 'Bifurcation in the presence of small noise', *Syst. & Control Lett.*, **1**, pp. 159–167

SASTRY, S. S. and VARAIYA, P. (1980): 'Heirarchical stability and alert state steering control of interconnected power systems', *IEEE Trans.*, **CAS-27**, pp. 1102–1112

SASTRY, S. S. and VARAIYA, P. (1981): 'Coherency for interconnected power systems', *IEEE Trans.*, **AC-26**, pp. 218–226

SAUER, P. W., AHMED-ZAID, S. and PAI, M. A. (1984): 'Systematic inclusion of stator transients in reduced order synchronous machine models', *IEEE Trans.*, **PAS-103**, pp. 1248–1255

SCHAAR, R. (1976): 'Singularly perturbed conservative systems', *Rocky Mountain J. Math.*, **6**, pp. 711–723

SCHAUER, M. and HEINRICH, R. (1983): 'Quasi-steadystate approximations in the mathematical modelling of biochemical reaction networks', *Math. Biosci.*, **65**, pp. 155–170

SCHUMACHER, J. M. (1984): 'Almost stabilizability subspaces and high gain feedback', *IEEE Trans.*, **AC-29**, pp. 620–628

SCHUSS, Z. (1980): 'Theory and applications of stochastic differential equations' (Wiley, New York)

SEBALD, A. V. (1979): 'Toward a computationally efficient optimal solution to the LQG discrete-time dual control problem', *IEEE Trans.*, **AC-24**, pp. 535–540

SEBALD, A. V. and HADDAD, A. H. (1978): 'State estimation for singularly perturbed systems with uncertain perturbation parameter', *IEEE Trans.*, **AC-23**, pp. 464–469

SEBALD, A. V. and HADDAD, A. H. (1979): 'On the performance of combined detection state estimation in convex uncertain spaces', *IEEE Trans.*, **AC-24**, pp. 651–653

SESAK, J. R., LIKINS, P. W. and CORADETTI, T. (1979): 'Flexible space craft control by model error sensitivity suppression', *J. Astronautical Sci.*, **27**, pp. 131–156

SESHADRI, S. R. (1976): 'Higher-order wave interaction in a periodic medium', *Appl. Phys.*, **10**, pp. 165–173

SESHADRI, S. R. (1977a): 'Asymptotic theory of mode decoupling in a space-time periodic medium-part—I: Stable interaction', *Proc. IEEE*, **65**, pp. 996–1004

SESHADRI, S. R. (1977b): 'Asymptotic theory of mode decoupling in a space-time periodic medium-part—II: Unstable interactions', *Proc. IEEE*, **65**, pp. 1459–1469

SHAKED, U. (1976): 'Design techniques for high feedback gain stability', *Int. J. Control*, **24**, pp. 137–144

SHAKED, U. and BABROVSKY, B. (1981): 'The asymptotic minimum variance estimate of stationary linear single output processes', *IEEE Trans.*, **AC-26**, pp. 498–504

SHENSA, M. J. (1971): 'Parasities and the stability of equilibrium points on nonlinear networks' *IEEE Trans.*, **CT-18**, pp. 481–484

SHEPHERD, J. J. (1978a): 'On the asymptotic solution of the Reynolds equation', *SIAM J. Appl. Math.*, **34**, pp. 774–791

SHEPHERD, J. J. (1978b): 'Asymptotic solution of a nonlinear singular perturbation problem', *SIAM J. Appl. Math.*, **35**, pp. 176–186

SHI, Y. Y. and ECKSTEIN, M. C. (1968): 'Application of singular perturbation methods to resonance problems', *Astron. J.*, **73**, pp. 275–289

SHIEH, L. S. and TSAY, Y. T. (1984): 'Algebraic-geometric approach for the model reduction of large-scale multivariable system', *IEE Proc. Pt. D.*, **131**, pp. 23–36

SHIMIZU, K. and MATSUBARA, M. (1985): 'Singular perturbation for the dynamic interaction measure', *IEEE Trans.*, **AC-30**, pp. 790–792

SHINAR, J. (1981): 'Solution techniques for realistic pursuit-evasion games' *in* LEONDES, C. T. (Ed.): 'Advances in control and dynamic systems, Vol. 17' (Academic Press, New York) pp. 63–124

SHINAR, J. (1983): 'On applications of singular perturbation techniques in nonlinear optimal control', *Automatica*, **19**, pp. 203–211

SHINAR, J. (1985): 'Zeroth-order feedback strategies for mediam-range interception in a horizontal plane', *J. Guidance, Control & Dynamics*, **8**, pp. 9–15

SHINAR, J. and FARBER, N. (1984): 'Horizontal variable speed interception game solved by forced singular perturbation technique', *J. Opt. Theory & Appl.*, **42**, pp. 603–636

SHINAR, J., MERARI, A., BLANK, D. and MEDINAH, M. (1980): 'Analysis of optimal tuning maneuvers in the vertical plane', *J. Guidance & Control*, **3**, pp. 69–77

SHINAR, J. and NEGRIN, M. (1983): 'An explicit feedback approximation for medium-range interceptions in a vertical plane', *Optimal Control: Appl. & Methods*, **4**, pp. 303–323

SILJAK, D. D. (1972): 'Singular perturbations of absolute stability', *IEEE Trans.*, **AC-17**, p. 720

SILJAK, D. D. (1983): 'Complex dynamic systems: dimensionality, structure, and uncertainty', *Large Scale Syst.*, **4**, pp. 279–294

SILVA-MADRIZ, R. (1986): 'Feedback systems and multiple time-scales', *Int. J. Control*, **43**, pp. 587–600

SILVA-MADRIZ, R. and SASTRY, S. S. (1984): 'Input-output description of linear systems with multiple time-scales', *Int. J. Control*, **40**, pp. 699–722

SILVA-MADRIZ, R. and SASTRY, S. S. (1986): 'Multiple time scales for nonlinear systems', *Circuits, Systems and Signal Processing*, **5**, pp. 153–169

SINGH. G. and YACKEL, R. A. (1973): 'Optimal regulator with order reduction of power system', *Computers in Elect. Engg.*, **1**, pp. 425–452

SINGH, R. P. (1982): 'The linear-quadratic-gaussain problem for singularly perturbed systems', *Int. J. Syst. Sci.*, **13**, pp. 93–100

SINGH, Y. P. (1982): 'Multiple time analysis of coupled nonlinear systems', *Int. J. Control*, **36**, pp. 99–107

SINGH, Y. P. (1986): 'Two-time analysis of jump behaviour in interconnected systems', *Int. J. Syst. Sci.*, **17**, pp. 141–149

SINGH, N. P., SINGH, Y. P. and ASHON, S. I. (1986): 'An iterating approach to reduced-order modeling of synchronous machines using singular perturbation', *Proc. IEEE*, **74**, pp. 892–893

SINGH, N. P., SINGH, Y. P. and AHSON, S. I. (1986): 'Near-optimal controllers for synchronous machines with prescribed degree of stability via iterative separation of time scales', *Proc. IEEE*, **74**, pp. 1466–1468

SKELTON, R. E. (1980): 'Cost decomposition of linear systems with applications to model reduction', *Int. J. Control*, **32**, pp. 1031–1055

SKINNER, L. A. (1981): 'Note on the Logerstrom singular perturbation models', *SIAM J. Appl. Math.*, **41**, pp. 362–369

SMITH, D. R. (1975): 'The multivariable method in singular perturbation analysis', *SIAM Rev.*, **17**, pp. 221–273

SMITH, D. R. (1985): 'Singular perturbation theory: An introduction with applications' (Cambridge University Press, Cambridge)

SOBOLEV, V. A. (1984): 'Integral manifolds and decomposition of singularly perturbed systems', *Syst. & Control Lett.*, **5**, pp. 169–179

SOLHEIM, C. A. (1980): 'On the use of low-order Riccati equations in the design of a class of feedback controllers and state estimation', *Modeling, Identification & Control*, **1**, pp. 231–246

SOLIMAN, M. A. and RAY, W. H. (1979*a*): 'Nonlinear state estimation of packed-bed tubular reactors', *Amer. Inst. Chem. Engg. J.*, **25**, pp. 718–720

SOLIMAN, M. A. and RAY, W. H. (1979*b*): 'Nonlinear filtering for distributed parameter systems having a small parameter', *Int. J. Control*, **30**, pp. 757–772

SOLOMON, L. P. and COMSTOCK, G. C. (1973): 'Two-time methods applied to under water acoustics', *J. Acoust. Soc. America*, **54**, pp. 110–114

SPEYER, J. C. and GUSTAFSON, D. E. (1979): 'Estimation of phase processes in oscillatory signals and asymptotic expansions', *IEEE Trans.*, **AC-24**, pp. 657–659

SPRIGGS, J. H., MESSITER, A. F. and ANDERSON, W. J. (1969): 'Membrance flutter paradox-an explanation by singular perturbations methods', *AIAA J.*, **7**, pp. 1704–1709

SRIDHAR, B. and GUPTA, N. K. (1980): 'Missile guidance laws based on singular perturbation methodology', *J. Guidance & Control*, **3**, pp. 158–165

STAVROULAKIS, P. and TZAFESTAS, S. (1982): 'Singularly perturbed large-scale distributed parameter control systems-an application to nuclear reactor control', *Math. Computers & Simulation*, **24**, pp. 303–313

STEIN, G. (1979): 'Generalized quadratic weights for asymptotic regulator properties', *IEEE Trans.*, **AC-24**, pp. 559–566

STEINBERG, A. M. (1986): 'Stabilization of nonlinear singularly perturbed systems with a periodic input', *Int. J. Control*, **43**, pp. 651–661

STEINMETZ, W. J. (1974): 'On a nonlinear singular perturbation problem in a gas lubrication theory', *SIAM J. Appl. Math.*, **26**, pp. 816–827

STEWARTSON, K. (1976): 'A note on perturbation series in fluid dynamics', *Qrt. J. Mech. Appl. Math.*, **24**, pp. 377–379

STINEMAN, M. (1985): 'Digital time-domain analysis of systems with widely separated poles', *J. Assoc. Comput. Machinery*, **12**, pp. 286–293

SUBRAMANIAN, R. and KRISHNAN, A. (1979): 'Nonlinear discrete-time systems analysis by multiple time perturbation techniques', *J. Sound & Vibration*, **63**, pp. 325–335

SUZUKI, M. (1981): 'Composite controls for singularly perturbed systems', *IEEE Trans.*, **AC-26**, pp. 505–507

SUZUKI, M. and MIURA, M. (1976): 'Stabilizing feedback controllers for singularly perturbed linear constant systems', *IEEE Trans.*, **AC-21**, pp. 123–124

SYRCOS, G. P. and SANNUTI, P. (1983): 'Singular perturbation modelling of continuous and discrete physical models', *Int. J. Control*, **37**, pp. 1007–1022

SYRCOS, G. P. and SANNUTI, P. (1984): 'Near optimum regulator design of singularly perturbed systems via Chandrasekhar equations', *Int. J. Control*, **39**, pp. 1083–1102

TANYI, G. E. (1982): 'Energy and biological evolution—I: The equilibrium states of biochemical processes', *Bull. Math. Biol.*, **44**, pp. 501–535

TAVANTZIS, J., REISS, E. L. and MATKOWSKY, B. J. (1978): 'On the smooth transition to convection', *SIAM J. Appl. Math.*, **34**, pp. 322–337

TENEKETZIS, D. and SANDELL, N. R., Jun. (1977): 'Linear regulator design for stochastic systems by a multiple time-scale method', *IEEE Trans.*, **AC-22**, pp. 615–621

TIKHONOV, A. N. (1952): 'Systems of differential equations containing small parameters multiplying some of the derivatives', *Mat. Sb.*, **31**, pp. 575–586

TIKHONOV, A. N., VASILEVA, A. B. and SVESHNIKOV, A. G. (1984): 'Differential equations' (Springer–Verlag, Berlin)

TRAN, M. T. and SAWAN, M. E. (1983a): 'Reduced order discrete models', *Int. J. Syst. Sci.*, **14**, pp. 745–752

TRAN, M. T. and SAWAN, M. E. (1983b): 'Nash strategies for discrete-time systems with slow and fast modes', *Int. J. Syst. Sci.*, **14**, pp. 1355–1371

TRAN, M. T. and SAWAN, M. E. (1983c): 'Singularly perturbed systems with low sensitivity to model reduction', *J. Astronom. Sci.*, **31**, pp. 329–333

TRAN, M. T. and SAWAN, M. E. (1984a): 'Low order observers for discrete systems with slow and fast modes', *Int. J. Syst. Sci.*, **15**, pp. 1283–1288

TRAN, M. T. and SAWAN, M. E. (1984b): 'On the well-posedness of discrete-time systems with slow and fast modes', *Int. J. Syst. Sci.*, **15**, pp. 1289–1294

TRAN, M. T. and SAWAN, M. E. (1984c): 'Decentralized control of two-time-scale discrete systems', *Int. J. Syst. Sci.*, **15**, pp. 1295–1300

TSAI, C. P. (1978): 'Perturbed stochastic linear regulator problem', *SIAM J. Control*, **16**, pp. 394–410

TSE, E. C. Y., MEDANIC, J. Y. and PERKINS, W. R. (1978): 'Generalized Wessenberg transformations for reduced order modelling of large scale systems', *Int. J. Control*, **27**, pp. 493–512

TUPCHIEV, V. A. (1962): 'The asymptotic behaviour of the solution of the boundary value problem for a system of first order differential equations containing a small parameter attached to the derivative', *Dokl. Akad. Nauk, SSSR*, **143**, pp. 1296–1299

TZAFESTAS, S. G. (1984): 'Optimal control of -coupled and singularly perturbed distributed-parameter systems', *Math. & Comput. in Simulation*, **26**, pp. 27–38

TZAFESTAS, S. G. and ANAGNOSTOU, R. E. (1983): 'Stabilizing controllers for singularly perturbed time-invariant bilinear systems', *Foundations of Control Engg.*, **8**, pp. 93–104

UTKIN, V. I. (1977a): 'Sliding modes and their applications to variable structure systems' (Mir Publishers, Moscow)

UTKIN, V. I. (1977b): 'Variable structure systems with sliding modes: A survey', *IEEE Trans.*, **AC-22**, pp. 212–222

UTKIN, V. I. (1978): 'Application of equivalent control method to the system with large feedback-gain', *IEEE Trans.*, **AC-23**, pp. 484–486

UTKIN, V. A. (1984): 'Variable structure systems: Present and future', *Aut. & Remote Control*, **44**, pp. 1105–1120

UTKIN, V. A. and UTKIN, V. I. (1984): 'Design of invariant systems by the method of separation of motions', *Aut. & Remote Control*, **44**, pp. 1559–1566

VAN DYKE, M. (1964): 'Perturbation methods in fluid mechanics' (Academic Press, New York)

VAN DYKE, M. (1975): 'Computer extension of perturbation series in fluid mechanics', *SIAM J. Appl. Math.*, **28**, pp. 720–734

VASILEVA, A. B. (1963): 'Asymptotic behaviour of solutions to certain problems involving nonlinear ordinary differential equations containing a small parameter multiplying the highest derivatives', *Russian Math. Surveys*, **18**, pp. 13–84

VASILEVA, A. B. (1968): 'Investigation of the asymptotic properties of a differential equation, appearing in some problems in kinetics', *Diff. Urav.*, **4**, pp. 397–408

VASILEVA, A. B. (1976): 'The development of the theory of ordinary differential equations with a small parameter multiplying the highest derivatives in the years 1966–76', *Russian Math. Surveys*, **31**, pp. 109–131

VASILEVA, A. B. and BUTUZOV, V. F. (1973): 'Asymptotic expansions of solutions of singularly perturbed differential equations' (Izdat. 'Nauke', Moscow) (in Russian)

VASILEVA, A. B. and BUTUZOV. V. F. (1978): 'Singularly perturbed equations in critical cases' (Izdat. Moscow University, USSR) (in Russian)

VASILEVA, A. B. and FAMINSKAYA, M. V. (1981): 'An investigation of a non-linear optimal control problem by the methods of singular perturbation theory', *Sov. Math. Dokl.*, **21**, pp. 104–108

VASILEVA, A. B. and STELMAH, V. G. (1977): 'Singularly perturbed systems in the theory of semiconductors', *USSR Comput. Math. & Math. Phys.*, **17**, pp. 48–58

VERGHESE, G. C., LEVY, B. C. and KAILATH, T. (1981): 'A generalized state space for singular systems', *IEEE Trans.*, **AC-26**, pp. 811–831

VERHULST, F. (1975): 'Asymptotic expansions in the perturbed two-body problem with application to systems with variable mass', *Celestial Mech.*, **11**, pp. 95–129

VERHULST, F., (Ed.) (1979): 'Asymptotic analysis: From theory to applications'. Lecture Notes Mathematics, Vol. 711 (Springer–Verlag, Berlin)

VIDYASAGAR, M. (1978): 'Nonlinear system analysis' (Prentice Hall, Englewood Cliffs)

VIDYASAGAR, M. (1984): 'The graph metric for unstable plants and robustness estimates for feedback stability', *IEEE Trans.*, **AC-29**, pp. 403–418

VIDYASAGAR, M. (1985): 'Robust stabilization of singularly perturbed systems', *Syst. Control Lett.*, **5**, pp. 413–418

VIKTOROV, B. V. (1977): 'Use of singular perturbation method to investigate automatic control systems', *Dokl. Akad. Nauk. SSSR*, **236**, pp. 296–299 (in Russian)

VISHIK, M. I. and LYUSTERNIK, L. A. (1957): 'Regular degeneration and boundary layer for linear differential equations with a small parameter', *Usp. Mat. Nauk*, **12**, pp. 3–122 (in Russian)

VISSER, H. G. and SHINAR, J. (1986): 'First-order corrections in optimal feedback control of singularly perturbed nonlinear systems', *IEEE Trans.*, **AC-31**, pp. 387–393

WASOW, W. (1965): 'Asymptotic expansions for ordinary differential equations' (Wiley–Interscience, New York)

WASYNOZUK, O. and DECARLO, R. A. (1981): 'The component connection model and structure preserving model order reduction', *Automatica*, **17**, pp. 619–626

WEINSTEIN, M. B. and SMITH, D. R. (1975): 'Comparison techniques for over damped systems', *SIAM Rev.*, **17**, pp. 520–540

WENDEL, J. G. (1950): 'Singular perturbation of a Van der Pol equation, *Annals Math.*, pp. 243–290

WESTON, A. R., CLIFF, E. M. and KELLEY, H. J. (1983): 'Altitude transitions in energy climbs', *Automatica*, **19**, pp. 199–202

WHITMAN, A. M. (1980): 'Asymptotic theory of freight car hunting', *Trans. ASME, J. Dyn. Syst. Meas. & Control*, **102**, pp. 190–193

WILDE, R. R. and KOKOTOVIC, P. V. (1972a): 'Stability of singularly perturbed systems and networks with parasitics', *IEEE Trans.*, **AC-17**, pp. 245–246

WILDE, R. R. and KOKOTOVIC, P. V. (1972b): 'A dichotomy in linear control theory', *IEEE Trans.*, **AC-17**, pp. 382–383

WILDE, R. R. and KOKOTOVIC, P. V. (1973): 'Optimal open- and closed-loop control of singularly perturbed linear systems', *IEEE Trans.*, **AC-18**, pp. 616–626

WILLEMS, J. C. (1981): 'Almost invariant subspaces: an approach to high gain feedback design-part—I: Almost controlled invariant subspaces', *IEEE Trans.*, **AC-26**, pp. 235–252

WILLEMS, J. C. (1982): 'Almost invariant subspaces: an approach to high-gain feedback design-part—II: Almost conditionally invariant subspaces', *IEEE Trans.*, **AC-27**, pp. 1071–1084

WILLOUGHBY, R. A. (Ed.) (1974): 'Stiff differential systems' (Plenum, New York)

WILLSKY, A. S., BELLO, M. G., CASTANON, D. A., LEVY, B. C. and VERG-HESE, G. C. (1982): 'Combining and updating of local estimates and regional maps along sets of one dimensional tracks', *IEEE Trans.*, **AC-27**, pp. 794–813

WINKELMAN, J. R., CHOW, J. H., ALLEMONG, J. J. and KOKOTOVIC, P. V. (1980): 'Multi-time-scale analysis of a power system', *Automatica*, **16**, pp. 35–43

WINKELMAN, J. R., CHOW, J. J., BOWLER, B. G., AVRAMOVIC, B. and KOKOTOVIC, P. V. (1981): 'An analysis of inter area dynamics of multi-machine systems', *IEEE Trans.*, **PAS-100**, pp. 754–763

WOLFE, R. J. (1978): 'Temporal oscillation in kinetic model of the Belousov–Zhabotinskii reaction', *Arch. Rat. Mech. Anal.*, **67**, pp. 225–250

WOLFE, R. J. (1979): 'Periodic solution of a singularly perturbed differential system with applications to the Belousov–Zhabotinskii reaction', *J. Math. Anal. Appl.*, **68**, pp. 488–508

WOLLKIND, D. J. (1977): 'Singular perturbation techniques: A comparison of the method of matched asymptotic expansions with that of multiple scales', *SIAM Rev.*, **19**, pp. 502–516

WOLLKIND, D. J. and LOGAN, J. A. (1978): 'Temperature-dependent predator-prey mite ecosystem on apple tree foliage', *J. Math. Biol.*, **6**, pp. 265–283

WOLLKIND, D. J., LOGAN, J. A. and BERRYMAN, A. A. (1978): 'Asymptotic methods for modelling biological processes', *Researches on Population Ecology*, **20**, pp. 79–90

WOMBLE, M. E., POTTER, J. E. and SPEYER, J. L. (1976): 'Approximations to Riccati equations having slow and fast modes', *IEEE Trans.*, **AC-21**, pp. 846–855

XU, P. Y. (1985): 'Singular perturbation theory of horizontal dynamic stability and response of aircraft', *Aeronaut. J.*, **89**, pp. 179–184

YACKEL, R. A. and KOKOTOVIC. P. V. (1973): 'A boundary layer method for the matrix Riccati equation', *IEEE Trans.*, **AC-18**, pp. 17–24

YOUNG, K. D. (1978*a*): 'Multiple time scales in single input-single output high gain feedback systems', *J. Franklin Inst.*, **306**, pp. 293–301

YOUNG, K. D. (1978*b*): 'Design of variable structure model-following control systems', *IEEE Trans.*, **AC-23**, pp. 1074–1085

YOUNG, K. D. (1982*a*): 'Near sensitivity of linear feedback systems', *J. Franklin Inst.*, **314**, pp. 129–142

YOUNG, K. D. (1982*b*): 'Disturbance decoupling by high gain feedback', *IEEE Trans.*, **AC-27**, pp. 970–971

YOUNG, K. D. (1985): 'On near optimal decentralized control', *Automatica*, **21**, pp. 607–610

YOUNG, K. D. and KOKOTOVIC, P. V. (1982): 'Analysis of feedback-loop interaction with actuator and sensor parasitics', *Automatica*, **18**, pp. 577–582

YOUNG, K. D., KOKOTOVIC, P. V. and UTKIN, V. I. (1977): 'A singular perturbation analysis of high gain feedback systems', *IEEE Trans.*, **AC-22**, pp. 931–938

YOUNG, K. D. and KWATNY, H. G. (1982): 'Variable structure servomechanism design and applications to overspeed protection control', *Automatica*, **18**, pp. 385–400

ZAID, S. A., SAUER, P. W., PAI, M. A. and SARIOGLU, M. K. (1982): 'Reduced order modeling of synchronous machines using singular perturbation', *IEEE Trans.*, **CAS-21**, pp. 782–786

ZINOBER, S. I., GHEZAWI, O. M. E. EL. and BILLINGS, S. A. (1982): 'Multivariable structure adaptive model reference following control systems, *IEE Proc. Pt.D.*, **129**, pp. 6–12

Subject index

www.ingramcontent.com/pod-product-compliance
Lightning Source LLC
Chambersburg PA
CBHW050523190326
41458CB00005B/1641